Data Scientists at Work

Sebastian Gutierrez

Apress®

For Madeleine and Hannah.

Contents

Foreword

In 2008, Google's Chief Economist, Hal Varian, stated:[1]

> I keep saying the sexy job in the next ten years will be statisticians .... The ability to take data—to be able to understand it, to process it, to extract value from it, to visualize it, to communicate it—that's going to be a hugely important skill in the next decades, not only at the professional level but even at the educational level for elementary school kids, for high school kids, for college kids.

Varian makes it clear he was really talking about what we now call a "data scientist," not a traditional statistician. So what is this skill set? How is a data scientist different from a statistician, mathematician, or computer scientist? And why does Varian call it sexy?

Recently I was discussing this very issue with some friends from different professions. We decided we could compare our respective fields by working on understanding the same problem—a theoretical economic marketplace that had a set of simple rules. Persi Diaconis, a mathematician, got out his favorite tool: a pencil and pad of paper. He was able to prove some convergence results about the game. Susan Holmes, a statistician (who was once thrown out of a math department for wanting to use computers to do data analysis) used her tool, the R statistical computing package, to run a simulation for 10 agents over 100 time steps, and plotted the results. The plots had quite a bit of random noise, so she had to employ some theoretical knowledge to come to an understanding of the situation. And I, as a computer/data scientist, had a variety of tools at my disposal, but I chose iPython Notebook with matplotlib to create a simulation that was very similar to Susan's, but with 5,000 agents over 25,000 time steps. Since I had 125,000 times more data, my plots were much smoother—the signal dominated the noise, and it was easier to see what was happening. In the end we all arrived at a similar level of understanding but came to it by different paths.

[1] Hal Varian, interview by James Manyika, "Hal Varian on How the Web Challenges Managers," McKinsey & Company Insights & Publications, October 2008 (transcript published January 2009). http://www.mckinsey.com/insights/innovation/hal_varian_on_how_the_web_challenges_managers.

One of the first explorers to tread all these paths was Leo Breiman, the statistician who wrote the influential "Statistical Modeling: The Two Cultures" paper in 2001.[2] In Breiman's view, most statisticians of that time belonged to the data modeling culture, which starts with the assumption that there is some underlying stochastic model that is generating the data, and the analyst's job is to measure the fit of a model to the data. Interpretability of the model is a primary concern. A minority of statisticians in 2001 and a majority of data scientists today belong to a culture of algorithmic modeling—one that recognizes that the data may derive from a complicated combination of unknown factors, and thus one that will resist characterization by a simple model. However, it is still possible to use the data to make predictions about new, unseen data, even without a model that fully characterizes the system. The primary concern is the accuracy of predictions, not interpretability. Breiman concludes his piece with the warning:

> We are in a period where there has never been such a wealth of new statistical problems and sources of data. The danger is that if we define the boundaries of our field in terms of familiar tools and familiar problems, we will fail to grasp the new opportunities.

Another statistician who was eager to grasp new opportunities was George Box, who wrote, "All models are wrong, but some are useful."[3] His career as a statistician deeply integrated with engineers led him to understand that "your model is wrong" is not a criticism, but rather an acceptance of the inherent complexity of the real world. Models are judged by their empirical utility, not by some elusive Platonic rationalist ideal.

Now for the final question: Why is data science sexy? It has something to do with all that grasping. And the begetting: so many new applications and entire new industries come into being from the judicious use of copious amounts of data. Examples include speech recognition, object recognition in computer vision, robots and self-driving cars, bioinformatics, neuroscience, the discovery of exoplanets and an understanding of the origins of the universe, and the assembling of inexpensive but winning baseball teams. In each of these instances, the data scientist is central to the whole enterprise. He or she must combine knowledge of the application area with statistical expertise and implement it all using the latest in computer science ideas. In the end, sexiness comes down to being effective. In an enterprise that

---

[2]Leo Breiman, "Statistical Modeling: The Two Cultures." *Statistical Science* 26 (2001): 199-231.
[3]George E. P. Box and Norman R. Draper, *Empirical Model-Building and Response Surfaces* (New York: John Wiley & Sons, 1987).

has many complex moving parts—interacting with customers, suppliers, raw materials, manufacturing, and everything else—it is quite common for a data analyst to be able to improve efficiency by 10% or more just by manipulating bits, never touching atoms. Sometimes a 10% increase is a nice bonus, and sometimes it's the difference between success and failure. In this book, you will see how some of the world's top data scientists work across a dizzyingly wide variety of industries and applications—each leveraging her own blend of domain expertise, statistics, and computer science to create tremendous value and impact.

—Peter Norvig

Director of Research, Google

Mountain View, California

About the Author

Sebastian Gutierrez is a data entrepreneur who has founded three data-related companies: DataYou (data science and data visualization consulting and education), LetsWombat (data-driven product sampling), and Acheevmo (athletic performance statistics). He was formerly an emerging markets risk manager at Scotia Capital and an FX options trader at JP Morgan and Standard Chartered Bank.

Gutierrez provides training in data visualization and D3.js to a diverse client base, including corporations such as the New York Stock Exchange, the American Express Company, and General Dynamics, universities, media agencies, and startups. He leads the 1,600-member New York City D3.js Meetup Group and is co-editor of *Data Science Weekly*, a weekly newsletter providing curated articles and videos on the latest developments in data science. He is a frequent speaker at meetups and conferences, such as Strata and Hadoop World in New York and Barcelona. He is a cross-disciplinary instructor at General Assembly. Gutierrez holds a BS in Mathematics from MIT and an MA in Economics from the University of San Francisco.

Acknowledgments

Many heartfelt thanks to the interviewees and to the friends and colleagues who made interviewing them possible. Words cannot possibly describe the inspiration each one of you has given me, so I will just say that you have made my life better and for that I am eternally grateful.

A thousand thanks to my family: Mom and Dad for passing on their love of books and learning; Laura for being an inspiration; Madeleine for bringing wonder into my life; Liz and Chris for your love and support; and Hannah for being my better half and partner in adventures.

A very special thanks to Apress for their support throughout the process: to Robert Hutchinson for his thoughts, advice, and care; to Rita Fernando for her patience and guidance; and to Kristen Ng for her perfect ear and encouragement.

Finally, thank you to the readers of this book. I sincerely hope that you see further by standing on the shoulders of these giants.

Introduction

Data is the new oil!

—Clive Humby, dunnhumby[1]

By 2018, the United States will experience a shortage of 190,000 skilled data scientists, and 1.5 million managers and analysts capable of reaping actionable insights from the big data deluge.

—McKinsey Report[2]

The emergence of data science is gathering ever more attention, and it's no secret that the term *data science* itself is loaded with controversy about what it means and whether it's actually a field. In *Data Scientists at Work*, I interview sixteen data scientists across sixteen different industries to understand both how they think about it theoretically and also very practically what problems they're solving, how data's helping, and what it takes to be successful.

Mirroring the flux in which data science finds itself, the sample of data scientists polled in this book are all over the map about the significance and utility of the terms *data science* to refer to a coherent discipline and *data scientist* to refer to a well-defined occupation. In the interests of full disclosure, I fall into the camp of those who believe that data science is truly an emerging academic discipline and that data scientists as such have proper roles in organizations. Moreover, I believe that each of the subjects I interviewed for this book is indeed a data scientist—and, after having spent time with all of them, I couldn't be more excited about the future of data science.

---

[1]Michael Palmer, "Data Is the New Oil," ANA Marketing Maestros blog, November 3, 2006. http://ana.blogs.com/maestros/2006/11/data_is_the_new.html.
[2]Susan Lund et al., "Game Changers: Five Opportunities for US Growth and Renewal," McKinsey Global Institute Report, July 2013. http://www.mckinsey.com/insights/americas/us_game_changers.

Though some of them are wary of the hype that the field is attracting, all sixteen of these data scientists believe in the power of the work they are doing as well as the methods. All sixteen interviewees are at the forefront of understanding and extracting value from data across an array of public and private organizational types—from startups and mature corporations to primary research groups and humanitarian nonprofits—and across a diverse range of industries—advertising, e-commerce, email marketing, enterprise cloud computing, fashion, industrial internet, internet television and entertainment, music, nonprofit, neurobiology, newspapers and media, professional and social networks, retail, sales intelligence, and venture capital.

My interviewing method was designed to ask open-ended questions so that the personalities and spontaneous thought processes of each interviewee would shine through clearly and accurately. My aim was to get at the heart of how they came to be data scientists, what they love about the field, what their daily work lives entail, how they built their careers, how they developed their skills, what advice they have for people looking to become data scientists, and what they think the future of the field holds.

Though all sixteen are demonstrably gifted at data science, what stuck out the most to me was the value that each person placed on the "people" side of the business—not only in mentoring others who are up and coming, but also in how their data products and companies interface with their customers and clients. Regardless of the diversity of company size and stage, seniority, industry, and role, all sixteen interviewees shared a keen sense of ethical concern for how data is used.

Optimism pervades the interviews as to how far data science has come, how it's being used, and what the future holds not just in terms of tools, techniques, and data sets, but also in how people's lives will be made better through data science. All my interview subjects believe that they are busy creating a better future.

To help recruit future colleagues for this vast collective enterprise, they give answers to the urgent questions being asked by those who are considering data science as a potential career—questions about the right tools and techniques to use and about what one really needs to know and understand to be hired as a data scientist. The practitioners in this book share their thoughts on what data science means to them and how they think about it, their suggestions on how to join the field, and their wisdom won through experience on what a data scientist must understand deeply to be successful within the field.

Data is being generated exponentially and those who can understand that data and extract value from it are needed now more than ever. So please enjoy this book, and please take the hard-earned lessons and joy about data and models from these thoughtful practitioners and make them part of your life.

Chris Wiggins

The New York Times

Chris Wiggins is the Chief Data Scientist at The New York Times *(NYT) and Associate Professor of Applied Mathematics at Columbia University. He applies machine learning techniques in both roles, albeit to answer very different questions.*

In his role at the NYT, Wiggins is creating a machine learning group to analyze both the content produced by reporters and the data generated by readers consuming articles, as well as data from broader reader navigational patterns—with the overarching goal of better listening to NYT consumers as well as rethinking what journalism is going to look like over the next 100 years.

At Columbia University, Wiggins focuses on the application of machine learning techniques to biological research with large data sets. This includes analysis of naturally occurring networks, statistical inference applied to biological time-series data, and large-scale sequence informatics in computational biology. As part of his work at Columbia, he is a founding member of the university's Institute for Data Sciences and Engineering (IDSE) and Department of Systems Biology.

Wiggins is also active in the broader New York tech community, as co-founder and co-organizer of hackNY—a nonprofit organization that guides and mentors the next generation of hackers and technologists in the New York innovation community.

Wiggins has held appointments as a Courant Instructor at the New York University Courant Institute of Mathematical Sciences and as a Visiting Research Scientist at the Institut Curie (Paris), Hahn-Meitner Institut (Berlin), and the Kavli Institute for Theoretical Physics (Santa Barbara). He holds a PhD in Physics from Princeton University and a BA in Physics from Columbia, minoring as an undergraduate in religion and in mathematics.

Wiggins's diverse accomplishments demonstrate how world-class data science skills wedded to extraordinarily strong values can enable an individual data scien-

tist to make tremendous impacts in very different environments, from startups to centuries-old institutions. This combination of versatility and morality comes through as he describes his belief in a functioning press and his role inside of it, why he values "people, ideas, and things in that order," and why caring and creativity are what he looks for in other people's work. Wiggins's passion for mentoring and advising future scientists and citizens across all of his roles is a leitmotif of his interview.

Gutierrez: Tell me about where you work.

Wiggins: I split my time between Columbia University, where I am an associate professor of applied mathematics, and The New York Times, where I am the chief data scientist. I could talk about each institution for a long time. As background, I have a long love for New York City. I came to New York to go to Columbia as an undergraduate in the 1980s. I think of Columbia University itself as this great experiment to see if you can foster an Ivy League education and a strong scientific and research community within the experiment of New York City, which is full of excitement and distraction and change and, most of all, full of humanity. Columbia University is a very exciting and dynamic place, full of very disruptive students and alumni, myself included, and has been for centuries.

The New York Times is also centuries old. It's a 163-year-old company, and I think it also stands for a set of values that I strongly believe in and is also very strongly associated with New York, which I like very much. When I think of The New York Times, I think of the sentiment expressed by Thomas Jefferson that if you could choose between a functioning democracy and a dysfunctional press, or a functioning press and a dysfunctional democracy, he would rather have the functioning press. You need a functioning press and a functioning journalistic culture to foster and ensure the survival of democracy.

I get the joy of working with three different companies whose missions I strongly value. The third company where I spend my time is a nonprofit that I cofounded, called hackNY,[1] many years ago. I remain very active as the co-organizer. In fact, tonight, we're going to have another hackNY lecture, and I'll have a meeting today with the hackNY general manager to deal with operations. So I really split my time among three companies, all of whose mission I value: The New York Times and the two nonprofits—Columbia University and hackNY.

Gutierrez: How does data science fit into your work?

---

[1] http://hackNY.org

Wiggins: I would say it's an exciting time to be working in data science, both in academia and at The New York Times. Data science is really being birthed as an academic field right now. You can find the intellectual roots of it in a proposal by the computational statistician Bill Cleveland in 2001. Clearly, you can also find roots for data scientists as such in job descriptions, the most celebrated examples being DJ Patil's at LinkedIn and Jeff Hammerbacher's at Facebook. However, in some ways, the intellectual roots go back to writings by the heretical statistician John Tukey in 1962.

There's been something brewing in academia for half a century, a disconnect between statistics as an ever more and more mathematical field, and the practical fact that the world is producing more and more data all the time, and computational power is exponentiating over time. More and more fields are interested in trying to learn from data.

My research over the last decade or more at Columbia has been in what we would now call "data science"—what I used to call "machine learning applied to biology" but now might call "data science in the natural sciences." There the goal was to collaborate with people who have domain expertise—not even necessarily quantitative or mathematical domain expertise—that's been built over decades of engagement with real questions from problems in the workings of biology that are complex but certainly not random. The community grappling with these questions found itself increasingly overwhelmed with data.

So there's an intellectual challenge there that is not exactly the intellectual challenge of machine learning. It's more the intellectual challenge of trying to use machine learning to answer questions from a real-world domain. And that's been exciting to work through in biology for a long time.

It's also exciting to be at The New York Times because The New York Times is one of the larger and more economically stable publishers, while defending democracy and historically setting a very high bar for journalistic integrity. They do that through decades and centuries of very strong vocal self-introspection. They're not afraid to question the principles, choices, or even the leadership within the organization, which I think creates a very healthy intellectual culture.

At the same time, though, although it's economically strong as a publisher, the business model of publishing for the last two centuries or so has completely evaporated just over the last 10 years; over 70 percent of print advertising revenue simply evaporated, most precipitously starting around 2004.[2] So although this building is full of very smart people, it's undergoing a clear sea change in terms of how it will define the future of sustainable journalism.

---

[2]www.aei-ideas.org/2013/08/creative-destruction-newspaper-ad-revenue-has-gone-into-a-precipitous-free-fall-and-its-probably-not-over-yet/

The current leadership, all the way down to the reporters, who are the reason for existence of the company, is very curious about "the digital," broadly construed. And that means: How does journalism look when you divorce it from the medium of communication? Even the word "newspaper" presumes that there's going to be paper involved. And paper remains very important to The New York Times not only in the way things are organized—the way even the daily schedule is organized here— but also conceptually. At the same time, I think there are a lot of very forward-looking people here, both journalists and technologists, who are starting to diversify the way that The New York Times communicates the news.

To do that, you are constantly doing experiments. And if you're doing experiments, you need to measure something. And the way you measure things right now, in 2014, is via the way people engage with their products. So from web logs to every event when somebody interacts with the mobile app, there are copious, copious data available to this company to figure out: What is it that the readers want? What is it that they value? And, of course, that answer could be dynamic. It could be that what readers want in 2014 is very different than what they wanted in 2013 or 2004. So what we're trying to do in the Data Science group is to learn from and make sense of the abundant data that The New York Times gathers.

Gutierrez: When did you realize that you wanted to work with data as a career?

Wiggins: That happened one day at graduate school while having lunch with some other graduate students, mostly physicists working in biology. Another graduate student walked in brandishing the cover of *Science* magazine,[3] which had an image of the genome of *Haemophilus influenzae*. *Haemophilus influenzae* is the first sequenced freely living organism. This is a pathogen that had been identified on the order of 100 years earlier. But to sequence something means that you go from having pictures of it and maybe experiments where you pour something on it and maybe it turns blue, to having a phonebook's worth of information. That information unfortunately is written in a language that we did not choose, just a four-letter alphabet, imagine ACGT ACGT, over and over again. You can just picture a phonebook's worth of that.

And there begins the question, which is both statistical and scientific: How do you make sense of this abundant information? We have this organism. We've studied it for 100 years. We know what it does, and now we're presented with this entirely different way of understanding this organism. In some ways, it's the entire manual for the pathogen, but it's written in a language that we didn't choose. That was a real turning point in biology.

---

[3]www.sciencemag.org/content/269/5223/496.abstract

When I started my PhD work in the early 1990s, I was working on the style of modeling that a physicist does, which is to look for simple problems where simple models can reveal insight. The relationship between physics and biology was growing but limited in character, because really the style of modeling of a physicist is usually about trying to identify a problem that is the key element, the key simplified description, which allows fundamental modeling. Suddenly dropping a phonebook on the table and saying, "Make sense of this," is a completely different way of understanding it. In some ways, it's the opposite of the kind of fundamental modeling that physicists revered. And that is when I started learning about learning.

Fortunately, physicists are also very good at moving into other fields. I had many culture brokers that I could go to in the form of other physicists who had bravely gone into, say, computational neuroscience or other fields where there was already a well-established relationship between the scientific domain and how to make sense of data. In fact, one of the preeminent conferences in machine learning is called NIPS,[4] and the N is for "neuroscience." That was a community which even before genomics was already trying to do what we would now call "data science," which is to use data to answer scientific questions.

By the time I finished my PhD, in the late 1990s, I was really very interested in this growing literature of people asking statistical questions of biology. It's maddening to me not to be able to separate wheat from chaff. When I read these papers, the only way to really separate wheat from chaff is to start writing papers like that yourself and to try to figure out what's doable and what's not doable. Academia is sometimes slow to reveal what is wheat and what is chaff, but eventually it does a very good job. There's a proliferation of papers and, after a couple of years, people realize which things were gold and which things were fool's gold. I think that now you have a very strong tradition of people using machine learning to answer scientific questions.

Gutierrez: What in your career are you most proud of?

Wiggins: I'm actually most proud of the mentoring component of what I do. I think I, and many other people who grow up in the guild system of academia, acquire a strong appreciation for the benefits of the way we've all benefited from good mentoring. Also, I know what it's like both to be on the receiving end and the giving end of really bad and shallow mentoring. I think the things I'm most proud of are the mentoring aspects of everything I've done.

---

[4]http://nips.cc

Here at the data science team at The New York Times, I'm building a group, and I assure you that I spend as much time thinking hard about the place and people as I do on things and ideas. Similarly, hackNY is all about mentoring. The whole point of hackNY is to create a network of very talented young people who believe in themselves and believe in each other and bring out the best in themselves and bring out the best in each other. And certainly at Columbia, the reason I'm still in academia is that I really value the teaching and mentoring and the quest to better yourself and better your community that you get from an in-person brick-and-mortar university as opposed to a MOOC.

Gutierrez: What does a typical day at work look like for you?

Wiggins: There are very few typical days right now, though I look forward to having one in the future. I try to make my days at The New York Times typical because this is a company. What I mean by that is that it is a place of interdependent people, and so people rely on you. So I try throughout the day to make sure I meet with everyone in my group in the morning, meet with everyone in my group in the afternoon, and meet with stakeholders who have either data issues or who I think have data issues but don't know it yet. Really, at this point, I would say that at none of my three jobs is there such a thing as a "typical day."

Gutierrez: Where do you get ideas for things to study or analyze?

Wiggins: Over the past 20 years, I would say the main driver of my ideas has been seeing people doing it "wrong". That is, I see people I respect working on problems that I think are important, and I think they're not answering those questions the right way. This is particularly true in my early career in machine learning applied to biology, where I was looking at papers written by statistical physicists who I respected greatly, but I didn't think that they were using, or let's say stealing, the appropriate tools for answering the questions they had.

And to me, in the same way that Einstein stole Riemannian geometry from Riemann and showed that it was the right tool for differential geometry, there are many problems of interest to theoretical physicists where the right tools are coming from applied computational statistics, and so they should use those tools. So a lot of my ideas come from paying attention to communities that I value, and not being able to brush it off when I see people whom I respect who I think are not answering a question the right way.

Gutierrez: What specific tools or techniques do you use?

Wiggins: My group here at The New York Times uses only open source statistical software, so everything is either in R or Python, leaning heavily on scikit-learn and occasionally IPython notebooks. We rely heavily on Git as version control. I mostly tend to favor methods of supervised learning rather than unsupervised learning, because usually when I do an act of clustering, which is generically what one does as unsupervised learning, I never know if I've done it the best. I always worry that there is some other clustering that I could do, and I won't even know which of the two clusterings is the better.

But with supervised learning, I usually can start by asking: How predictive is this model that we've built? And once I understand how predictive it is, then I can start taking it apart and ask: How does it work? What does it learn? What are the features that it rendered important?

That's completely true both at The New York Times and at Columbia. One of the driving themes of my work has been taking domain questions and asking: How can I reframe this as a prediction task?

Gutierrez: How do you think about whether you're solving the right problem?

Wiggins: The key is usually to just keep asking, "So what?" You've predicted something to this accuracy? So what? Okay, well, these features turned out to be important. So what? Well, this feature may be related to something that you could make a change to in your product decisions or your marketing decisions. So what?

Well, then I could sit down with this person and we could suggest a different marketing mechanism. Now you've started to refine and think all the way through the value chain to the point at which it's going to become an insight or a paper or product—some sort of way that it's going to move the world.

I think that's also really important for working with junior people, because I want junior people always to be able to keep their eyes on the prize, and you can't do that if you don't have the prize in mind. I can remember when I was much younger—a postdoc—I went to see a great mathematician and I talked to him for maybe 20 minutes about a calculation I was working on, as well as all of the techniques that I was learning. He sat silently for about 10 minutes and then he finally said, "What are you trying to calculate? What is the goal of this mathematical manipulation you're doing?" He was right, meaning you need to be able to think through toward "So what?" If you could calculate this, if you could compute this correlation function, or whatever else it is that you're trying to compute, how would that benefit anything? And that's a thought experiment or a chain of thinking that you can do in the shower or in the subway. It's not something that even requires you to boot up a computer. It's just something that you need to think through clearly before you ever pick up a pencil or touch a keyboard.

John Archibald Wheeler, the theoretical physicist, said you should never do a calculation until you know the answer. That's an important way of thinking about doing mathematics. Should I bother doing this mathematics? Well, I think I know what the answer's going to be. Let me go see if I can show that answer. If you're actually trying to do something in engineering, and you're trying to apply something, then it's worse than that, because you shouldn't bother doing a computation or collecting a data set or even pencil-and-paper work until you have some sense for "So what?" If you show that this correlation function scales to $T^{7/8}$, so what? If you show that you can predict something to 80-percent accuracy on held-out data, so what? You need to think through how it will impact something that you value.

Gutierrez: What's an interesting project that you've worked on?

Wiggins: One example comes from 2001 when I was talking to a mathematician whom I respect very much about what he saw as the future of our field, the intersection of statistics and biology, and he said, "Networks. It's all going to be networks." I said, "What are you talking about? Dynamical systems on networks?" He said, "Sure, that and statistics of networks. Everything on networks."

At the time, the phrase "statistics of networks" didn't even parse for me. I couldn't even understand what he was saying. He was right. I saw him again at a conference on networks two years later.[5] Many people that I really respected spoke at that conference about their theories of the way real-world networks came to evolve.

I remember stepping off the street corner one day while talking to another biophysicist, somebody who was coming from the same intellectual tradition that I had with my PhD. And I was saying, "People look at real-world networks, and they plot this one statistical attribute, and then they make up different models—all of which can reproduce this one statistical attribute." And they're basically just looking at a handful of predefined statistics and saying, 'Well, I can reproduce that statistical behavior.' That attribute is over-universal. There are too many theories and therefore too many theorists saying that they could make models that looked like real-world graphs. You know what we should do? We should totally flip this problem on its head and build a machine learning algorithm that, presented with a new network, can tell which of a few competing theorists wins. And if that works, then we're allowed to look at a real-world network and see which theorist has the best model for some network that they're all claiming to describe."

That notion of an algorithm for model testing led to a series of papers that I think were genuinely orthogonal to what anybody else was doing. And I think it was a good example of seeing people whom I respect and think are very smart people but who were not using the right tool for the right job, and then trying to reframe a question being asked by a community of smart people as a prediction problem. The great thing about predictions is that you can be wrong, which I think is hugely important. I can't sleep at night if I'm involved in a scientific field where you can't be wrong. And that's the great thing about predictions: It could turn out that you can build a predictive model that actually is just complete crap at making predictions, and you've learned something.

---

[5] http://cnls.lanl.gov/networks

Gutierrez: How have you been able to join that point of view with working at a newspaper?

Wiggins: It's actually completely the same. Here we have things that we're interested in, such as what sorts of behaviors engender a loyal relationship with our subscribers and what sorts of behaviors do our subscribers' evidence that tends to indicate they're likely to leave us and are not having a fulfilling relationship with *The New York Times*. The thing about subscribers online is that there are really an unbounded number of attributes you can attempt to compute. And by "compute," I really mean that in the big data sense. You have abundant logs of interactions on the web or with products.

Reducing those big data to a small set of features is a very creative and domain-specific act of computational social science. You have to think through what it is that we think might be a relevant behavior. What are the behaviors that count? And then what are the data we have? What are the things that can be counted? And, of course, it's always worth remembering Einstein's advice that not everything that can be counted counts, and not everything that counts can be counted. So you have to think very creatively about what's technically possible and what's important in terms of the domain to reduce the big data in the form of logs of events to something as small as a data table, where you can start thinking of it as a machine learning problem.

There's a column I wish to predict: Who's going to stick around and who's going to leave us? There are many, many attributes: all of the things that computational social science, my own creativity, and very careful conversations with experts in the community tell me might be of interest. And then I try to ask: Can I really predict the thing that I value from the things that the experts believe to be sacred? And sometimes those attributes could be a hundred things and sometimes that could be hundreds of thousands of things, like every possible sequence element you could generate from seven letters in a four-letter alphabet. Those are the particular things that you could look at.

That is very much the same here as it is in biology. You wish to build models that are both predictive and interpretable. What I tell my students at Columbia is that as applied mathematicians, what we do is we use mathematics as a tool for thinking clearly about the world. We do that through models. The two attributes of a model that make a model good are that it is predictive and interpretable, and different styles of modeling strike different balances between predictive power and interpretability.

A few Decembers ago, I had a coffee with a deep learning expert, and we were talking about interpretability, and he said, "I am anti-interpretability. I think it's a distraction. If you're really interested in predictive power, then just focus on predictive power." I understand this point of view. However, if you're interested in helping a biologist, or helping a businessperson, or helping a product person, or helping a journalist, then they're not going to be so interested in .08

error on held-out data. They're going to be interested in the insights and identification of the interesting covariates, or the interesting interactions among
the covariates revealed to you.

I come from a tradition in physics that has a long relationship with predictive
interpretability. We strive to build models that are as simple as possible but
not simpler, and the real breakthroughs, the real news-generating events, in the
history of physics have been when people made predictions that were borne
out by experiment. Those were times that people felt they really understood
a problem.

Gutierrez: Whose work is currently inspiring you?

Wiggins: It's always my students. For example, I have a former student, Jake
Hofman, who's working with Duncan Watts at Microsoft Research. Jake was
really one of the first people to point out to me how social science was birth-
ing this new field of computational social science, where social science was
being done at scale. So that's an example of a student who has introduced me
to all these new things.

I would also say that all of the kids who go through hackNY are constantly
introducing me to things that I've never heard of and explaining things to me
from the world that I just don't understand. We had a hackNY reunion two
Friday nights ago in San Francisco. I was out there to give a talk. We organized
a reunion, and the Yo app had just launched. So a lot of the evening was me
asking the kids to explain Yo to me, which meant explaining the security flaws
in their API and not just how the app worked. So that's the benefit of working
with great students. Students are constantly telling you the future of technology, data science, and media amongst other things, if you just listen to them.
Former students and postdocs of mine have gone on to work at BuzzFeed,
betaworks, Bitly, and all these other companies that are at the intersection of
data and media.

I have also benefited greatly from really good colleagues whom I find inspiring. The way I ended up here at The New York Times, for example, was that,
when I finally took a sabbatical, I asked all my faculty colleagues what they did
with their sabbaticals, because I had never taken one. My friend and colleague
Mark Hansen did the "Moveable Type" lobby art here in the New York Times
Building. So if you go look at the art in the lobby, Mark Hansen wrote the
Python to make the lobby art "go", and he did that in 2007 when they moved
into this building. So he knew many people at The New York Times, and he
introduced me to a lot of people here and was somebody who explained to
me—though he didn't use these words—that The New York Times is now in
a similar state to the state that biology was in 1998. That is, that it's a place
where they have abundant data, and it's still up for grabs what the right way is
to use machine learning to make sense of those data.

Mark Hansen is a good example of somebody who's done great work. In fact, although he won't admit it, he was using the phrase "data science" throughout the last 12 years. He's been writing for years about what he often called "the science of data." He's been somebody who's been really thinking about data science as a field much longer than most people. Actually, he worked with Bill Cleveland at AT&T. Bill Cleveland, in turn, had worked with Tukey, so there is a nice intellectual tradition there. There's a reason why data science resonates so much with academics. I feel it's because there's been an academic foundation there in the applied computational statistics community for half a century.

David Madigan, who's the former chair of stats at Columbia, is also inspiring. He is somebody who's done a great job showing the real impact of statistics—good and bad—on people's lives. All the people I respect are people who share my value for community. Mark Hansen is trying to build a community of data journalists at the journalism school. His PhD was in statistics, but now he's a professor of journalism who is trying to build a community of data journalists. David Madigan similarly—he was the chair of statistics and now he's the Executive Vice President for Arts and Sciences at Columbia.

The people I find the most inspiring are the people who think about things in this order: people—in terms of how you build a strong community; ideas—which is how you unite people in that community; and things that you use to build the community that embodies those ideas.

But mostly, I would say my students—broadly construed, at Columbia and at hackNY—who inspire me.

Gutierrez: What was it that convinced you to join The New York Times and try to make a difference when you did your sabbatical?

Wiggins: It was clear to me by the end of my first day here that we should build a predictive model for looking at subscriber behavior. I spent some time interviewing or meeting everyone around here in the company who I felt was likeminded. I found some good collaborators, worked on this project, and it was clear from the way people reacted to it that no one had done that before. I did that without a real clear sense of whether or not I was reinventing a common wheel.

I got the impression from the way people reacted that people had been sort of too busy feeding the goat, meaning doing their daily obligations of running a company, even worse in journalism. In journalism, you have constant deadlines, but even on the business side, there's a business to run. Nobody has time to do a two-month research project. I think that's what convinced me that there really was a lot to be learned from the data that this company is gathering and curating.

Gutierrez: What do you look for in other people's work?

Wiggins: Creativity and caring. You have to really like something to be willing to think about it hard for a long time. Also, some level of skepticism. So that's one thing I like about PhD students—five years is enough time for you to have a discovery, and then for you to realize all of the things that you did wrong along the way. It's great for you intellectually to go back and forth from thinking "cold fusion" to realizing, "Oh, I actually screwed this up entirely," and thus making a series of mistakes and fixing them. I do think that the process of going through a PhD is useful for giving you that skepticism about what looks like a sure thing, particularly in research. I think that's useful because, otherwise, you could easily too quickly go down a wrong path—just because your first encounter with the path looked so promising.

And although it's a boring answer, the truth is you need to actually have technical depth. Data science is not yet a field, so there are no credentials in it yet. It's very easy to get a Wikipedia-level understanding of, say, machine learning. For actually doing it, though, you really need to know what the right tool is for the right job, and you need to have a good understanding of all the limitations of each tool. There's no shortcut for that sort of experience. You have to make many mistakes. You have to find yourself shoehorning a classification problem into a clustering problem, or a clustering problem into a hypothesis-testing problem.

Once you find yourself trying something out, confident that it's the right thing, then finally realizing you were totally dead wrong, and experiencing that many times over—that's really a level of experience that unfortunately there's not a shortcut for. You just have to do it and keep making mistakes at it, which is another thing I like about people who have been working in the field for several years. It takes a long time to become an expert in something. It takes years of mistakes. This has been true for centuries. There's a quote from the famous physicist Niels Bohr, who posits that the way you become an expert in a field is to make every mistake possible in that field.

Gutierrez: What's been the biggest thing you've changed your mind about?

Wiggins: That's a tough choice. There are so many things that I've changed my mind about. I think probably the biggest thing I've changed my mind about is the phrase that you can't teach an old dog new tricks. I think if you really care about something, you'll find a way. You'll find a way to learn new tricks if you really want to.

The other thing that I've changed my mind about is that I grew up, like most academics, with the sense that scientists somehow functioned with some orthogonal value system that was different than the world. I think one thing that I've changed my mind about in this area—but this is over like a 20-year period since I was not yet a PhD, is that scientists are human beings too, whether they know it or not. And that science is done by scientists, and scientists are human beings. And so all the good and the bad about humans, and how they make their choices, and what they value, carries over to the scientific and academic enterprise. It's not different. It's a lovely guild and it's functioned fantastically for centuries, and I hope it continues to function for a long time because I think it has been very good for the species, but we shouldn't believe that scientists are somehow not subject to the same joys and distractions as every other human being.

So that's part of what I learned—that science is somehow not a qualitatively different enterprise than, let's say, technology or any other difficult human endeavor. These are very difficult human endeavors, and they take planning, and attention, and care, and execution, and they take a community of people to support it. Everything I just said is completely true of academic science, writing papers, winning grants, training students, teaching students, as well as forming a new company, doing research, or using technology in a big corporation that's already been established. All of those things are difficult and require a community of people to make it happen. As they say: "people, ideas, and then things, in that order." That's true in any science and that's also true in the real world.

Gutierrez: What does the future of data science look like?

Wiggins: I don't see any reason for data science not to follow the same course as many other fields, which is that it finds a home in academia, which means that there becomes a credentialing function, particularly around professional subjects. You'll get master's degrees and you'll get PhDs. The field will take on meaning, but it will also take on specialization. You see this already with people using the phrases "data engineering" and "data science" as separate things. My group here at The New York Times is the Data Science group, which is part of the Data Science and Engineering larger group. People are starting to appreciate how a data science team involves data science, data engineering, data visualization, and data architecture.

Data Product is not sort of a thing yet, but certainly, if you look at how, say, data science happened at LinkedIn—data science reported up through the product hierarchy. At other companies, data science reports through business; or it reports through engineering. Right now I'm located within in the engineering function of The New York Times, separate from the product, separate from marketing, and separate from advertising. Different companies are locating data science in different arms.

So I think there'll be credentialing. I think there will be specialization. New fields are born—I wouldn't say all the time, because by real-world standards, nothing ever happens in academia—but there are new departments born at universities every few years. It happens, and the way that it happens is part of the creation of new fields. I'm old enough that I had the benefit of watching, say, systems biology be born as a field, synthetic biology be born as a field, and even nanoscience be born as a field in the time that I've been a practicing academic. My first research project in the 1980s was in chaos, which at that time was being born as a new field. There's a famous book on this by James Gleick, at that time writing for The New York Times, called *Chaos: Making a New Science*.[6] It's not that new fields aren't created in academia. It's just that it's so damn slow compared to the pace of the real world, which I think is really for the best. There are young people's futures at stake, so I think it's actually not so bad.

So I think the future of data science is for it to become part of academia, which means a vigorous, contentious dialog among different universities about what is really data science. You're already starting to see work in this direction. For instance, at Columbia, a colleague of mine, named Matt Jones, who's an historian, is writing a book about the history of machine learning and data science. So you're already starting to see people appreciate that data science wasn't actually created from a vacuum in 2008. Intellectually, the things that we call data science had already been sort of realized—that is, that there was a gap between statistics and machine learning, that there was sort of something else there. So I think there will be a greater appreciation for history.

Part of what happens when a field becomes an academic field is that three main things occur—an academic canon is set, a credentialing process is initiated, and historical study provides the context of the field. An academic canon is the set of classes that we believe are the core intellectual elements of the field. The credentialing process, which is another separate function from academia, which can be unbundled, is initiated so you can get master's and PhD degrees. Lastly, historical study occurs to appreciate the context: Where did these ideas come from?

As the names and phrases people use become more meaningful, then you get the possibility of specialization, because what we have now is that when people say "data science" they could mean many things. They could mean data visualization, data engineering, data science, machine learning, or something else. As the phrases themselves become used more carefully, then I think you'll get to see much more productive specialization of teams. You can't have a football team where everybody says, "I'm the placekicker." Somebody needs to be the placekicker, somebody needs to be the holder, and somebody needs to be the linebacker. And as people start to specialize, then you can pass. You can have meaningful collaborations with people because people know their roles and know what "mission accomplished" looks like. Right now, I think it's still up for grabs what a win in data science really looks like.

---

[6]James Gleick, *Chaos: Making a New Science* (Viking, 1987).

Again, I come from a very old field. Physics is a field where the undergraduate curriculum was basically canonized by 1926. Years ago I picked up a book at the Book Scientific bookstore called *Compendium of Theoretical Physics*.[7] It had four chapters: classical mechanics, statistical mechanics, quantum mechanics, and E&M—electricity and magnetism. Those are the four pillars on which all of physics stands. And physics has a pretty rich intellectual tradition, with some strong clear wins behind it, but it's really built on those four pillars. You can see that it has a strong canon. Most fields don't enjoy that. I think you really need to have a well of a mature field for you to be able to say, "Here are the four classes that you really need to take as an undergraduate."

Gutierrez: What does the academic canon at the Institute for Data Sciences and Engineering at Columbia cover?

Wiggins: I'm on the education committee for the Data Science Institute at Columbia, so we've created a canon of four classes: Probability and Statistics, Algorithms for Data Science, Machine Learning for Data Science, and EDAV, which is short for Exploratory Data Analysis and Visualization. The three letters, *EDA*, are taken directly from John Tukey.

Tukey had a book in the 1970s called *Exploratory Data Analysis*—which was basically a description of what Tukey did without a computer, probably on the train between Princeton and Bell Labs, whenever somebody gave him a new data set.[8] The book is basically a description of all the ways he would plot out the data, histograms, Tukey boxplots, Tukey stem-and-flower plots—all these things that he would do with data. If you read the book now, it looks like, "man, this guy was kooky. He should have just opened up R. He should have just opened up matplotlib."

Around the same time, he was co-teaching a class at Princeton with Edward Tufte. If you pick up the book *Visual Display of Quantitative Information,* look at whom it's dedicated to.[9] It's dedicated to Tukey. Again, there's a very old academic tradition on which many of the data science ideas lie. People have been thinking in academia for a long time about what the visual display of quantitative information is. How do we meaningfully "do" data visualization? What do we do when someone hands us data and we just have no distribution? The world doesn't hand you distributions. It hands you observations.

---

[7]www.wachter-hoeber.com/Books.html?bid=002
[8]John W. Tukey, *Exploratory Data Analysis* (Pearson, 1977).
[9]Edward R. Tufte, *The Visual Display of Quantitative Information* (2nd ed.) (Graphics Press, 1983).

Much of what we do in physics or mathematical statistics organizes our worldview around what the appropriate model is. Is this the time when I should treat it as statistical mechanics and, if so, what terms do I put in my Hamiltonian? Is it the case that this is a quantum mechanical problem? If so, what terms do I put in my Hamiltonian? Is this a classical mechanics problem? If so, what terms should I put in my Hamiltonian?

The world's like that. The world doesn't hand you models. It doesn't come to you with a model and say, "Diagonalize this Hamiltonian."[10] It comes to you with observations and a question usually being asked by the person who gathered those data. So that's the tradition that I thought was important enough that we make one of our four pillars of data science at Columbia. We want students to think about how we explore data before we decide that we're going to model it using some particular distribution or some particular graphical model. How do you explore a data set that you've been handed?

Gutierrez: What are the most exciting things in data science for you?

Wiggins: The things that are most exciting to me are not new things. The most exciting thing to me is realizing that something everybody thinks is new is actually really damn old. That's why I like Tukey so much. There's a lot of excitement about this new thing called "data science." I think it's really fun to go see really old papers in statistics that are even older than Tukey. For instance, Sewall Wright was using graphical models for genetics in the 1920s.[11] The things that really capture my excitement are not the newfangled things. It's particularly around the ideas, not so much things, because, again—people, ideas, and things in that order. The things change. It's fun when we think we have a new idea, but usually we then realize the idea is actually very old. When you have an understanding of that, it's a really frickin good idea.

Stochastic optimization and stochastic gradient descent, for example, has been a huge, huge hit in the last five years, but they descend from a paper written by Robbins and Monro in 1951.[12] It is a good idea, but the fact that I think it's a good idea means somebody really thought through it very carefully with pencil on paper a long time back. Trying to understand the world through data and your computer is a very good idea. That's why Tukey was writing about it in 1962 when he was ordering everybody to reorient statistics as a professional discipline and a funding line for the NSF organized around computation and data and data analysis. He wrote an article in 1962 called "The Future of Data Analysis."[13] And he wasn't the last, right?

---

[10]http://vserver1.cscs.lsa.umich.edu/~crshalizi/reviews/fragile-objects/
[11]Wright, Sewall. "Correlation and causation." *Journal of Agricultural Research* 20.7 (1921), 557-585.
[12]Herbert Robbins and Sutton Monro, "A Stochastic Approximation Method": *Ann. Math. Statist.*, Volume 22, Number 3 (1951), 400-407.
[13]John W. Tukey, "The Future of Data Analysis": *Ann. Math. Statist.*, Volume 33, Number 1 (1962), 1-67.

Leo Breiman all throughout the 1990s was writing to his community of statisticians, "Let us get with data, statistics community!" He was writing papers in the late 1990s telling all his colleagues to start going to NIPS.[14] It was like he had gone into the wilderness and come back and said to everybody at Berkeley, which was one of the first mathematical statistics departments, "You guys need to wake up because it's on fire. You guys are proving theorems. It's on fire out there. Wake up!"

So I think there was a strong tradition of people understanding how powerful and how different it was to understand the world through data. The "primacy of the data" was a phrase that one of the mathematical statisticians at Berkeley used a long time back for Tukey's emphasis.[15] This strong tradition carried on through this sort of heretical strain of thought from John Tukey through Leo Breiman to Bill Cleveland in 2001. All of them saw themselves as orthodox statisticians, though they were people who were sufficiently heretical. It's just that as statistics kept doubling down on mathematics every five years because of their origin from math that made statistics a bona fide field, you found this strain of heretics who were saying, "No, you should really try to get with data." That's what I think is most exciting in terms of people, ideas, and things—don't be distracted by today's things but find the people and their ideas that are actually much older.

[14]https://www.stat.berkeley.edu/~breiman/wald2002-1.pdf
[15]Erich L. Lehmann, *Reminiscences of a Statistician: The Company I Kept* (NY: Springer Science+Business Media, 2008), 198.

Caitlin Smallwood

Netflix

Caitlin Smallwood is the VP of Science and Algorithms at Netflix, a provider of on-demand Internet media, streaming both TV programs and movies. Netflix has over fifty million subscribers in more than forty countries and is expanding rapidly, with members able to watch as much as they want, anytime, anywhere, on nearly any Internet-connected screen without commercials or commitments. Netflix stands at the forefront of Internet television and has further bolstered its leadership position by starting to develop and produce its own programming. As the amount of content available and consumed grows rapidly (more than one billion hours of TV shows and movies are currently streamed per month), the predictive engines and data infrastructure required to effectively personalize recommendations and ensure instantaneous content delivery become ever more complex.

To tackle these challenges at Netflix, Smallwood draws on her broad technical expertise in experimentation, analytics, and recommendations acquired over the course of more than twenty years of experience in Internet data products, including spells at Intuit, Yahoo!, and Hyperion Solutions. She has also held various prior roles at Netflix, most recently as Director of Consumer Science and Analytics, and worked as an analytical consultant for PwC and SRA International. She holds an MS in Operations Research from Stanford University and a BS in Mathematics from the College of William and Mary.

Smallwood's career has spanned the evolution of big data, analytics, experimentation, and recommendations from the infancy of the Internet to the ever-connected, data-rich world we live in today. Her remarkable perspective comes through as she shares her thoughts on analytics pre-Internet, her excitement at first encountering massive data at Yahoo! and her first data set at Netflix, and her views on the importance of culture and team in data-centric organizations. Smallwood's interview exudes wisdom, experience, and leadership.

Sebastian Gutierrez: What is it like to work at Netflix?

Caitlin Smallwood: It's been a riveting and exciting experience on many levels. Though I feel very fortunate to have worked in a lot of great places, I've been amazingly happy at Netflix.

Mission-wise, the company is transforming the television business. The dawn of Internet TV is fascinating and has been a great thing to participate in. People-wise, I love working in a culture like ours. We have this public culture document called "Netflix Culture: Freedom & Responsibility" that's posted on our jobs site, which defines our core values. What's great is that the culture is actually exactly what you read in the document. It really is. You can see it from the moment you interview. It's powerful to be in a place that really does have so much freedom and has amazing talent that you get to participate in. Personally and professionally it's great to reap the benefits of working in this culture.

Data-wise, being in the world of data and working at a company like Netflix is just outstanding because throughout the company, from the executives down, data is prioritized so heavily. This perspective makes it a real treat to work with any aspect of data, whether it is data engineering, analytics, or the more hardcore mathematical modeling. It's great because you don't fight battles about things like, "Can we capture this kind of data?" or, "Will somebody put effort into structuring this data that we really want to do great things with?" It's not hard to convince people that those projects are important. And then there's so much debate and public sharing of data and results that it makes it really exciting.

Gutierrez: Where does your team fit into the organization?

Smallwood: I report up through our product organization and my manager is Neil Hunt, the chief product officer. Though I'm part of the product org, my team works more like a centralized team that supports the whole company's modeling needs, be it predictive modeling or other modeling. The people in the product org are really fantastic, as each person, no matter what his or her level, is really strong at what they do.

Companywide, the level of innovation is inspiring to be around all the time. For example, in the product org we frequently have strategy meetings with great agendas lined up that Neil Hunt, all the VPs, and people at all other levels attend. At these meetings, somebody will present a new idea for the product, backed up by data or analysis, or results from an A/B test, and everybody will debate that idea in public, which is a really great way to learn and get more and more ideas flowing. So that piece of it is just really energizing.

Gutierrez: Who drives the data culture and was it present when you joined?

Smallwood: The data culture was present when I came in and it continues to get stronger. Regarding who drives it, I would say there's no one person or organization. It's really a collective effort. You see how much it energizes everyone to participate in that, so it just feeds off of itself. That said, our CEO, Reed Hastings, is a big reason we have such a strong data culture. He believes in causality over correlation and is really strong on making decisions with as much data as possible rather than just judgment. Good judgment is important, of course, but trying to back things up with data and analytics has always been a belief of his, and it's just spread from there.

Gutierrez: How open is the company internally with its own data?

Smallwood: Internally, we're very open. In fact, people will be publicly called out if they are not being open with their data, as it's not seen as a good thing to hide information that you know. It's encouraged that you think about the company strategies in different areas of the business, and if you know something, or if you have an insight or a piece of data that would help some other organization, you're expected to share that in a way that is useful to the company. Of course, you also want to do it without drowning people in too much information, right? There's judgment there, as well as around how much to share.

Gutierrez: What have you been working on this year?

Smallwood: My first thought is that a Netflix year is like a dog year. It's such a productive company because of the low level of process. This means that we get a tremendous amount done in a year, so I'll just rattle off a list of some of the things that we've been working on in the last year. We do a lot of work on the personalization algorithms and recommender systems and algorithms that contribute to that space, so this is an area of ongoing continuous optimization. That's always a big investment, as we're always trying to improve that.

We've also been working on search, as it's a really interesting area within the Netflix product. We've been trying to make the search experience really nimble for people. For example, with the search autocomplete—when you type the letter A, we'd like to instantly know what movie or TV show you're searching for, based on your history and based on the population usage of Netflix.

We've developed a big culture of experimentation on the product side, and now we're trying to bring that same level of rigor to other parts of the business, like our marketing organization and other parts of Netflix. This has been interesting because other parts of the organization have a little bit of a different mindset and approach toward decision-making and exploring ideas. So that's been a big area for us.

A key internal thing we've really been focused on is little, nimble user interfaces for people to explore the output of models so that it's not a thing that's hard to access. For instance, what would our ranking algorithm show you of your ranking of all the titles in Netflix? Can we show you that as an internal tool for us to help you? This type of tool helps innovation be faster, because then everybody at Netflix can access model outputs to study and understand them. This then feeds into their thinking about what they might be working on. So we've been trying to really expose more of the outputs of these complex models in ways that more people can leverage.

Gutierrez: What does it mean to have a low level of processes, and how does it allow you to accomplish these great products?

Smallwood: It means that we really don't have formalities that cause you to jump through hoops to get things done. I don't have to get approval to work on things, nor do I have to go through some approval chain, let's say, to get a piece of software. It's just that we're expected to use good judgment and not waste the company's money. Every little thing like that adds up. It's everywhere. It's in purchasing, it's in travel, it's in everything.

This low level of process goes for projects as well. It's not like we plan out a project and formally acquire a resource from the different teams that would have to contribute to get that project done. In the strategy meetings, it's clear if something's a great idea, and so somebody will take the lead on that initiative and go to the different teams that need to be involved. We trust one another's judgment, and therefore people figure out how and when to get it done. This makes things very nimble.

Gutierrez: What makes Netflix move so fast that it feels like you fit many years into one year?

Smallwood: For us, the speed is driven by our passion for our business and our excitement about the positives of changing the TV viewing experience and letting it be much more in the hands of the customer. Helping the customer figure out what they want to watch, how much of it they're going to watch, and having that be a seamless, easy experience is something we're really excited about. We can't help ourselves.

We hire very motivated people. You just can't really move slowly when you've got a whole company full of super-motivated people excited about what they're doing. It's just not in your DNA. Of course, as competitors enter the market, there's also a legitimate business need of moving fast if we really want to keep our awesome business thriving.

Gutierrez: How would you describe your work to a data scientist?

Smallwood: I would say we're a team that does all kinds of statistical modeling. We really focus and output three things as a team. We work on predictive models using all of the techniques that people in this field would be familiar with—regression techniques, clustering techniques, matrix factorization, support vector machines, et cetera, both supervised and unsupervised techniques. A second thing is algorithms, which I would say are obviously closely related to models, except that they're embedded in some sort of ongoing process, like our product. And then the third thing is experimentation and all the scientific methodology behind that, which we leverage, as well as all the analytics that go with each experiment that we run.

Gutierrez: How would you describe your work to a non–data scientist?

Smallwood: I would say that we collect the data on all of our customers about how and what they're watching. Then we hunt for patterns in the data, which we can then leverage to recommend things to them that they might want to watch and essentially improve the service for everyone. So we leverage the information across the whole population to really make each of our customers happier.

Gutierrez: How did you come to your current position at Netflix?

Smallwood: My first job was as a programmer with a consulting company. I wasn't even really doing anything intense algorithmically. Through this work, I happened upon an optimization problem and that's what got me interested in operations research or OR. The interest was high enough that I went back to graduate school to study OR. After grad school, I went to work for a smaller company focused entirely on building OR kinds of models for different industries and companies. The company was later purchased by PwC. And that was great, but it was right around the time when the Internet boom started, and so that changed everything in terms of how opportunities just exploded.

I feel like I just kind of lucked into this career that happened to coincide with the Internet. Suddenly it was like, hey, there's all this data that wasn't available before, whole new opportunities of types of things you could build models for, and whole new problems that need to be solved. Things you used to have to "model" did not need models anymore, because there was so much data. All you needed to do was figure out the median of some particular dimension value. So that changed the whole world of opportunities.

As this boom happened, I became really interested in working on a product. This led me into the space of creating analytic products, like personalization engines or other components of products. That was a pivotal point in my transition from operations research into algorithms and data. Really being able to connect data to people and actions that people might take brought me a new sort of thrill that I hadn't had before. To me, discovering things about people through their data was a really a cool thing. One thing led to another and I worked on more and more analytic products and components.

Eventually Netflix called me. The very first time someone called me from Netflix, I thought, "Oh my gosh. I'll bet their data is amazing! I would love to work on Netflix data." So I didn't have to think too hard about joining Netflix.

Gutierrez: What was the specific aha! moment where you thought personalization models made sense?

Smallwood: I think it was really during my time at Yahoo!, where for the first time I had massive data at my fingertips. It's just so exciting to see how much variety there is in the world. When you start looking at user-level profiles of information—of pretty much any kind of user-generated data—and you're aggregating at the user level to try to understand the head and the tail and the incredible diversity within the human population, it's very obvious how different people are. That to me is fascinating. How can you build things that can satisfy the whole population? That's an exciting problem.

Gutierrez: When you came to Netflix, what was the first data set you worked with?

Smallwood: The first data I worked with, about four years ago, was our viewing data, which was our largest data. At that point in time, my role was slightly different than it is now. It was tilted slightly more toward the data engineering side than it is now. The project involved an overhaul of the viewing data and the data engineering behind it. At that point, even though the data was much smaller than it is now, we could see the trajectory that we were on and we knew we needed to redesign how that data was represented so that it could both scale but still be granular enough for the things we wanted to study then and in the future.

The project involved understanding all the data we were collecting at the log level and what it included. This data set includes every segment of every stream at every bit rate. So while you are watching a movie, the bit rate is flapping around and we're serving you many, many streams that come together to form one view that you see as the customer. So, it's just a tremendous amount of data. We also collect every action you take—like pausing, rewinding, and at what bit rate you are watching. This data set also includes your bandwidth changes, network congestion, rebuffer events, or whatever else may happen while you are trying to watch something on the service. As you can imagine, this volume and detail makes things both daunting and fun.

Gutierrez: Internet entertainment as an industry is a little bit hard to nail down. What are the main types of problems the industry is tackling?

Smallwood: You're right that it's hard to exactly place our industry—even internally we joke about it. If you're down in our Beverly Hills office, you feel like you're working for an entertainment company. But if you're up here in our Los Gatos headquarters, you feel like you're working for an Internet technology company. We really are both.

I would say that the key problems people are trying to tackle are what do consumers want to watch and how do they want to watch? How people watch is a big problem now that there are so many more choices of things to watch. There's so much flexibility in what Netflix—and now others—are starting to offer that it's really important that we figure out how to set things up in a way that's best for consumers. For instance, there have been various public articles about Netflix and how people want to watch Internet entertainment. These articles have focused on binge viewing, whether that is a good thing, and whether people like that or not.

Anecdotally, I've had a few people tell me that they cancel Netflix for a couple of months at a time because they feel they're spending their weekends on the couch. So they exercise self-discipline and cancel Netflix for a few months, intending to come back later. You can see this is a tough area, because offering more choices and flexibility in how and when to watch leads some people to cancel their accounts for a few months.

The other problem is determining what people want to watch. We've observed in broadcast television the whole transition to reality TV. And, just like everything else on the Internet, content from all over the world is becoming more accessible, and so now you have more choices and flexibility around what you can watch. With all of these new types of things people can watch, it becomes harder for us to be able to provide what they really want down the road. At an industry and company level, we have to figure out where we should be focusing to be able to provide what's best for our customers and their watching choices.

Broadcast TV and other players are also thinking about these same kinds of things. Everyone is really focusing in on what kind of content should be developed. For instance, people consume much more serialized programming than they do movies. Does that mean that they're happier with serialized television than movies? Or do they prefer a mixture? So there are all sorts of nebulous things like that that we really want to understand.

Gutierrez: Who's driving the thought process behind understanding what people want to watch?

Smallwood: I actually think that in this particular issue, Netflix is doing more than others. This is partially because we have a lot more data than a lot of other companies have. Most of our competitors have the data just from their area, so it tends to be data on a smaller set of content than what we have. The combination of having so many viewers and such a variety of content from many different studios gives us a much richer data set in order to study these things. That said, of course, Amazon has tremendous creativity and innovation, and they also have a lot of data from similar kinds of digestion mechanisms, so they're doing some really great innovation in this space as well. Hulu is an interesting player because they've got their own Internet data, as well as the channel, and so I think they have some more opportunities there.

Again, it's hard to pinpoint how you would characterize our industry. I would say even companies like Facebook are interesting in this space, because we're all competing for someone's entertainment time online, essentially electronic entertainment time. They're doing a ton of interesting things in data science with regard to understanding social relationships and how people want to spend their time online.

Gutierrez: You've amassed a data set big enough that you see it as a competitive advantage. What other types of data, outside the ones just mentioned, do you use?

Smallwood: A key component of the data is that we also have a lot of metadata about the titles themselves. We do a lot of work in terms of tagging movies. We have folks who watch every single piece of content on Netflix, and then intricately tag those with a predetermined set of tags that we get folks to use consistently. So we have that data, and then we also compile external data related to the content we have. Some of that data we can access and some of it we don't touch because we don't want to mess with sources that we shouldn't, as we're very conservative about not overstepping our bounds.

Gutierrez: Given all this data, can you tell how old someone is by their viewing patterns?

Smallwood: We can guess, though we'd rather ask people, because as I've said before, people have such individual tastes that it's fascinating. Sometimes we'll do qualitative research where we'll look at the data to find the people we want to talk to, and then we'll talk to them under a certain set of assumptions that we've drawn from the data. And we'll learn that, oh, actually, that's not the case. So if we want to use information on whether this is a kid's or an adult's profile to help optimize people's experience, we'd rather ask people to tell us in their profile that it's a kid's profile. Once they tell us, we can then adapt accordingly.

Gutierrez: If someone wanted to learn more about data science for entertainment, where should they look?

Smallwood: A great source is the amazing number of articles out there these days. Just do a Google search for entertainment optimization, or predictive models for entertainment, or related searches, and a lot of interesting things will pop up. I would also say that talking to people, of course, is a great way to do it. Go on LinkedIn and find people who are working in those kinds of companies and reach out to them. Most people are happy to have a half-hour phone call, which can really give you a lot of information as well.

Gutierrez: What in your career are you most proud of?

Smallwood: I would say of my whole career I'm most proud of the team that I lead right now. I've hired a bunch of great people and the fact that we're such a tight team and super fun to work with is incredibly gratifying to me.

If you were looking for more of an individual output thing, I'm probably most proud of some work I did at Intuit prior to it acquiring Mint. We had a scrappy little team of four people doing an internal startup-like project. I had the chance to lead the creation of a personalization system. It was Mint-like in that we were using a recommendation engine to match a couple hundred advertisers we signed up and who had coupons to people based on people's spending behaviors. It was super exciting to build a whole recommendation system from scratch that actually worked quite well. It contributed to Intuit's decision to acquire Mint, because the project was sort of a proof of concept that we could do it and make it work.

Gutierrez: What is a typical Netflix day for you and your team?

Smallwood: It would be quite different for me versus my team, so I'll talk about my team. Across the team we do a collection of things, like the strategy meetings where we debate ideas and output. We also do that internally as a team. For example, we have a brainstorming meeting every week with a specific agenda for everybody who works on experimentation. Somebody comes with a topic to present and discuss, and then at the meeting we try to figure out things like, "What should we do about this problem?" or, "I had this great idea. Do you guys think it's great? I think it's great. Or does it suck?" We do that internally as a team, as well as with different parts of the organization.

Another chunk of time is spent on sharing results, ideas, and data with other teams where there are dependencies across the organization.

A fair chunk of time is on the source data. We always have more things we want out of the data, so we work closely with other teams who either log data, or do the data warehousing, or do the business logic around the data. Working with those teams is always a big part of our team's work.

We also spend time studying model results and iterating on models. This includes the modeling piece as well as the tactical piece of whether a model is working. And if it's not working, what are some of the ideas that are around that we could try differently? So studying results here as well.

Similarly, we spend time studying results of experiments that we do across various topics. Experimentation and getting your statistical results is the easy part. The hard part is interpreting those results when you know it's still a world of uncertainty and you might have metrics that are telling different stories about the same test. How do you interpret those results and try to translate what the test was testing, what the change in the product was, and imagining all the reasons why you might be getting these inconsistent metrics?

Gutierrez: How do you help new team members develop this way of thinking about interpreting outputs and results?

Smallwood: A great deal of individual one-on-one talks. I focus on asking them, "How do you read this result? What are you seeing in these numbers?" Once they share their answers, then I'll share what I'm seeing in those numbers. It goes both ways, so that we are both learning from each other. Sometimes they have a great idea that I didn't think of and sometimes it's the other way around. A lot of the one-on-one is spent really sharing and understanding how you're seeing the world.

And then a lot of it is people working with each other in the team. We have so many opportunities in these debate sessions to hear other people's ideas and how they're seeing the world. This exposure also allows for practicing on one situation after another, getting through a variety of models, and a variety of A/B tests. As people work through various situations, people naturally start to see patterns and are able to learn about different things going on and how to interpret them.

Gutierrez: Do you tape the debates for further instruction?

Smallwood: We don't really tape them. They happen often enough that it's just an ongoing learning opportunity. In fact, we also have a weekly analytics meeting where we present analytics results. The analytics meeting is similar to the strategy meeting, except it's less about a specific project and it's more about questions like, "How should we measure people streaming?" We also put a lot of effort in these meetings to talk about interpretation. It's really useful to get the project leaders and everyone thinking the same way, or hearing about the same gotchas or problems that are common problems, and seeing examples of those, and then talking through them together.

Gutierrez: How do you measure success at work?

Smallwood: We're lucky that we're in an experimentation-heavy culture, because it's very easy to say that something is a success when it's a winning test. Our closest reliable measure of customer satisfaction is customer retention, which is determining if people stay month after month. When we increase customer retention in an A/B test, then you get to remove all judgment and people's egos and everything else. The data speaks for itself. That's the easiest measure of success. We're lucky that a lot of the algorithms that we work on are amenable to tests. So we'll build version A and set it up versus version B of, say, our ranking algorithm, and then we'll just test it to find out if it is better or not better in the eyes of our customer. It's awesome that we have that opportunity.

The harder situation is when our models are used as input to human decisions. A great example of this is when we try to predict customer demand for titles we don't have yet. The information the predictive models generate is used for decision making in our Beverly Hills office, where they're negotiating with studios about what content to acquire. For instance, if they have the choice of five different titles, which one should they pick for Netflix given limited amounts of money?

That's a very hard problem, as it's a particularly hard space to model because it's so dependent on what other titles we have in our catalog and many factors about our member base and their interests in different kinds of content. In this situation, it's harder to measure the success of a model, because you never know at the point when you're making the decision whether the model is right or wrong. The content acquiring model has been a very successful model, but that doesn't mean it's accurate for every single content decision.

We monitor how successful this type of model is in aggregate over time. The way we measure success for this type of model is more about how the company internally views the usefulness of the model. We monitor things like: Are they using it and do they mostly trust it? Do they ask for feature enhancements? Do they want more out of it? Do they come to our team looking for more modeling to solve other problems? So it's a softer measure of success for that kind of modeling.

Gutierrez: Is there a number of models you need to create per year?

Smallwood: This goes back to the previous point of low process. There isn't a requirement on my team to push out a certain number of models per year. We have very few, if any, requirements. What we do have are responsibilities. For me personally, it's my responsibility to figure out how to evolve the team and our work to keep pace with all the innovation going on across the company.

For example, I see that with the movement toward more originals at Netflix, a different set of content decisions are going to need to be made. From this I know that my team needs to beef up our talent and level of resources in that area, so that we can help them make some really difficult decisions. I need to understand what the company strategy is in all the different parts of the business.

It's fortunate that I've worked in a lot of different industries and have a lot of years of experience under my belt because I kind of know how much resource it takes to do what I envision we will need to do in all the different areas. And it's always a moving target. It never goes backwards. I'm never worried about overhiring.

Gutierrez: What's a specific project that you have worked on recently?

Smallwood: I'll talk about a project that pertains to both experimentation and predictive modeling. Experimentation is a core part of how Netflix evolves our product. We really believe in this approach toward our product innovation, and so it's important that we are firm on how we measure those experiments, because there are many of them and they're precious to everyone.

One of the key metrics that we leverage for measuring our experiments is the number of viewing hours. How many hours did a particular customer spend watching Netflix over, let's say, a 28-day period, because that's really the key measure of engagement. If you're spending time on Netflix, it's extremely highly correlated with retention. Though retention is our core metric, I'm going to talk about hours in relation to this project.

For a long time we just measured total hours and percentage of customers that stream more than n hours at a couple of different thresholds. It's a really good way to think about hours because you get a sense of the distribution. However, it's much more impactful if we get people who are streaming only 3 hours a month to stream 4 hours a month than it is if we get people who are already streaming 30 hours a month to stream 31 hours a month. We really thought about this and we weren't happy with treating all customers equally when we were assessing hours.

We wanted a way to compare statistically how algorithm A versus algorithm B is impacting the lighter streamers, but we didn't want to do it at just one threshold. We wanted to do it at all the thresholds. To solve this, we essentially built a predictive model that has to do with the relationship between hours and retention at all parts of the distribution.

We came up with a metric that we call the "streaming score." The streaming score is related to both the number of hours that a customer streamed and how long they've been with Netflix, as well as the relationship with retention. The streaming score also takes into account viewing region, because people in the US are very different than people in Latin America in terms of how much TV they watch. So we built this model and now we have this metric, the streaming score, with test statistics around it that let us get a much more granular, impactful, and essential view of the streaming.

Gutierrez: How was it built?

Smallwood: First, we did a lot of research on different ways we could model it. It's a little bit complicated because customers get billed once a month, whereas the streaming is happening daily. So you have this marker on the day you decide to cancel your Netflix account. You may stop streaming but you also have service until the end of your billing cycle, so there's a little bit of a nebulous relationship there. This relationship made us think about doing a survival model.

We ended up implementing a custom model for this particular metric. But we did a bunch of research, and then we built a couple different versions of the model. We studied both versions of the model for both accuracy [AUC] and performance on A/B test results. And, again to the point of learning from each other, as a group we debated the results and decided what we thought was a good version to model.

Gutierrez: How did you set out the research path?

Smallwood: A lot of the research ideas came from different people on the team. Across the team there are enough people with different methodologies that they're familiar with that we have a lot of ideas of how to do things. Many folks on the team are great at keeping up with research in the industry, so they'll bring new techniques back to us as a team and share them with everybody. This means we've got good coverage of new techniques that are popping up everywhere, both within our team and the research that is happening elsewhere.

So first we brainstormed about the classes of techniques that would be good here. The different techniques depend on the characteristics of the data, so we made sure to really understand all the key dimensions that we cared about and how the data was distributed. We also worked to really understand which dimensions were actually important to this problem. As this process evolved, one person on the team—she's this brilliant statistician—was really leading the effort, so she started hammering away at the problem.

Gutierrez: How did she take the lead? Was she assigned, or did it just evolve naturally?

Smallwood: A lot of things just naturally fell into place, as she was very passionate about this problem and had the time to work on it. She and I really saw eye-to-eye that this was an important thing to solve for Netflix, especially since we both agreed that we really hadn't quite been looking at it the right way before. We also both understood that it would be a little bit of a hard sell, even at Netflix, because it's been the way of looking at tests for so long. So not only did she have the perfect kind of brain for this particular kind of problem, she also understood the implications of tackling and leading the project.

Gutierrez: How do you balance time, projects, and priorities?

Smallwood: Within my group, we have several teams that are focused on different parts of the business as their primary focus. It's not the only thing people in that team are working on, but at least they have a primary focus so that they can have all the domain knowledge that goes with that part of the business.

Obviously the marketing organization has totally different strategies that they're thinking about and totally different kinds of data they're looking at than the product organization. So we've found that it's really important to have people on our team who completely understand that part of the business. Otherwise, how would you do that translation in your head of what you're trying to model? This partial specialization also helps us know how much load is coming in from the different areas, and then we are able to readjust as needed on an ongoing basis.

Ongoing, there are things that pop up that you didn't expect, and suddenly you feel like you have fifty percent more work to do as a team and wonder how you are going to do it. Since I pretty much understand most of what's going on at Netflix, and folks within the teams understand their areas as well, we're pretty good at figuring out what to de-prioritize in order to let the super-fire-drill thing come in the door.

Gutierrez: How do you know you're solving the right problem?

Smallwood: I think it really comes down to understanding the business goals and understanding what the priority of the business is for the thing that you're trying to solve. Is it a curiosity, or is it something that's actually going to be used for an important decision, or for an important process or product? Or is it something that's operational in nature? And is there already something there that's pretty good and you're just looking to optimize it? Or is it something where there's nothing's there, and it's just a glaring issue where you know you can improve things dramatically?

There's a judgment call there that comes from understanding the business priorities, as well as understanding what you know you could offer from a modeling or algorithm standpoint. I do think it has to do with understanding the data at hand, what's possible with the data and models, as well as what's important from a business perspective.

Gutierrez: How do you know you're capturing the right data?

Smallwood: That is a really fascinating question for me, because I feel like through the process of my career at the different companies I've worked with, I've seen it all. At Yahoo!, all data under the sun was captured and it was great. When you wanted to study something, you knew you could find the data somewhere. However, it was challenging because the data was a mess, which meant that the ratio of the time you spent massaging the data and getting it to the point where you could actually use it for something, compared with the stuff you're actually doing—was high.

At Netflix, it's quite the opposite, from the standpoint that we are very intentional about what data we capture and how we capture it. We try to embed any critical business logic that we know we're going to want to apply, no matter what we do with the data, at the point of capture. This makes just about everything much easier. Sometimes we get that wrong. And more work falls to the engineering teams when that's the case because then they have to go back, detangle what we did, and rework something new, which can be painful.

But you're going to have that no matter what at some layer of the data stack, and we have found that it's easier to do that at the source where you can. Now, you can't always do that, because you want granular data that you can aggregate in as many ways as you want to later when you think of new ideas. But there are certain things you know that you really don't need. We try to weed *out* as strongly as possible in things, but you're never one hundred percent right. You're occasionally going to have to go back and rework things. You just want to try to minimize that.

Gutierrez: How do you think about the technology selection for the data stack?

Smallwood: This is a hard one because technology, especially in the data space, evolves more quickly than most companies can evolve. This is true especially at the data warehousing level, whether it's in the cloud or dedicated warehousing. There are so many different broad mechanisms, and once you've built a lot of infrastructure within your company, it's incredibly expensive to switch over to some new technology.

We use Teradata for a large part of our data warehousing. If we wanted to move from Teradata to some other data-center–oriented warehousing system, we would have so much to move that it would be a year's worth of work for the entire data organization. Perhaps not quite that much, but it would be a lot of work. So the farther upstream you are in your stack, the harder it is, I think, to change technologies.

That said, we have been, like many companies, moving more and more toward cloud-based analytics. When I first started at Netflix, pretty much all of our data was in Teradata and we had a little bit of data in the cloud. Netflix was just moving toward serving our whole product off of the cloud, so as you can imagine, that meant we started having more and more data in the cloud. With this evolution of the product occurring, what has worked for us is doing parallel development. We've moved from having the majority of our analytics in Teradata to now having the majority of our analytics in the cloud.

We do still have a formidable amount of data in Teradata, but we've switched our philosophy. We have aggregate data we use for ongoing reporting in Teradata. We have granular data we use more for the modeling in the cloud, and all sorts of analytics go in both places. But we have more data in the cloud, as it's closer to the point of capture. And it's worked really nicely for us to have those systems in parallel. Maybe eventually we'll get completely away from the data center. But it's a matter of efficiency and investment in what works best for our needs.

Gutierrez: How do you think about the technology selection for general use within your team?

Smallwood: It's a lighter thing when thinking about technology selection for the team rather than for the data infrastructure. We want to do analysis and then have some visualization on top of that analysis. For this work, we experiment with all different types of tools, and the technology choice essentially comes down to whatever the passion is of the person who's working on it. We use all sorts of tools at that layer of the stack. For analytics, we heavily use R. Any open source software for the most part is preferable to licensed software, so we are heavily open source–oriented. We use a ton of R, Python, and things that are easy for people to pick up, learn, and then do all sorts of visualization things with as well.

Gutierrez: Whose work is currently inspiring you?

Smallwood: My colleagues are very inspiring to me, both on my own team but also other colleagues across Netflix. We have an amazing guy leading the algorithms from the product innovation side—Carlos Gomez-Uribe. He's an outstanding, brilliant guy, and his whole team is super strong and inspiring. I feel silly saying this over and over, but it really is true: We have amazing people at Netflix!

Outside of Netflix, on the experimentation side, I always enjoy Ronny Kohavi, who is a great speaker and excellent at experimentation. He really gets the power of it and is great at conveying that in his talks. In terms of algorithms and predictive models, so many people doing great stuff that it's less about following particular people than it is about noticing particular papers or applications coming out the companies that are doing interesting things in experimentation or in the predictive algorithms and models space.

I've always thought that Amazon has done and continues to do super-amazing things with data. LinkedIn also continuously does interesting things with data. Some of it is that they have interesting data, but they also have like a long history in data science with DJ Patel, and progressing from there. I've always been a fan of LinkedIn and their output.

Gutierrez: What inspired you to get involved as an advisor to HiQ Labs, a company that does predictive modeling on employee churn?

Smallwood: I was drawn to it really because of the application. I'm always looking for interesting things you can do with data. To me, what they are doing is super interesting to think about. As I've said, as my career has progressed, I feel really cognizant of how awesome it is to have a great team. And once you have a great team, you want to keep that team. So to me it's very interesting to think about the factors that cause people to move on. Especially here in the Valley, where there are so many interesting opportunities and where anybody on my team could jump ship and find another super-interesting position somewhere else very easily. So why they stay at Netflix, and how I can keep them engaged and wanting to stay rather than jumping into something else— I'm really interested in understanding those factors.

I think it's really hard, because people in this day and age—in the Valley anyway—are sort of trained to move every couple of years. That's how you progress your career. It's by doing some other new, exciting thing. So, it's been interesting working on this side project as an advisor and studying the problem from a data perspective. It's essentially using the LinkedIn API to study and try to understand at what point people churn, and whether you can predict that from things like their LinkedIn data.

Gutierrez: Are there any interesting insights that you've been able to apply to your team here at Netflix?

Smallwood: A few. Whether the model is right or wrong about somebody— predicting that somebody is a risk has caused me to have conversations with my own team that I might not have otherwise had. It's probably those conversations that mattered the most. It really helps to make sure you check in with people and ask, "Are you happy and engaged? If you're not, let's talk about it, because I can't promise you we can change it, but we can certainly talk about it and try." So I think it's very easy in a fast-paced business to lose track that a whole year has gone by and you don't know if people are happy or not. That's a dangerous thing to let happen if you want to make sure your team's engaged.

Gutierrez: You're incredibly proud of the team and really enjoy the camaraderie that you have. Is this something you've instilled in your team, or is it something that you look for when you hire?

Smallwood: I think it's both. It certainly has a lot to do with the characteristics of the people that you hire. And it's also about having common goals and things you're excited about as a team. Having people feel like they can really trust one another and be inspired by one another is a big part of it as well.

I also think some of it is my own experience from decades of working in this space. I've learned to appreciate things like team camaraderie more than I might have earlier. When I was five years into my career, I probably didn't nearly appreciate the depth of what it is to have a great team as much as I do now.

Gutierrez: When did your appreciation change?

Smallwood: I think it changed after making it through the early experience of leading my first few teams. I made all of the standard rookie management mistakes in learning how to manage, and I made hiring mistakes and had to live with those. After having to figure out how to deal with all of those things, you kind of learn.

Gutierrez: What do you look for in people when in hiring?

Smallwood: I would say the top things are hunger and insatiable curiosity. You imagine a data set and you salivate at just thinking about that data set. Those are the top qualities, because people who always want to dig more, mine the data, and learn new things from the data are the people who are happiest in this kind of job. Obviously, the technical skills are important. But that's always the easiest thing to interview for because it's straightforward to ask those technical questions.

However, it's not as straightforward to try to get a feel for how curious you are. You can't ask someone, "How curious are you?" But you can tell by how many questions they ask. And if you describe to them a data set and ask, "What would you do with that data set?", people either can't stop talking about idea after idea, or they're like, "Oh, I don't know. Maybe I would look at the average minutes"—or something inconsequential like that. So I obviously look for the technical skills and the curiosity.

The last important facet I look for is tenacity. You're really never done with data and algorithms, so tenacity is a core thing. Passion for our business and what we're trying to achieve is important as well.

Gutierrez: Do you look for tool-set knowledge when you're hiring, or is it more of a matter of, "We need the thinking to be in place, and we can teach you what you may be missing in your tool arsenal"?

Smallwood: It's more the latter, though it's important that people know some tools. If you've never worked with any kind of data querying, you're not going to understand data that well, and you're going to have a long learning curve to get to the point where you're efficient. So that would be a showstopper if somebody didn't know some tools at each layer of the data stack. But the specific tool doesn't matter that much. Generally, once people know any tool at some layer of the stack, they understand that layer and they're able to pick up another tool quite easily.

Gutierrez: What's the biggest thing you've changed your mind about with respect to using data?

Smallwood: I would say hands down it's A/B testing. When I was first exposed to it many years ago as "web analytics", I didn't have much of a level of respect for it. It seemed very straightforward and trite to me. I didn't really get the power of it until I came to Netflix. What's interesting to me is that people's intuition is wrong so often, even when you're an expert in an area. Even when you have a lot of domain and background in an area, your intuition to answer any one question is still wrong sometimes.

Similarly, with predictive models we get excited when we see an AUC of 0.75. We're like, "Oh my gosh! That's a great model!" But it's still wrong a lot of the time, and it's only through experimentation that you can actually get a causal read on something. And it is so fun and fascinating to watch. We get to do both of these things at Netflix: We get to develop the alternate algorithm, and then we get to test it. We may initially think, "This algorithm will be so much better than what we had before! It has an amazing AUC." Or whatever we measured offline—perhaps MRR [Mean Reciprocal Rank]. And then we test the model and it's instead worse than what we had before. There will be one surprising result after another. And to me, it's the power of getting that real causality that just completely rocked my world of thinking about data.

Gutierrez: Do you codify what you learn from each A/B test?

Smallwood: We definitely try to look at themes of things we learned across the tests, but the focus is more on where else we can do testing that we're not doing yet. We would love to test in the content space to learn more about the titles and catalog makeup that are most important to our customers, but we don't want to test things that are a negative experience for customers. So we haven't and won't do that. We've debated minimal experiments like, "What if we just took one title out of our library and tried to see if we could measure an impact?" Still, not only do we have contractual agreements with the studios, but we also don't want to degrade the experience for our customers. But we do always think about whether there is anything we could do to experiment in the content space to help inform those decisions.

Gutierrez: Do you do A/B testing in your personal life?

Smallwood: Not quite A/B testing, but when I was young, I used to build little models for my personal decisions. Dorky little models to make personal decisions. I'd figure out what all the attributes were about decision A versus decision B, and come up with my weights on how important each of those attributes were.

Gutierrez: You have mathematicians, operations researchers, statisticians, and data scientists in your group. How do you classify them?

Smallwood: Titles are always a challenging thing. We didn't even use the word "data science" a decade ago. So titles across the industry are a challenging thing. In terms of what people do, it's more self-selecting. People tend to be either more interested in the product and experimentation kind of work, or more interested in algorithms. Even in the algorithms space, some people have very specialized interest in particular types of algorithms. So I think, in general, we try to group people in ways and in areas that they're most excited about so that you're getting both the business benefit and people are happy. We're pretty flexible on titles at Netflix. People can call themselves what they feel like calling themselves.

Gutierrez: What does the future of data science look like?

Smallwood: I think it's going to continue to explode and it will become quite ubiquitous. Kind of like how trend charts are everywhere. I think it will be the same thing for models, algorithms, and the deeper data sciences. The techniques are already amazing and they'll continue to get better. Even where we stand today, they've really evolved amazingly with the depth and sophistication of the math and the different kinds of techniques. And it's due to having so much data available that's let people really investigate different nuanced techniques. So I think those techniques will continue to explode. And what will become more challenging for data science is being aware of so many different techniques, being good at knowing how to formulate a problem with the proper kinds of techniques, and not misusing the techniques.

I think that there will be a little bit of a danger involved as well. I think it will be very easy to develop bad models because of open source availability, how easy it is to program these techniques, as well as the vast array of complicated techniques. It is already easy to develop bad models, and unfortunately I think that will become even easier.

I think this is true in particular for companies that are trying to start a data science team where they didn't have one before. I think it's a bit of a danger zone for them. It will be very easy to hire someone who's built one regression model and uses fancy terms in their interviews like "support vector machine," and the executives will go, "Wow! Come in and build our data science team." For these companies, there's no way to gauge whether the person they hire knows what they're doing, because a great model versus a bad model is still a model. They both spit out the same type of result and you have some measure of whether that's a good result or not, but it's impossible to really know whether it's a good model or not. In my mind, it's all about the quality of the people building those models, and I think it will be hard for inexperienced companies to discern that.

Gutierrez: How can people discern the good people from the rest?

Smallwood: I think it's important to look for experience—especially if you're starting a new team. You can't take someone without experience, even if they were valedictorian of their PhD program at MIT. I still wouldn't take that person as my only or first data scientist, because they haven't worked enough. I think the education is great—don't get me wrong, education is fantastic. But in today's world, where people are working on more practical problems, the actual experience of wrestling through one model after another matters for the only or first data scientist hire. Especially if the experience happened under very different data circumstances, different distributions of the underlying data, and different data characteristics. You also want to see experience with missing data, duplicate data, and all the challenges that you actually face with raw collected data. And that's just on the data side.

On the modeling side, you also want to see experience in thinking about whether you're solving the right thing, and then learning from the business perspective that it's completely impractical and you solved the wrong thing. You have to go through all of those experiences to really build up the beefier level of experience.

I think if you've found someone with that experience—even just one great person like that, you're set. Because then you can hire some junior-level people and teach them along the way. But you've got to have at least one person who's actually experienced.

Gutierrez: What is one problem you think data scientists need to fix?

Smallwood: That is a really hard question, as I feel like all of the important problems that can be tackled with data are already starting to be tackled by people. In the future, perhaps, there may be other new important problems, but at this point people are looking at the current important problems. But I will say that to me the most valuable things to improve are things where data is shared more openly across the industry, regardless of what industry you're talking about.

If I were to pick one thing that needs fixing in this space, it is that health care data sharing is moving way more slowly than it should. If we could somehow figure out how to speed up privacy-protected information sharing across all that great data that's collected, I'm sure we could make more progress on diagnosing things, even self-diagnosis. Especially with more ubiquitous open sharing of information about symptoms and what they can be connected to health-wise.

Anything where you're sharing data across the whole consumer and business industry within the same industry, I think is super beneficial. I love what's been done in microphilanthropy. The work in this area is very cool. There's a ton of environmental work that's going on that's also really cool. So I think that there's real opportunity everywhere with today's ability to collect data and then share it.

Gutierrez: What is something a small number of people know about that you think is going to be huge in the future?

Smallwood: One thing I think will grow exponentially is anything in the world of motion, text, and language analytics. Basically anything in the non-numerical data world will grow very fast. Specifically to text, I find what Twitter is doing to be really interesting, especially the skills around taking that super-fuzzy text data and being able to identify important patterns. I think this is an area that will really explode.

Gutierrez: What is driving the growth in text analytics?

Smallwood: More and more mechanisms are dealing with text data because we're able to handle larger volumes. You used to have a multiple-choice question where you had to answer a, b, c, or d, and then that data would get encoded. Well, now people can just answer open textings and we have ways to interpret those text blobs mathematically.

It's still a complex problem because the world of ontologies and text interpretation and disambiguation. This brings a lot of really hard challenges. However, products are evolving in the right direction because more and more collection mechanisms are collecting that data and are able to handle those volumes. So companies that make products as easy as possible for the customer are going to evolve things in the right direction, which often means voice or text inputs and analytics.

Gutierrez: Does Netflix have anything in the pipeline around voice?

Smallwood: We have an experimental talking UI named Max on the PS3. Max talks and tries to be conversational with the things that he's trying to ask you like, "What kind of mood are you in? What would you like to watch?" We haven't yet started doing things in the other direction, where customers are speaking. We do have pointer devices, so that's one step in the motion direction.

Gutierrez: What advice do you have for people to do great work?

Smallwood: I think it's huge to embrace your curiosity. You should never quite feel satisfied with an answer and you should always essentially have more questions than answers. Enjoying that as a sort of way of life really goes a long way. Being able to connect your own work to things that you're passionate about also goes a long way as well.

Personally, I really love trying to understand people through their data, especially if I can have a way to cycle the understanding back to them with the hope of helping to improve their world. So for me, anything where I get to try to understand people and improve their world as a result is especially gratifying. And that's different for every person. So I think its super important to find whatever the connection is for you that makes you feel excited about what you're doing.

In the world of data science, being somewhat egoless is really important. There are some industries where your role allows you to have a big ego, and that works great for having a big impact. However, in data science, you get humbled over and over again as you try to do things. You think you've come up with some brilliant idea, and then you build some model or you build some metric, even just a simple metric, and you look at the results and you're like, "Oh, that's terrible." It's so disappointing. Or the reverse can happen too. You do some dinky little thing that you hardly put any thought into and it turns out to have a massively positive impact. And so being somewhat egoless in this field is important.

If you do have an ego, not only will you be disappointed all the time, but you also will not be open-minded enough about what the data can say to you, because you'll be too stuck in your one mindset. You have to be really flexible with your mindset, thinking in tiny, tiny details, and then all the way up to super-elevated levels about the forces that are going on across the data.

You also have to be open-minded about different techniques. Let's say you build a regression model and you find that one signal you added to it really made your regression model successful. Then you try a totally different technique and that signal was diminished to nothing in the other technique. You have to be open-minded enough to potentially throw away the signal. You can't get too attached to that signal because maybe you haven't learned that the signal you got all excited about wasn't that important. This type of open-minded, flexible, egoless kind of attitude is important.

Gutierrez: Is egoless attitude something that can be taught?

Smallwood: I think that's something that people can learn, but I don't know that you can teach it person to person. I think people learn it more from experiences.

Gutierrez: What should someone starting out try to understand deeply?

Smallwood: I'm a big believer in understanding probability distributions. Understanding all the different types of distributions and what those characteristics look like in your data really goes a long way toward understanding how to build different types of models. If you only know the normal distribution, you're not going to be nearly as effective as if you know Poisson distributions and all the other different kinds of distributions. Knowing and understanding the distributions really help guide how you think about modeling things.

Also important is studying a variety of techniques: clustering techniques, regression techniques, tree-based techniques, and others. Try to get experience with a gamut of different kinds of techniques, because then over time you realize there are subtle similarities across them. Sometimes you can learn about a problem by coming at it from all of those angles and see what comes out as the commonalities across all those approaches. As we talked about earlier, experience with different models, different data sets, and wrinkles with data sets is hugely important.

Gutierrez: How does someone develop the skill to know how to choose the right technique to apply to a problem?

Smallwood: It's about trying a lot of the different techniques and learning some of the common pitfalls that you would come across with the different techniques. I also think there's also a lot to be said for working in a collaborative environment where you can show your approach to someone else and hear their feedback on questions like: Why was that a good idea? Why was that not a good idea?

It's hard to know if you're working in isolation on a model. You would have a hard time knowing whether you built the right kind of model or not, because the model will output something regardless if you modeled it correctly or not. If you're cocky or full of ego, you'll just believe you did the right thing and not stop to think about whether you actually did the right thing. It comes back to being egoless and open-minded. So I think it's really hard to learn how to choose the right technique to apply to a problem without getting feedback from multiple people in the space who have experience as well. The more people whom you can get feedback from over time, the better. I really think that's a great way to progress.

Gutierrez: What advice is helpful for people moving into the field?

Smallwood: I would say to always bite the bullet with regard to understanding the basics of the data first before you do anything else, even though it's not sexy and not as fun. In other words, put effort into understanding how the data is captured, understand exactly how each data field is defined, and understand when data is missing. If the data is missing, does that mean something in and of itself? Is it missing only in certain situations? These little, teeny nuanced data gotchas will really get you. They really will.

You can use the most sophisticated algorithm under the sun, but it's the same old junk-in–junk-out thing. You cannot turn a blind eye to the raw data, no matter how excited you are to get to the fun part of the modeling. Dot your i's, cross your t's, and check everything you can about the underlying data before you go down the path of developing a model.

Another thing I've learned over time is that a mix of algorithms is almost always better than one single algorithm in the context of a system, because different techniques exploit different aspects of the patterns in the data, especially in complex large data sets. So while you can take one particular algorithm and iterate and iterate to make it better, I have almost always seen that a combination of algorithms tends to do better than just one algorithm.

Similarly, I would say that if you are just developing one model— let's say, a predictive model—it really is great to try a variety of techniques. You'll learn a lot. It's like the art and the science. There's the science, and that's actually the easier part of the whole thing. The art is imagining all the possible signals you could try as inputs to your model. And trying the different techniques and seeing what themes arise will get your far. It will help you understand what classes of signals always seem to pop no matter what technique you use. Lastly, along with trying different techniques, is to A/B test whenever and wherever you can, as it's really great to add that to your rigor and understanding of the different techniques and themes.

Gutierrez: What do you love about data science?

Smallwood: What I love about it is that it's incredibly creative and innovative. If someone's just dipping their toes into the field—come on in and learn more, as it is a fascinating field. If you think, "Oh, it's just going to be some boring math field," it's not that at all. It's incredibly challenging and creative. And that's been a constant surprise to me through the years, and continues to be. It continues to be more and more creative the longer you're in it, because you have more tools at your disposal and more intuition, right or wrong, at different times about the data. And so it just gets more exciting and creative over time.

Yann LeCun

Facebook

Yann LeCun *is the Director of AI Research at Facebook, the world's largest social networking service. Facebook's core business is to facilitate communication among people and between people and the digital world. The technology required to support this mission is immense given the sheer scale of data involved. As of 2014, Facebook had over 1.3 billion active users (with more than 150 billion connections among each other) and 829 million unique daily logins. On an average day, these users uploaded around 350 million photos, shared around 4.75 billion items, and sent around 10 billion messages. The availability of these streaming data sets is so large that, for the most part, Facebook's systems only have time to look at any piece of data just once. Such activity levels bring with them a unique set of challenges: how best to make sense of and understand all the data, and how then to use this intelligence to make decisions.*

Prior to joining Facebook, LeCun was and continues to be the Silver Professor of Computer Science, Neural Science, and Electrical and Computer Engineering at New York University, where he is the founding director of the NYU Center for Data Science. After his postdoc work in Geoff Hinton's group at the University of Toronto developing the theory of artificial neural networks with back-propagation, he joined AT&T Bell Labs, where he later became the head of the Image Processing Research Department. LeCun then worked briefly as a Fellow of the NEC Research Institute in Princeton before joining NYU in 2003. Over his career to date, he has published more than 180 technical papers and book chapters on machine learning, computer vision, handwriting recognition, image processing and compression, and neural networks. He is particularly well known for his work on deep learning methods, which are used by companies to understand images, video, documents, human-computer interactions, and speech.

LeCun is a peerless example of a data scientist with a transformational vision—in his case, using deep learning to teach machines to perceive the world—who strives to actuate that vision in both academic and industrial research laboratories. LeCun's indefatigable pursuit of his vision, from the early days of AI to the current days of Internet-scale machine learning, is the dominant theme of his stories about publishing the MNIST data set and how it's frequently used for testing out new machine learning methods, the ups and downs in the vogue for neural networks in academic circles, and his own evolving beliefs about the comparative merits of supervised and unsupervised learning. LeCun's interview is a testament to his passion for machines that can learn and his belief in their future.

Sebastian Gutierrez: Tell me about where you work.

Yann LeCun: I'm the director of AI Research at Facebook. Part of this role involves data science, though there are other groups doing data science at Facebook. AI Research can be thought of as the more advanced side of data science if you want. I'm also a professor at NYU part time, which is conveniently located just across the street from my Facebook lab. Though I'm now a university professor, most of my career has been in industry research. Early on I worked at Bell Labs in a group that was, at the time, working on machine learning and neural nets and similar projects. Then I became a department head at AT&T Labs, which was the name of AT&T's research lab after the company split up in 1996. I joined NYU in 2003, so I've been here a little over 11 years. I joined Facebook at the end of 2013.

Gutierrez: What excited you about the opportunity at Facebook?

LeCun: The main thing is that I was given the opportunity to create a world-class research lab from scratch. Facebook did not have a tradition of being active in research, so this was a bit of a new experiment for Facebook. It's not every day that someone comes to you and says, "We'd like you to create a research lab and hire the best people in the world." You can't refuse that. Our mission is very ambitious: understand intelligence and make machine more intelligent. That's what attracted me here.

Gutierrez: What is the makeup of your team?

LeCun: There are about 35 people in the team as of November 2014. It's a mix of research scientists and people who are more on the engineering side. It's about three-quarters research scientists and one-quarter research engineers. That 35 number will change soon, as we're going to grow really quickly. Right now, our growth rate is actually limited by the number of good people who are available, particularly since every other company is trying to hire in this same area. We are in a big competition for talent with Google, Microsoft, Baidu, Yahoo!, IBM, and a bunch of others.

Gutierrez: What are the short-term and long-term goals of the lab?

LeCun: I'll start with the long-term goal. It's going to sound incredibly arrogant, but it's basically to solve AI. Of course, a lot of people have tried to do this and failed, so we're not claiming we're going to succeed, at least we're not giving a timeline. The long-term goal is to really understand intelligence, be it natural or artificial, and try to make machines more intelligent so that they can help people in a better way all around—help people communicate and help people deal with the digital world. So that's the long-term goal.

The shorter-term goal is to understand content. People upload hundreds of millions of images onto Facebook every day and also make billions of posts and comments. One of the main things that Facebook has to do is essentially decide what to show to users. Every single day users could be shown roughly between 1500 to 2000 items. Users don't have time for that, so Facebook's algorithms basically selects a couple hundred of those that the user does have time to see every day. To do a good job at picking what to show users, we have to essentially match the content to users' interest. Being able to figure out what's in an image, what a post is about, what topic it is, and whether it's a topic that a particular user is likely to be interested in allows us to better match content to users' interests. So understanding content is really essential for Facebook.

Gutierrez: How much deep learning is involved in this process?

LeCun: A lot of things we do are based on deep learning, but not everything. Deep learning is very useful for us, particularly for things like images and some text representation and some topic extraction, because deep learning shines in situations where we have lots of data. Of course, Facebook certainly does have lots of data, so that's why we use it. In a lot of situations, it beats the approach of handcrafted features followed by classifiers just because of the amount of data we have.

Gutierrez: Looking back at how your career has progressed, were you following a path?

LeCun: Though there is no clear path in terms of institutions I've worked in, there has certainly been a path in terms of the technical problems I've been interested in. In fact, what I'm interested in has been pretty constant except for short time period. I've always been really fascinated by AI and related subjects since I was a kid. While I was undergrad in the late 1970s through early 1980s, I studied electrical engineering. During this time, I did a bunch of projects trying to figure out if we could make machines learn. I was always convinced that the only way to make intelligent machines was to get into learning, because every animal is capable of learning. Anything with a brain basically learns.

I approached the problem by searching the literature for machines that could learn and realized that, at least in the early 1980s, nobody was working on these types of problems. The only literature I could find was from the 1960s and some of it from the 1970s, but mostly from the 1960s. It was the old work on the sort of neural net version 1.0, from the 1950s. Things like the perceptron and other techniques like this and then the statistical pattern recognition literature that followed in the early 1970s. But by the time I started to take an interest in this research area, the field had been pretty much been abandoned by the research community. This time period is sometimes referred to as the "neural net winter."

I graduated—though my specialty was not actually machine learning, as there was no such thing as machine learning back then. In fact, in France at that time, there wasn't even such thing as computer science. The specialties I graduated with were VLSI integrated circuit design and automatic control. After undergrad, I went to grad school. Unfortunately, I had a hard time finding people who were interested in what I wanted to do, as I already knew exactly what I wanted to work on. I had already realized by the time I was an undergrad that the thing that people were after back in the 1960s and could never solve was basically the idea of multi-layer neural nets and deep learning.

Maybe two years before I started grad school, I started experimenting with various algorithms. I came up with something that eventually became what we now call the back-propagation algorithm—which we use every day at Facebook on a very, very large scale—independently from David Rumelhart, Paul Werbos, David Parker, Geoff Hinton, and others. I had a very hard time getting senior people in grad school to help me because the field had been abandoned. Luckily, I had a very nice advisor, Maurice Milgram, and I had my own funding, which was mostly independent from my advisor. The very nice advisor, who wasn't really working at all in anything I was doing, basically told me that he would sign the paper, as I seemed like a smart guy, but he couldn't help me.

Gutierrez: What made you so sure that this was the right direction?

LeCun: I don't know. I was just so convinced. It was just so obvious to me that learning in multi-layer neural nets was the key. Then in the early 1980s, I discovered people like Geoff Hinton, as well as Terry Sejnowski, and realized they were interested in exactly the same question. So Geoff Hinton was the guy I wanted to meet. He was the only guy that I could meet who could understand what I was saying basically. It just so happened that he was invited to one of the first conferences on neural nets back in 1985 in France. I met him at the conference and we immediately clicked. He invited me to a summer school the next year, and then basically invited me to do a postdoc with him. Geoff Hinton has been and continues to be a big mentor to me. So I was driven. I knew what I was doing. It's not like I went to see an advisor who told me, "Why don't you work on this algorithm?" It was completely self-determined.

Gutierrez: What initially sparked your interest in AI?

LeCun: I was born in 1960. So by the time I was nine, you had rockets flying out in space, people landing on the moon, and *2001: A Space Odyssey* came out, where you had space and intelligent computers in it. Science fiction was the spirit of the time. I've always been interested in science. When I was a kid, I thought and hoped I would become a scientist. I hesitated—not for a very long time, unfortunately—between things like astrophysics, paleontology, neuroscience, or AI. But I'm really an engineer. I got this from my dad who is a mechanical engineer, I like to build stuff.

So what I thought about, as I thought about doing science, was: What are the big scientific questions of our time? One question is: What is the universe made of? Which things like astrophysics and fundamental physics try to answer. Another question is: What's life all about? Which biology and so on try to answer. Another question is: How does the brain work? And this question is a big, big scientific mystery. If you are a young scientist who has not yet realized your limitation, you go for the big thing. And understanding intelligence is a great big question.

As an engineer, I think of the brain as a very complex system. Intelligence is something that is very abstract, which maybe can be modeled by mathematics, and so we can use an engineering approach to figuring out how the brain works by trying to build intelligent machines to validate the designs or the conceptual ideas that we have. A great deal of things have been said, some very abstract, about how the brain actually works. But how do you know they are right until you build a system that actually works? So at least there you have most of the ingredients that are necessary. So that's the kind of scientific question that has interest for me.

Of course, not only did I have to satisfy my desire to build stuff, I also had to get jobs where I could develop good technology and do great work. It's strange for me to say this, but it was never quite clear for me that I would ever become an academic. I have—and maybe I should have become one earlier— but industry research was kind of a perfect environment for me for a long time. And so I'm sort of coming back to that now, though keeping a foot in academia, and the two worlds are really complementary in this way I find. And so I'm in this incredibly privileged situation where I can have one-and-a-half feet in industry and half a foot in academia, which allows me to take advantage of the complementarity between the two. In academia you can do things like computational neuroscience and theory, while in industry you build ambitious things that are difficult to do in academia.

Gutierrez: What was the first data set that you worked with?

LeCun: The first real data set I worked on was a medical data set that I worked with while I was doing my PhD. The data set came from a medical study of patients that come to an emergency room with abdominal pain. It turns out that deciding whether or not to operate was a very difficult diagnosis to make for surgeons based on just abdominal pain. There are on the order of 20 different basic diagnoses you can make based on abdominal pain. Some of these diagnoses require very quick surgery, like appendicitis, for example. And so I was given a fairly large data set for the time, with thousands of samples, with basic descriptions of patients, with missing values, and things like that, which you would expect.

The people I talked to who had collected this data set had tried naïve Bayes and similar approaches on it. I tried neural nets. Neural nets didn't exist yet, but I basically tried this newfangled thing on it—back propagation—and I got some pretty decent results. This helped me come up with the idea of tailoring the architecture of the system so that it would be able to identify syndromes and things like this, which are collections of symptoms, so as to reduce the number of free parameters in the system, because we knew, even back then in 1986, that overfitting was a big issue.

Gutierrez: Was there a specific aha! moment when you grasped the power of data?

LeCun: It was never about data for me. For me, data was and is a means to an end. For me, it's always been about the power of the model you can train, and so it's about learning algorithms. The wide availability of data came way, way, way later—like 20 years after I started working on these questions. We started having large data sets or decent-sized data sets for things like handwriting recognition or speech recognition in the 1990s. In fact, I published one of those data sets—the MNIST data set, which is used very frequently for handwriting recognition. Now it's not considered big at all, but at the time it was.

The availability of data sets so large that you don't even have time to look at any piece of data more than once because you have streaming data coming at you is a very recent phenomenon. A lot of the methods that I am interested in happen to scale very well in those situations, because I have always been a believer in things like stochastic gradient descent and similar techniques. These are things people use now after a hiatus of 10 years. People used other methods that didn't scale very well because they weren't confronted with this flow of data, and now that they have data of this size, they're now coming back to these techniques.

Gutierrez: What conferences, papers, or books in AI research would you recommend to someone just starting out?

LeCun: There are two different things I am interested in. One is AI or ambitious machine learning, and the other one is what I'll call data science. However, it's not the industry meaning of data science. What I mean by data science in this context is really the general problem of extracting knowledge from data, whether that is done automatically or semiautomatically, and whether we're talking about the methods or the tools or the infrastructure, and whether the data has to do with things like business or science or social science in particular. Those are two of my slightly different interests. In both cases, I believe things like deep learning will probably have a big impact on the practice of data science in the near future.

If you're interested in AI or machine learning, the main conferences are NIPS and ICML, and also conferences like AI Stats, UAI, and KDD, which is more data science–oriented actually, but there are quite a lot of methods papers now. And then there are lots of other good smaller conferences. In fact, there's one that's really important to me, not just because I run it, called ICLR, which stands for the International Conference on Learning Representation. This is a deep learning conference. It's relatively small, but it's very focused and very interesting. So those are conferences for machine learning–type things.

There are other areas of machine learning that I'm not very active in. Things like reinforcement learning, which is actually very important for industry and is used in things like ad placement, so there's AAAI and similar conferences. I also have a foot in computer vision. In computer vision, the main conferences are CVPR, ICCV, and ECCV. CVPR is more into images, videos, and similar things.

Gutierrez: Is there an area today that you feel is somewhat analogous to deep learning when you started, in that you think it's going to be giant in the future but people just aren't looking at it right now?

LeCun: I think it goes in cycles. We have a new set of techniques that comes up, and for a while the technique is under the radar and then it kind of blows up, and everybody explores how you can milk this technique for a while until you hit a wall. Progress slows and becomes more boring. Then some new set of techniques comes up and the whole process starts over again.

In my area, back in 1986 and 1987, neural nets had been under the radar and sort of blew up in 1986. So a lot of interesting stuff happened, a lot of crazy things happened—and a lot of hype happened, as well—until the early 1990s, when, in my own lab at Bell Labs, the next wave came up. The next wave was support vector machines or kernel methods, which are very popular and work very well. That replaced, to some extent, a lot of the work on neural nets. So neural nets migrated at that time to other conferences, like NIPS.

The ICML conference was actually more focused on a completely other branch of machine learning that was more symbolic. That branch of machine learning has essentially disappeared now. The transition to "statistical" machine learning at ICML was completed around 2009, when numerical machine learning techniques basically wiped out symbolic machine learning, so there are barely any papers on these topics at ICML anymore.

Simultaneous with kernel method, graphical models came up in the mid 1990s. Graphical models are really orthogonal to things like neural nets, and can be viewed as more like a framework around other techniques. In conferences such as NIPS, neural nets started to be heard again in the mid 2000s. I had actually left the field for a few years, not because people were not interested in neural nets, but because I had other interests that I wanted to pursue. I worked on image compression between 1996 and around 2002.

Then I came back to machine learning in the early part of 2000. Geoff Hinton, Yoshua Bengio, and I started what you could call a conspiracy—the deep learning conspiracy, basically—where we attempted to rekindle the interests of the community in learning representations as opposed to just learning classifiers. For the first few years, it was very difficult for us to get any papers published anywhere, be it computer vision conferences or machine learning conferences. The work was labeled "neural nets" and basically not interesting for that reason. People just didn't seem interested in digging past the title essentially. It's only around 2007 or so that things started to take off.

For deep learning, it was still a bit of a struggle for a while, particularly in computer vision. In computer vision, the transition to deep learning happened just last year. In speech recognition, it happened about three years ago, when people started to realize deep learning was working really well and it was beating everything else, and so there came a big rush to those methods. But it was a struggle for almost 10 years.

Gutierrez: So you've done vision and audio. What's next?

LeCun: Natural language is what's next. At Facebook, we have quite a lot of effort going on with deep learning for natural language. That's kind of obvious though, right? Google also has pretty big efforts in that direction. After natural language, there's video, and then after that there is the combination of all of the above. For in video, you very frequently also have audio. People then make comments on images and videos. So what you'd like to be able to do is to represent all of those different pieces of content in the same space, so that things that are talking about the same thing on similar topics end up in the same region of that space. This is called "embedding."

Gutierrez: What in your career are you most proud of so far?

LeCun: The thing I'm most proud of is that back in the early 1990s at Bell Labs, there were a bunch of very smart people working together, and together we built a check-reading/check-recognition system. But it wasn't just a check-reading system; it was an entire process for doing end-to-end image recognition. The data that was available, and the limitation of the computers of the time drove us to apply these ideas to check recognition. That was basically one of the practical applications for which we had data, and there were people willing to actually do the development, commercialize it, and everything.

People still have to catch up with the technology that we used to solve this problem. The kind of techniques that we used there—the integrated deep learning convolutional nets in particular, with what we now call "structure prediction"—I'm willing to bet people are going to reinvent over the next few years. I have this paper that I coauthored with Léon Bottou, Patrick Haffner, and Yoshua Bengio that basically describes convolutional nets as well as the full end-to-end system we built. The paper is called "Gradient-Based Learning Applied to Document Recognition." It's from 1998 and it was published in the *Proceedings of the IEEE*. It's a very long paper and it was written a couple years after we finished the system.

The check-recognition system we built started as a research prototype that then got turned into a product. A company named NCR, which at the time was a subsidiary of AT&T, deployed the product in their machines. It was first deployed in 1996. And in 1996 AT&T split itself up and basically ended the project because the research group stayed with AT&T while the development group went to Lucent Technologies. The product group, NCR as a company, was spun off, so the whole project was disbanded right after it became incredibly successful. The only disappointing thing about the project was that we never really received internal credit for the success except that I was made a manager! It was depressing for me that at this great moment of success, all of a sudden the whole company that made it possible decided to break itself up.

It took us a couple of years to write a long paper that described the entire system. So the first half of it is basically: Here is what convolutional nets are all about, here is how you implement them, and then here is everything else you need to know about the technique. Then the second half of the paper is how you integrate this with things like language models or language interpretation models. For instance, when you're reading a piece of text, if it's English text, you have grammar for English, so you want a system on top of it that extracts the most likely interpretation that is part of the language. And what you'd like to be able to do is to train the system to simultaneously do the recognition and the segmentation, as well as provide the right input for the language model. We managed to figure out how to do this.

Since then the technique has been reinvented in different forms multiple times for different contexts of natural language processing and for other things. There are models called CRF—conditional random fields—as well as structured perceptron, and then later things such as structured SVMs, which are very much in the same spirit except they're not deep. Our system was deep. So the second half of that paper talks about how you do this. Sadly, it seems like very few people ever read the second half!

I'm extremely proud of our work. It's probably the thing I'm most proud of. The convolutional net, which is what I'm known for, has had a huge impact in recent years. There was nothing conceptually complicated about it. What was difficult to do at the time was to actually make it work, particularly given the computational resources we had at the time and the software tools that we had to build ourselves. I'm very proud of some of the early work on back prop and other techniques, some of which I never published properly. I am also proud of the DjVu system, which is not machine learning; it's a digital document compression system.

Gutierrez: You're a professor and one of the founders of the NYU Center for Data Science. You're also leading a research lab at Facebook. What does a typical workday look like?

LeCun: I spend a little bit of time at NYU. If you had asked me this question a year ago, the answer would have been very different because I was basically running the Center for Data Science and putting together some of those programs. This entailed a good deal of administrative work. But I don't do this anymore because other people at NYU do this now. Whenever I'm at NYU, which is a relatively small amount of time, I talk to my students and postdocs. So it's mainly getting an update on their projects, giving them ideas on what to do next, and things of that nature.

When I'm at Facebook, there is also a good bit of technical meetings. Several meetings are more organizational, so these entail meeting with people in the company who have AI or ML problems that perhaps my group could help with—hiring, talking to candidates, interviewing candidates, and related things. A good deal of my energy is actually devoted to hiring. And then there is the daily work. A lot of what we do all of us are trapped in—like doing email and procedural meetings. Then, of course, there is some work that is of service to the community, like running the ICLR conference, doing peer paper reviews, writing recommendation letters, and related activities. Some of these are more science-related.

Gutierrez: How do you view and measure success?

LeCun: For scientists in academia and industry, the criteria are slightly different but not that different. In each area, the real criterion is impact. There are three kinds of particular impact in industry. One impact is the intellectual impact on the world at large, or on science and technology. So this is what you do by publishing papers, giving talks, and helping to change the way people do things.

You would think that a company like Facebook would not have any reasons to publish the research they do internally because it would be telling competitors what they do. But, in fact, there is a big incentive to do this. The incentive is that for projects that are very upstream of applications, but even for projects that are very close to applications, the best way to measure the quality of what you're doing is to test your methods on some standard data set or measure yourself through the kind of formal crowd-sourcing that peer review is. And for projects that are far from products, it's a much more accurate process than internal evaluation despite all its shortcomings.

Submitting yourself to the scrutiny of the scientific community is a much better way of evaluating the quality of a piece of work than internal evaluation. It's very difficult for a small number of people who may not be specialists to critically evaluate a piece of work within a company. This type of evaluation essentially leads to internal hype. You see this in some types of industries that tend to be more secretive, such as the defense industry. There are a lot of things going on there that are not so great but, because they can't publish, they really don't know. So it's useful for a company to have its scientists actually publish what they do. It keeps them honest. It's also very good for the prestige of the company. Publishing allows them to attract high-quality people, who generally want to brag about what they do, and need to talk about what they do if they want to be an integral part of the research community. So that's the first type of particular industry impact—intellectually strong impact.

The second type of impact is focused on research projects that are deemed important internally. These types of projects are ones where we say, "We know we want to work on this project." We don't know what ultimate application it will have, but we know it's important. A bunch of people will work on it and, if you have an impact on it, internally we can evaluate whether that impact is important.

The third type of impact, of course, is impact on things that are deployed. So if you work on something that ends up being deployed or you improve something that's already deployed in some measurable way, then that's a very direct way of measuring impact. There are organizations that claim to be research labs, but they only measure the third kind of impact and so they're not really research labs. They're just focusing on the short-term opportunities.

Gutierrez: How do you get ideas for things to study or analyze?

LeCun: That's actually a very important question that determines how a research lab should be organized. There are several kinds of research scientists. As a first step, you need people with vision. These are research scientists who have a long-term vision. They may or may not be that good at actually implementing it. They might associate with other people who do the implementation. Then there are people who are good at keeping their eyes on a long-term goal that will have a long-term impact and are good at ignoring fashions. Then there are people who are excellent problem solvers, who may not necessarily have the long-term vision, but they do have an ability to solve complex problems that other people just can't. And finally, you have people who can actually implement things and get them to work. In a research lab, you need all of those people.

Yes, you need all those people, but it is very essential that the people in the management of the research have vision, which means that they have to be respected in their field. The idea that somehow you can put a bunch of research scientists together and then put some random manager who's not a scientist directing them doesn't work. I've never ever seen it work. You need to have someone that has vision, who also has some standing in the community, directing a research lab or directing a group within a research lab. Management skills are a little overrated in the sense that managing research scientists is like herding cats. You basically have to make sure the litter is clean every morning so that people can do what they're best at and then you get out of the way.

Gutierrez: What specific tools do you use?

LeCun: I'm very special in this regard. I've had a long history of building my own tools, some of which were really exotic. Even before I joined Bell Labs, I built a tool with my friend Léon Bottou that at the time was called SN. Eventually it turned into project called Lush. The Lush project started in 1987, so this was before the days of MATLAB. It was basically a MATLAB-like environment with an interpreted language at the front end and with a numerical library in the back end. The front-end language was a Lisp dialect that we wrote. We continued using it once I joined Bell Labs. Then, over the years and the decades at Bell Labs, people like Léon Bottou, Patrice Simard, and others, developed all kinds of tools around it, including a compiler. It became a very useful tool for us at Bell Labs because eventually we were able to write research prototypes in Lush, compile them, and then turn them into product directly, without having to transcribe the code into something else.

The first version of our check reader, in fact, was prototyped in Lush and then translated into C++ by a team of developers. The result of this, unfortunately, was that now our research environment was different from our production environment. In these types of setups, you can't really improve the production environment, as it becomes a black box because it's not instrumented. After this experience, we decided that we had to use the same tool for prototyping and productization, and so that's why we wrote the Lush compiler. This compiler had tons of limitations but ended up being very useful. We managed to release this in open source by the time Léon and I left AT&T in 2002. We used Lush when I worked briefly at NEC Research Institute in Princeton. We also used Lush at NYU for quite a while for a lot of our projects until about 2010.

A few years ago, we switched to another similar environment called Torch 7. The philosophy of it is very similar. It's a flexible, dynamic language that's compiled. Torch 7 is built on top of LuaJIT, so it's the compiled version of the Lua programming language. Lua is a simple language that is very popular in the video game and computer game industry. It's a scripting language and so it's a very easy to extend language. Torch 7 is basically a numerical and machine learning extension to Lua. Torch has been really flexible for us.

We use Torch 7 at NYU in our research and we also use it at Facebook AI Research. We've actually released some of our Facebook source code related to Torch as open source. Google is also using Torch 7. They recently acquired a London company called Deep Mind. All of Deep Mind's code is essentially built around Torch. Ronan Collobert who is one of the originators and main developers of Torch, has joined Facebook. The other two main developers are both former students of mine. Koray Kavukcuoglu is at Deep Mind. Clément Farabet is at Twitter here in New York. There are other developers as well, but these are the main three characters. Many other companies are using Torch7 for deep learning besides Facebook, Google, and Twitter.

Gutierrez: What papers have been published on the work you're doing at Facebook?

LeCun: We had two papers at the CVPR 2014 conference. One of the papers was on a system called PANDA, which basically does human detection in images. The other paper was on a system called DeepFace that does face recognition with very good accuracy. Both of those systems use convolutional nets. Both of those projects, by the way, were started before I joined Facebook. We also had two papers at the EMNLP 2014 conference on topics such as text embedding—for things like hashtag prediction—and question answering. There are other projects that were started more recently that we have submitted papers on just in the last few weeks on essentially natural language-related tasks.

A natural language model is a system that can predict, upon seeing a piece of text, what word is going to come next in the text. It can't predict the exact word, but it can predict the distribution of the words and can then measure the accuracy of that prediction. So we've been working on basic techniques to be able to do this—not necessarily because we're interested in language models, but because we're interested in the underlying techniques that can be used for all kinds of different domains. Variations of recurrent neural nets, for example, is one of these techniques that we are exploring. A particularly interesting development is what we call "Memory Networks", which are recurrent nets augmented by a separate "short-term memory" that the neural net can use to store temporary states. Using a memory network, we have been able to build models of simple worlds, similar to the world of a text adventure game, from which the model can answer complex questions after being told a number of events. An impressive example is a paragraph that described the sequence of major events in Lord of the Rings. Then we can ask questions like "where is the ring?" (answer: Mount Doom), "where is Frodo?", etc.

Gutierrez: What lessons have you learned from being at Facebook and working with their data sets?

LeCun: Well, the data sets are truly gigantic. There are some areas where there's more data than we can currently process intelligently. Content analysis and understanding users are very important problems. And then there's the big question of matching users' interest with content. There is an interesting interplay in areas like image recognition and natural language understanding inside companies like Facebook, Google, and Microsoft. There are other players as well, of course, because it's a bit of a gold rush. It's not quite a land grab because you can't stop other people from doing the same thing you are doing, even though there are such things as patents. At Facebook we don't file patents to sue other people; we occasionally file them so that other people don't sue us.

Essentially, Facebook files patents so that we can say, "We came up with this idea, perhaps at the same time that maybe you did, but you can't sue us for using it." Another way of protecting ourselves like this is to publish papers, because that also establishes priority. So there is a bit of this interesting land grab. What's been very surprising to me is the amount to which there is now a sense in the industry that AI is going to revolutionize everything. AI all of a sudden or machine learning, particularly deep learning, went from some sort of obscure academic field of investigation to front and center at major successful companies like Google, Facebook, Microsoft, IBM, Yahoo!, Baidu, Yandex, and others. It's been a very recent, quick, and surprising phenomenon to me.

Gutierrez: How have you found this sudden change and attention?

LeCun: At some level it's not really a surprise, but it's not like you can count on this kind of stuff happening. As a researcher, you do research and you hope someone will pick it up. When it's Mark Zuckerberg picking it up, it's a different kind of person than the usual colleagues in academia or industry. A lot of people in Silicon Valley and technology are fascinated by the potential impact of AI in the future. And that's what's driving a lot of this growth and concentration in this academic field. I agree with this. Certainly, AI will have a big impact on society. And it's probably true that companies that can position themselves in the right way in AI are going to have some sort of position of advantage going into the future.

That's why, for example, Google is going into robotics. They believe that robotics is going to be the big thing over the next 10 years. Now, for some other companies with less financial strength, it might seem premature. Maybe the industry is not going to take off for the next 5 or 10 years, but Google has enough resources to wait it out and make sure that they're ready or actually help it happen. Facebook has a similar approach. Facebook's core business is to facilitate communication between people and also facilitate communication between people and the digital world. That's going to be mediated by AI. In fact, it's already mediated by machine learning.

Gutierrez: How do you know whether you're solving the right problem?

LeCun: The question is different depending on whether we're talking about research projects or more practical short-term things. It's always very difficult to know whether you're working in the right direction for a research project. This relates to the vision thing I was talking about earlier, that if you have some idea of where the world might be going, then it's easier to continue to work on it if you really believe in it. For example, to me, the motivation for deep learning is a complete no-brainer. It was completely obvious to me 25 years ago and it's surprising to me that it's taken so long for other people to realize it. But it's still surprising to me how fast they converted to it once they became convinced.

It's the fact that when you build your system, most of the time you've spent is in building the data analysis system or data mining system, machine learning system, whatever it is. Most of the time you spent was doing feature design, data cleaning first of all, and then feature design. Then you turn the crank on your favorite SVM or logistic regression or boosted trees or whatever you're using for classification and prediction. From this point of view, feature design is where all the time is spent. And how good of a job you do on feature design limits the ultimate performance of the system.

So, clearly, if you could use learning for that—and with enough data you can, you could use learning to design the feature extraction system. It would be a big win because of all of the manual labor that would disappear and perhaps your system would work better because the feature extractor would be tuned for the data you had at hand. That was the motivation behind deep learning. Of course, the danger of this is that the system now has too many parameters and overfits. So we have all of those related concerns. So you need a lot of data for that to work, and that's why people haven't picked up on this until recently.

That logic was obvious to me 25 years ago, and it's still obvious. It's surprising to me that it's taken so long for people to realize this. So that's one example of where having this sort of long-term vision helps you convince yourself, in the face of all your papers being rejected and nobody picking up on your work, that you're actually on the right track. There's also some limit to this type of belief in the absence of success. I mean it's not like I haven't been successful at all with this. Of course, the check-reading system was a big success. It's one of the few things that people remember from early success with neural nets. It's not like my work was completely ignored. However, there was certainly a winter period between the mid 1990s and mid to late 2000s.

Now, for more practical or short-term things, there are obvious measures of success. There are metrics. All of the web companies have metrics for how well they're doing, like how many clicks you get and what's the lift on whatever things you measure and things like this. Those are pretty obvious. And those things are being tested all of the time.

Gutierrez: Whose work is currently inspiring you?

LeCun: The people whose work inspires me are the people whom I really learn something from when I talk to them. They are old friends. So people like Geoff Hinton, Léon Bottou, and Yoshua Bengio. In the more industrial context, my bosses at Bell Labs, Larry Jackel and Larry Rabiner, were an incredibly good mentors. In neuroscience it's Sebastian Seung and Eero Simoncelli. These are people my age or slightly younger or older who I know are interested in the same questions and have slightly different takes on things and we always learn from each other. I cherish all the time I can spend with them. There are also lots of younger people whose work I find amazing. I've tried to hire them all at Facebook! In terms of people whose work I find inspiring, they're not necessarily in my field. They are people like Richard Feynman, Albert Einstein, or people with similar breakthroughs. What I find inspiring about them is their particular intellectual way of approaching problems, which I find fascinating, if not mysterious, and which I'd like to be able emulate.

Gutierrez: What do you look for in other people's work?

LeCun: The work that I tend to be interested in is innovative, creative work, which occasionally doesn't make it to the big conferences because it's too innovative and too creative. The review process for big conferences tends to focus on incremental improvements on mainstream models, which I find completely boring. I mean, it's useful, so don't get me wrong. I don't want to say people should stop doing this type of work, but it's not like I learn much from that. It's just that I am interested in the innovative creative work. I learn much more from that type of work.

Another type of work that interests me is the type that is very careful experimental work, which is very impressive. This is where the work that's been done is relatively straightforward and obvious, but it's great that people have done this work so exceedingly well that you now have a piece of data that you can point to and say, "This actually works if you do it right." In fact, to some extent, convolutional nets were like that to many of us.

I remember that I'd been out of my postdoc with Geoff Hinton for about a year and I gave a talk at NIPS in 1989 on convolutional nets. It was the first big talk on convolutional nets that I gave, and Geoff Hinton was in the audience. I had started thinking about things related to the topics I was speaking about when I was a postdoc with him in Toronto, so he knew a bit about my thoughts. At the end of my talk he said, "This is a very good talk you gave. Basically, the result is that if you do all the sensible things right, it actually works."

It was at the same time a compliment and not a compliment in that he was sort of saying, "This is not particularly innovative. We knew that this was the right thing to do, and it's nice that you've done it and shown that it's the right thing to do." I think there is something to be said for that, and I find this interesting in other people's work as well. Of course, these ideas weren't nearly as obvious to other people as they were to us. But the best ideas are always obvious a posteriori. Some people are more impressed by mathematical virtuosity—you know, really impressive technical work at a very high mathematical level. Some of that work is occasionally useful, though not always. I'm very much interested in theoretical work that sets the foundation for how to think about a problem, but I'm not impressed by vacuous theory, even if it's technically impressive.

Gutierrez: What's the biggest thing you've changed your mind about?

LeCun: Unsupervised learning. I've actually changed my mind two or three times about this, and I'm probably not done. Back in the old days, when I started working with neural nets in my undergrad and early grad student days, I worked on unsupervised learning and never published anything about it. I mean, not really. I did, but it wasn't that great. At that point, I thought that unsupervised learning was really ill-defined. You can measure performance, but there was really no objective way of saying, "Here is an objective function to minimize. And I know if I minimize it, my system will work well." And so, because I thought that it was so badly formulated, I decided it was useless.

So I started working on and became a big believer in discriminative, purely supervised learning. Convolutional nets are a result of this thinking. Basically, there were similar kinds of architectures that were used before. Fukushima's neocognitron is one, for example, which had a similar architecture to convolutional nets. It's not the same, but it's really similar. They were trying to use unsupervised learning mostly. I thought that that particular approach was insufficient, so convolutional nets are basically a supervised simplified version of Fukushima's neocognitron. At the same time, my friend Vladimir Vapnik came up with the theoretical argument that you should never try to solve a more complex problem than you have to. In this case, unsupervised learning, to some extent, is a more complex problem than, say, classification in the sense that it's like learning a density in a high-dimensional space. That's like the hardest thing you can imagine. So I was against unsupervised learning in some ways.

At the time, however, Geoff Hinton was actually a big advocate and believer of unsupervised learning. It's funny because we've been sort of out of phase in our beliefs about supervised learning versus unsupervised learning. So at the time he was trying to convince me that I should work on unsupervised learning, all the while I was telling him that the stuff that really works is actually supervised learning. Then came the early 2000s and my opinion changed completely. I realized that to really solve the deep learning problem with very deep networks, perhaps you would need unsupervised learning to do pre-training. Geoff came up with some idea on how to do this, and that was very inspiring.

So I started to get really interested in this work and unsupervised learning. I worked on a technique called "sparse auto-encoders," which are now relatively widely used, although not really for industrial applications. We got a bunch of interesting work done with this technique and saw some improvement on some data sets which were really, really interesting. Unsupervised learning is more biologically plausible, as it's clear that the brain is trained more in an unsupervised manner than a supervised manner. So this work had more connections with neuroscience. That was a way of getting papers published, because if you publish another paper on supervised convolutional nets, people say, "You talked about this in 1991. Why don't you do something else?" Whereas unsupervised learning was kind of new, so that's what actually got people interested in deep learning.

Then what happened is that the practical applications of the deep learning are all purely supervised. They're all basically just back prop on convolutional nets. And the funny thing is that Geoff was one of the people who changed their minds about this. I mean, he didn't change his mind in the sense that he still thinks unsupervised learning is the way to go in the long run, and I believe this, and Andrew Ng believes this, and Yoshua Bengio believes this. But in terms of practical applications, he changed his mind in the sense that he started working on purely supervised convolutional nets like me. He applied this to speech recognition, image recognition, and contributed to changing the opinion of the community about the whole idea of deep learning.

We've been sort of oscillating between the case of Geoff thinking unsupervised learning is the way to go in the long run, but sort of holding his nose and, you know, making supervised learning work because it actually works. And for me, it's been more of a change of opinion over the years. I'm more like Geoff now. I'm still convinced that in the long run for things like natural language and video, unsupervised learning will have to play a big role, but the stuff we use today in practice is all back prop.

Gutierrez: What does the field of data science look like in the future?

LeCun: What I say very often in regards to the future of data science is that one of most important things to notice is that the amount of data that's being collected and stored is growing exponentially. It either grows at the speed at which our communication network increases in bandwidth or at the speed at which our hard drives increase in capacity. It's always one of the two, depending on whether it is streaming data or it is stored data. And so that's an exponential with a pretty big rate. Currently, when you try to extract knowledge out of that data, there are humans in the loop. The amount of human brainpower on the planet is actually increasing exponentially as well, but with a very, very, very small exponent. It's very slow growth rate compared to the data growth rate.

What this means is that inevitably—in fact, this has already happened—there is a point where there are just not enough brain cells on the planet to even look or even glance at that data, let alone analyze it and extract knowledge from it. So it's clear that most of the knowledge in the world in the future is going to be extracted by machines and will reside in machines. It's probably already the case actually, depending on what your definition of knowledge is. For me, knowledge is some compilation of data that allows you to make decisions, and what we find today is that computers are making a lot of decisions automatically. That's not going to get any better in the future.

So if we look at future of data science, data science is not going away in the sense that the science and the technology—as well as the engineering around extracting knowledge from data—are going to be one of the big things of the future that societies are going to be relying on. It's already the case to some extent. The web relies on this already. But all society will rely on this. So this is not a fad; it's not going away. If you say that data science is a fad, it's as if you had said in 1962 that computer science was a fad. Look where we are now.

So my thinking around this phenomenon is that it will create—of course, it's created an industry, which we all know—a demand for people who are educated in this area. And it's also creating a need for an academic discipline that deals with this. This is something that some people aren't quite grasping at this point. For instance, if you are a statistician, you say, "Well, that's just statistics." If you're a machine learning person, you say, "Well, that's just machine learning." If you are a database person, you say, "Well, that's just a database, with a bit of machine learning and statistics on top." If you're an applied math person, you say, "Well, all of these techniques and methods use applied math."

All of those people are wrong. It's all of those things combined into one discipline: statistics plus applied math plus computation plus infrastructure plus the application areas, which are the things that those methods can be applied to, which require expertise. So techniques such as deep learning allow us to reduce or minimize the amount of human expertise required to attack a new problem, so that the machine does it by itself as much as possible. Of course, at this point, there are always humans in the loop. And things like data visualization make it easy for people to do things like this as long as there are still humans in the loop. Eventually, however, those models will essentially just build themselves.

I really believe in this concept of data science being a new academic discipline. At NYU, we helped start this trend because we were early with the creation of a Center for Data Science. We were also very early with the creation of the master's degree in data science, which is and has been a big success. We have had incredible support from the Moore-Sloan Data Science Environment Initiative, which is a big program by the Moore and Sloan Foundations. This initiative has grouped NYU, the University of Washington, and the University of California, Berkeley, together with the purpose of establishing data science as an academic discipline for the sciences.

Gutierrez: What types of academic scientific disciplines will benefit?

LeCun: The new way of doing genomics, astrophysics, neuroscience, and social science is through analyzing massive amounts of data. That's going to revolutionize all of these fields. In fact, it's already revolutionized genomics. Genomics itself as a field is actually entirely dependent upon data analysis. It didn't even exist before that. Genetics did, but not genomics.

In physics, a lot of the new results in astrophysics and high-energy physics actually rely very heavily on large data and complex statistical models. Things like the discovery of dark energy, for example, co-discovered by Saul Perlmutter, Nobel Prize winner, who is my counterpart of the Moore-Sloan Data Science Initiative at UC Berkeley, was made using massive statistical analysis. Also, a thing like the discovery of the Higgs boson was the result of massive statistical data analysis and results. Part of the system for this work was actually designed by my NYU colleague, Kyle Cranmer, who designed the integration for all the statistical models.

Data Science is also on its way to revolutionize social science. There is actually a big push from social scientists who would love to put their hands on Facebook's data. Facebook does not give its data away because it's private for users, and so that data is inaccessible for science. It's sad perhaps, but that's the way that Facebook operates.

Gutierrez: What is something that a smallish number of people know about now that you think 5 or 10 years from now will be huge?

LeCun: I think we'll have systems that do a much better job than they currently do at things like language translation. The difference will be that they will actually understand what it is that they are translating. Current systems that do translation don't actually understand what they are translating. It's just glorified statistical pattern matching. Eventually, however, these systems will understand more and more about the text that's being translated. The techniques for language understanding, in general, will have wide applications, not just for translation, but also for search indexing and intelligent agents. One of the things that people have been asking me about in the last few months is the movie *Her* and my thoughts on interaction with the intelligent agents. We're not going to have this within 5 or 10 years. This is way beyond what we can do, but intelligent systems that have some hint of common sense may appear in the next decade.

Gutierrez: What advice would you give to someone starting out?

LeCun: I always give the same advice, as I get asked this question often. My take on it is that if you're an undergrad, study a specialty where you can take as many math and physics courses as you can. And it has to be the right courses, unfortunately. What I'm going to say is going to sound paradoxical, but majors in engineering or physics are probably more appropriate than say math, computer science, or economics. Of course, you need to learn to program, so you need to take a large number of classes in computer science to learn the mechanics of how to program. Then, later, do a graduate program in data science. Take undergrad machine learning, AI, or computer vision courses, because you need to get exposed to those techniques. Then, after that, take all the math and physics courses you can take. Especially the continuous applied mathematics courses like optimization, because they prepare you for what's really challenging.

Gutierrez: What's something someone starting out should strive to understand deeply early on?

LeCun: It depends where you want to go because there are a lot of different jobs in the context of data science or AI. People should really think about what they want to do and then study those subjects. Right now the hot topic is deep learning, and what that means is learning and understanding classic work on neural nets, learning about optimization, learning about linear algebra, and similar topics. This helps you learn the underlying mathematical techniques and general concepts we confront every day.

Gutierrez: What do you look for when hiring others?

LeCun: I'm in a special situation because I'm building a research lab with world-leading scientists and so I hire research scientists at the junior level, mid-level, and senior level. What I look for is a track record in research, which means a strong publication record, not necessarily lots of papers, but papers with a particularly large impact that we know contain really interesting ideas. A large number of people that we hire tend to have been on our radar screen for a few years. Occasionally, someone shows up that wasn't on our radar, so we are constantly looking for great people as well.

There is another category of people that we recruit, but frankly, it tends to be more internal recruiting than external. We look for people with extraordinary programming skills combined with a good knowledge of things like machine learning or at least the ability to learn it really quickly. We're very fortunate at Facebook AI Research that some of the people in the group are essentially the most respected and top engineers at Facebook, which is amazing. These people are just astonishingly good. They're making things possible that we wouldn't otherwise be able to do and we couldn't have even approached.

So we look for mainly those two types of people—research scientists and exceptional engineers. You need a very wide spectrum of expertise. And don't forget, diversity of point of view is also a very important thing. You don't want to just hire clones of the same person, because then they will all want to explore the same things. You want some diversity.

Gutierrez: Where do you see the biggest opportunities for data science?

LeCun: If you are a scientist in an experimental science, particularly social science, I think there's a huge amount of opportunities at the boundary between the method side of data science and the domain science. This is going to revolutionize a lot of areas of science and so it's a very exciting place to be, particularly in social science. Other areas have already had a head start, like genomics and biology. But neuroscience, in particular, and social science are big areas of opportunity. If people are just starting out, I would suggest looking there for big interesting and exciting problems to tackle. And of course, if you are interested in methods, deep learning is where the action is.

Erin Shellman

Nordstrom

Erin Shellman *is a statistician and data scientist in the Nordstrom Data Lab—a small team of data scientists and software developers who build data-driven technologies like product recommendations. As of the time of writing, Nordstrom operates 289 full-line department stores across the United States and in Canada, and 151 Nordstrom Racks. Shoppers increasingly buy online and Nordstrom offers a variety of retail experiences to fit different preferences. For bargain-hunters, the flash sale site Hautelook offers daily deals for members, and for men who want a personal touch, Trunk Club provides a curated home try-on experience. In combination with* Nordstrom.com, nordstromrack.com, *and brick-and-mortar stores, these channels result in a rich data ecosystem that the Data Lab uses to inform business decisions and enhance the customer experience.*

Shellman's data science career began with an internship at the National Institutes of Health in the Division of Computational Biosciences. It was here that she initially learned and applied machine learning to uncover patterns in genomic evolution. Following her internship, she completed a Master of Science degree in biostatistics and a doctoral degree in bioinformatics both from the University of Michigan in Ann Arbor. While at the University of Michigan, Shellman collaborated frequently and analyzed many types of heterogeneous biological data including gene expression microarrays, metabolomics, network graphs, and clinical time-series.

A frequent speaker and teacher, Shellman has presented at conferences such as Strata and the Big Data Congress II, and also speaks regularly at meet-ups and gatherings in the Seattle technology community. She is also an instructor at the University of Washington where she teaches data mining in the continuing education program. As a community outreach leader, Shellman is a co-organizer of the Seattle chapter of PyLadies, an international mentorship group for women in the Python open-source community, and a mentor to young women pursuing education and careers in math through the Association for Women in Mathematics.

Shellman is an outstanding exemplar of the trend of physical scientists transitioning into the field of data science to enable companies and organizations to make better decisions with data. In her interview, Shellman reveals her love of biology, the application of scientific reasoning to the world of data science, and how, as a newcomer to the field, she is adapting her skills and mindset to meet the challenges of her new career. She discusses the application of record linkage to match similar products with high SKU turnover, how she thinks about what to build and which problems to tackle, and provides advice to undergrads and grads interested in transitioning into a career in data science.

Sebastian Gutierrez: Tell me about where you work.

Erin Shellman: I work at Nordstrom in the Nordstrom Data Lab. It's an exciting job because in retail, specifically fashion retail, I have the opportunity to work on data applications and problems in lots of areas, including transactional, business operations, clickstream, and seasonal trends in garment colors. Retailers the size and age of Nordstrom have access to unique data that many online retailers don't, because they don't have brick-and-mortar stores. The heterogeneity of the data we work with makes the job really fun and challenging.

Gutierrez: How is the Nordstrom Data Lab structured?

Shellman: We are a small, multidisciplinary team split nearly equally into data scientists and software developers, and we work closely together on all of our projects. Together, we become greater than the sum of our parts.

Gutierrez: How is it working in this type of multidisciplinary team?

Shellman: I love my team because everyone is awesome and there are so many opportunities to learn from one another. For example, we use a technique called pair programing. It's a mutual programming exercise where we have two monitors, keyboard, and mice tethered together so that we can work on the same program simultaneously. It's a great way to code review, learn and teach. That's really been my favorite part of working on a multidisciplinary team.

Gutierrez: In addition to pair programming, do you do pair data science?

Shellman: We don't formally pair on statistics or data science work. For these subjects we have standing discussions around the whiteboards that surround our open-plan office. For instance, yesterday we finished the day with a discussion of how a statistical model could be applied, what data would be needed, the limitations of the model, and the latency expected when using the model in a real-time application. So while we weren't pair programming, we were discussing behavior and expected results as a group. The great thing about our workspace is that these discussions happen in the open, so everybody can hear, chose to participate, and join in if they have something to contribute.

Gutierrez: How has it been working with developers?

Shellman: I've been an R programmer for a decade, but you don't develop general-purpose software in R, so I've had to learn a ton and think about things differently. We released Recommendo 2.0, the second version of our product recommendations API, earlier this year and I worked on a real-time component of that called "the scorer." In a nutshell, the scorer receives messages about customer actions like product page views or add-to-bag events. It then re-orders recommendations in real-time based on the customer behavior right now in the session. I initially wrote the scorer using the Python library pandas because we'd been using it for our nightly batch recommendations. Well, I learned that the conveniences of the pandas data frame that we were enjoying for batch jobs had subpar performance in real-time applications. Each time the scorer runs, a message needs to be parsed, scores need to be computed and updated, and the process requires multiple reads and writes to our Dynamo tables on AWS. In this situation the set-up costs of the data frame objects were too high, and I ended up having to re-write the whole thing without pandas.

In the past I didn't really have to worry much about the performance of my programs, so I didn't recognize the performance limitations of the scorer I'd written. It was really amazing to have talented developers around when I needed advice.

Gutierrez: What's been the progression from undergrad to Nordstrom?

Shellman: I started undergrad as an economics and philosophy major because I really liked thinking about complexity in human behavior and interactions. I was drawn specifically to economics because I liked the process of describing that complexity with math. As I got further along in my program, I realized that I couldn't see myself at the age of 40 still working in economics, and knew I needed a change. I was young and didn't realize that you don't have to do the thing you majored in for the rest of your life.

Part of what drove my shift from social science was reading Richard Dawkins' *The Selfish Gene*[1] and Robert Axelrod's *The Evolution of Cooperation*[2], which is a classic text about evolutionary game theory. I developed an admiration and love of applying econ-like math models to biology. So rather than continue to focus on economics, I decided that I wanted to study biological sciences while retaining the math focus. To really explore this area, I got an internship at the National Institutes of Health in the Division of Computational Biosciences. I worked for Jim Malley, a mathematician who taught me machine learning and

[1]Richard Dawkins, *The Selfish Gene, 2nd ed.* (Oxford University Press, 1990).
[2]Robert Axelrod, *The Evolution of Cooperation* (Basic Books, 2006).

how to apply it to biological problems. I did a lot of work using random forest for research in comparative genomics.

I loved the internship so much that I knew I wanted to do something with statistics and biology after undergrad. I started the master's program in biostatistics at the University of Michigan and later went on to do a PhD in bioinformatics, also at Michigan. When I finished, I decided that the traditional academic path wasn't for me, and I wanted to work on challenging problems in data science. Interestingly, even though academia was ahead of industry in applying the techniques of machine learning, it lags in the effective use of data technology. I thought that to really learn that technology, I needed to leave academia, and that's what I did.

Gutierrez: What advice do you have for current undergrads?

Shellman: My advice to undergrads is to study computer science, math, or statistics, and a combination of the three. It doesn't matter what else you study alongside them, if you have those three skills, you can do whatever you want, literally. I think the opportunities are endless, which means you don't actually have to commit to any industry. I can't advocate enough for the study of math in general and the maths more broadly because it's where you learn to reason and think critically. The really exciting industries that are experiencing a lot of growth all involve math, computer science, and statistics in some way.

Gutierrez: Now that you are in the industry, what are you studying?

Shellman: Right now I'm studying programming and computer science, and I'm hitting it hard. I'm going to a lot of meet-ups, speaking at a lot of meet-ups, and even organizing one. I'm a co-organizer of the Seattle chapter of PyLadies, which is an international mentorship group whose goal is to help women become active participants and leaders in the Python open-source community. I'm also revisiting some old projects from grad school and applying my new and improved data skills to see if I can revitalize them.

I'm at a point where I feel myself getting up this mental hill. I'm done learning basics and have a good sense for how all the pieces fit together and I'm building on that knowledge. Although I'm starting to realize I've gotten hazy on things I was solid on before, so I have to go back and review. I have to make sure I am maintaining a balance between programming and math.

Gutierrez: When did you realize you wanted to work with data as a career?

Shellman: I knew I wanted to go into science as a career after my internship at the NIH, and collecting and analyzing data is a huge part of scientific work. I don't think I was ever really aware of wanting to work with data _per se_, I just wanted to answer cool questions. Data's just the world making noises at you. I think I was more interested in the application first.

Gutierrez: Do you see yourself working in data science in your 40s?

Shellman: Well, the thing about data science is that it's almost a catchall to the point that it's meaningless. The reason that almost everybody starts a data science talk with a slide discussing "What does it even *mean?*" is that it almost means nothing. A data scientist to me is a person with a certain set of quantitative and computational skills that are applicable across different domains. So as a data scientist, even if I don't have the domain expertise I can learn it, and can work on any problem that can be quantitatively described. I can almost guarantee that I won't be in fashion retail in my forties, but I'm sure I'll be working on something that relies on data and using similar techniques and methodologies.

Gutierrez: How would you describe your work to a data scientist?

Shellman: I build the recommendation engines like the ones you're used to seeing all over the web, and sometimes I do it with really unique data, like transactions involving personal stylists in our brick-and-mortar stores or color trends from fabrics.

Gutierrez: What have you been working on recently?

Shellman: Over the last year and a half I've mostly worked on Recommendo, building new algorithms and the real-time scorer. For the last couple months we've been working on a follow-up to Recommendo that will offer customer segmentation as a service. We're calling it Segmento and we've already conducted some initial tests with the email marketing team.

Segmento is an internal tool for creating custom audiences partly inspired by the Facebook audience builder tool. For our initial test we constructed customer affinity scores for certain brands and categories of products from past purchase activity and product page views. I've been working on a classifier to assign similar customers to pre-defined groups based on demographics and purchase behavior.

Gutierrez: Why are recommendations important to Nordstrom?

Shellman: For starters, recommendations play to Nordstrom's strengths, because we have data other retailers don't. We've got those lovely brick-and-mortar stores! The first recommendation engine I worked on is called Our Stylists Suggest, and it works by analyzing transactions that occurred in our full-line department stores and were facilitated by a personal stylist.

The idea is that our personal stylists are the best recommenders we could ever hope to find and are fantastic at building cohesive outfits for our customers. Our goal was to emulate those stylists online and build a recommendation strategy to help our customers coordinate their wardrobe.

Our Stylists Suggest is a great example of a recommendation algorithm that's unique to fashion data and takes advantage of the deep expertise of our stylists. It's a competitive edge against strictly online retailers because it's a customer experience they can't easily replicate.

Gutierrez: What data did you work with when you joined Nordstrom?

Shellman: Our Stylists Suggest was the first project we tackled as a team, and that dealt primarily with in-store transactional data from our 117 full-line stores across the country. We started by examining what stylists were selling, specifically the composition of shopping baskets

Now that we've expanded the recommendations work into a full platform we work with data from many sources including clickstream and online purchases.

Gutierrez: How do you mix together the venerable Nordstrom brand with the innovative data science approach?

Shellman: It's an ongoing challenge, but we've developed a couple strategies that have been successful. The first is actually a company-wide open-door policy that invites people to come by and see us, and they do that quite a lot. Folks from all parts of the business will come and tell us how they've been approaching a problem and where they think we can help. Often they've been spinning their wheels, and are interested in our thoughts on how to approach the problem. Other times people just stop to chat and use us as a sounding board.

We also work hard to have good relationships with the people who are ultimately responsible for getting our work in front of customers—primarily the web team. So we keep in touch with them regularly and let them know what is going on from our side. We also make sure that when they find bugs that we respond right away—especially because in the enterprise world, they're used to having to file a service request and it taking a week for someone to respond to them. So when we can respond with, "I just pushed the change and in about ten minutes it will be live on the site," they obviously love that.

Finally, prototyping our products so that internal customers can use them early on has been crucial for our success. It doesn't even have to be something fancy. For instance, our recommendations preview tool doesn't have a particularly interesting visualization, but it's enough to show the result of our algorithms. Now we can shoot off a URL to internal customers and it allows them to sit at their desk and understand the behavior of our product and experiment with it, and provide feedback way before we're talking about getting it into production. This has been super helpful and has been a really great way to get people excited about what we're working on. Building and maintaining those relationships are just like anything else—you always have to be working on them. Nurturing those relationships are part of the job, and if you leave them unattended, they might not be there later.

Gutierrez: How do you interface with stylists in your full-line stores?

Shellman: Fortunately for us, Nordstrom loves to promote from within, so a lot of our collaborators used to be on the selling floor or in the stockroom, and that comes in really handy when you want to get in touch with people in the store.

One of my recommendation strategies analyzed beauty items that customers buy repeatedly, so that we could deliver an email reminder when we think you'll run out of your favorite product. Nordstrom has "beauty stylists"— personal stylists whose expertise is specifically in beauty products and cosmetics, and we used a collaborator's connections to go the flagship store to speak to the beauty stylist there. We took her out for coffee to get an idea of what her day was like and how she interacted with customers. We also spoke with her about a tool we were planning to build that would alert her when a customer was almost out of a beauty product. Then the stylist could give them a call to catch up and check in about the item.

However, through the course of our chat we learned that she was not interested in our tool because if a customer she chatted with replenished their product online, our stylist wouldn't be credited with the sale. It was great to find this out before we spent a lot of time developing a tool for her, and we wouldn't have known without going in and talking to her. This discovery process was crucial to the development of the final product, a personalized email campaign.

Gutierrez: What's an experiment you have run in the store?

Shellman: We ran an experiment that involved placing Pinterest-themed signs on top-pinned products in the store. Primarily these were small signs that said, "Top-pinned items this week." This was an interesting case of using online data in our physical stores to see what value it provided to our customers. Specifically we measured if the new Pinterest signage increased the sell-through rate of those popular products.

We tested a couple experimental conditions. For example, is it better to place all the top-pinned items in a central location so they're easy to see, or pin them in their normal spots to encourage discovery?

Though just a small experiment, it helped draw attention of customers to popular products and was a conversation starter for our stylists.

Gutierrez: What other companies are doing interesting work in the fashion retail and specialty retail space?

Shellman: Rent the Runway rents high-end garments and is a cool and innovative business. Birchbox has a neat monthly subscription model for grooming products for men and women. Then you've got your crate-based companies like Stitch Fix and of course Trunk Club, which was recently acquired by Nordstrom. I'm really excited about Trunk Club because they will have fun new problems to solve, like how to optimize product assortment, and even optimizing weight and cost of shipping. In the flash sale arena you've got Haute Look, also a Nordstrom company, and Zulily who are both innovating and doing a lot great work especially in the area of product recommendations. Flash sale sites have extremely high product turnover so they're dealing with the cold start problem for every product every day.

Gutierrez: Are there other types of data that you think are ripe for plucking given that you have the big department store model?

Shellman: There's lots of exciting new data coming from our recent acquisitions, HauteLook and Trunk Club. It's smart that Nordstrom recognizes that there isn't one type of customer and provides different shopping experiences through the full-line stores, the Rack, HauteLook, and Trunk Club. It's going to be really interesting to see how customers interact with all these channels and it's going to pose new challenges. Measuring behavior across all those channels will be difficult, but lends itself to so many opportunities to cross-promote.

The most interesting types are data are those collected for one purpose and used for another. For example, one of our developers, Jason Wilson, had a really cool idea to look at what was purchased when people asked for gift receipts. Then you could make recommendations for the most gifted products for an upcoming holiday. It's an approach in the same spirit as Our Stylists Suggest. Just let people do what they're doing and use that behavior to enhance our services.

Gutierrez: How could someone starting out learn more about the fashion retail industry from a data science point of view?

Shellman: Conferences are a great place to learn and meet people working on similar problems. Twitter, honestly, is a really great resource. It's really easy to find and follow people from Rent the Runway, Warby Parker, Zulily, and other companies to see what they're up to and what tools and technologies they're currently into. Another great place is company blogs. MailChimp and Etsy have really great blogs.

I would also suggest that when you're in a conversation and have the opportunity to learn, ask lots of questions and don't worry about if you're perceived as not knowing what's going on. Then ask follow-up questions. A good medium for this type of exchange is emails. I'm not against cold emails at all. Don't worry about not knowing the language. Coming from biomedical sciences, I didn't have any experience in retail, so there was a lot of stuff that I didn't know and I just asked people to explain it to me. If you go into every conversation with that attitude, people will tell you a lot of stuff, and it's just your job to absorb it.

Gutierrez: What's something you've been proud of from your work at Nordstrom?

Shellman: I'm really proud of what we've been able to accomplish in a short amount of time. In under a year we released a recommendations API that is used in so many different parts of the business. Recommendo is all over our website, in our marketing and transactional emails, and soon in the mobile app.

From a personal point of view, I'm also proud of how quickly I was able to learn and adapt to the industry after school. When I think about my skills and the technology I used as a bioinformatician, there's not a lot of overlap with the skills and technology I use now. I've learned so much and matured in my skills quite a lot and that's been really awesome. I don't feel like I'm done though, I don't even feel close to done, but I'm really excited to see where I'll be in another year's time.

Gutierrez: What does a typical workday look like?

Shellman: We start with a group stand-up where we discuss what we're working on for the day, any requests for help, or anything that needs attention. Our group uses a kanban board to keep track of our work, so we describe tasks on post-its and move them around from to-do, doing and done. Then generally, depending on what I'm doing, I spend most of the day programming. Yay!

Occasionally, I will have meetings, though our director does an excellent job of shielding us from the bulk of these. This allows us to spend a lot of the day doing productive work. The team also engages in a lot of ritualistic behavior. Group lunch is typical, as is coffee in the afternoon. I feel really lucky that all the people on my team are people I want to be around all day.

Gutierrez: How do you view and measure progress and success?

Shellman: We are all responsible for our own progress, and don't do sprint planning. We write specific tasks on the Post-it notes, like "add more tests to scorer" and work on those tasks until they're done. We don't necessarily have things to hit every week, and just try to work as fast as we can on the task-based projects we own.

In terms of measuring success, on Thursdays, the Nordstrom Innovation Lab hosts show-and-tell, which entails 5 five-minute lightning talks. The purpose is to show what you've been working on and get early feedback. It doesn't have to be remotely complete. It could even be something that's broken that you're stuck on and need advice.

For us, Thursday is an unwritten deadline because we all want to make sure we have something for show-and-tell. There's only 5 slots so it might be the case that you don't get to show your work, but the goal is to have something, even if it's a couple of figures. Transparency into one another's work—the progress and the obstacles—is really helpful.

Immediately following show-and-tell we "retro," short for retrospective, which for us is basically just happy hour. At retro we talk about how the week went in terms of productivity, what worked, what didn't, and how to be better next week. Regardless of what we talk about, we see it as a time to get together as a team to reflect, or just chat.

Gutierrez: How do you come up with ideas for things to study or analyze?

Shellman: We develop ideas internally or they come from our collaborators in the company. One of the coolest things about our group is that we collaborate with people from many different parts of business. For example another data scientist on the team, Elissa Brown, has been working in the area of store fulfillment. In addition to our warehouses, each Nordstrom store can operate as a mini fulfillment center. Elissa's work involves forecasting expected fulfillment requests so that they can be efficiently distributed across stores and hopefully reduce the time it takes to get products to the customer.

We also develop our own projects. For example, we created a recommendation strategy called My Color Trends that is an interactive visualization of a customer's Nordstrom wardrobe through the lens of color. The recommender is an interactive way to visualize your own palette and explore the colors in the products you bought. My Color Trends is also a tool for customers to find new products that are a precise match with selected colors, or items that complement. One of the features I built was a strategy that recommends products in complimentary colors. What is considered a complementary color is debatable, but I got around the issue all together by creating associations rules from the most commonly co-occurring colors in the fabrics themselves. The great thing about that strategy is that it's data-driven so I don't have to assign rules that are inflexible and arbitrary. When seasons and trends change, the association rules will adjust to reflect the shifts in taste.

Some projects that start out with collaborators have to evolve to find a good fit. For example, the beauty replenishment work I mentioned previously started out as tool for beauty stylists and evolved into a personalized e-mail campaign It's been great because we have the freedom to work on that and be able to change according to what we think is best. Fortunately, we've had really good initial success, which has made it easy for us to go out on a limb to create new things. We didn't mess it up on the projects we were asked to do, so now we have the latitude to do some crazier work.

Gutierrez: What specific tools are you using?

Shellman: I'm writing a lot of Python these days, it's what all our recommendation algorithms are written in. The Recommendo API is written in node and hosted on AWS. We use a lot of open source libraries in Python, like scikit-learn and pandas. As someone who used to work almost exclusively in R, pandas is great because it's cheating in a way. It makes Python a lot like R, so you get to code in Python but get a lot of the conveniences that we've all come to expect from R. Of course, you'll also make yourself insane trying to remember whether it's "len" or "length," and 0 or 1 indexed.

Gutierrez: What is a specific project you have worked on recently?

Shellman: It's been a little while, but the beauty replenishment project was a really fun one. The project started out as a tool for beauty stylists but evolved into a personalized e-mail campaign. Initially we thought it would be helpful for stylists to know when their clients were running low on product so they could give them a call and remind them to come into the store. After early feedback from the stylists that they likely wouldn't use a tool like that, we found a home for the beauty replenishment work in a personalized email. I started by analyzing active beauty customers, going through their beauty transaction histories to understand what they purchased and then estimate when they would be ready to replenish.

The biggest challenge was that beauty products have fast SKU [Stock Keeping Unit] turnover. For example, say four months ago I bought lotion, and now there's a new and improved formula. As a customer, when I replenish my lotion, the new and improved formula is the same product that I bought four months ago. However, from the manufacturer's perspective, it's a new SKU. The issue is that if I didn't account for that SKU ancestry in my analysis I'd miss a lot of replenishment purchases.

I used record linkage to solve the problem. Record linkage is a technique used to find duplicates in things like census data and medical records. In survey data it's typical to have typos and variations in name spellings and you want to link those separate records into a single entry. I was doing the same thing—only instead of names and addresses, I had brands, categories, and product descriptions. I forced matching on things like product type and brand, and then used fuzzy string matching to measure the similarity between product descriptions. My output was a probability that two items were the same "record" for each candidate product.

Going in I didn't know that SKU turnover would be such a large part of the project. I was green and not familiar with the product catalog and how SKUs evolve. That made the project fun and challenging.

The beauty replenishment emails were the first campaign personalized at the individual level that Nordstrom had launched, and they out-performed traditional marketing emails. This summer we had a very capable intern who picked up this work and automated everything into an ongoing weekly campaign. It's been really rewarding to see that work go from initial idea, to test, to automated, and now in production. I'm really proud that it's a Nordstrom product conceived and built in-house, and think it showcases Nordstrom's technology capabilities.

Gutierrez: How do you or the labs think about whether to undertake new projects or improve projects you've already done?

Shellman: We usually just have a simple cost-benefit conversation with the person asking for new features. It typically goes something like this: "I could build a new recommendation strategy that reads the mind of the customer and automatically ships products to their house, but I would have to stop working on the other features you requested." Not always, but most of the time the requested improvements are nice to have, but not essential, although a mind-reading recommender would probably be a good idea to pursue. We prioritize based mostly on the web teams' launch dates, so as long as we can hit those we have no problems adding new features.

Gutierrez: How do you think about whether you're solving the right problem?

Shellman: I typically start from the finish line. Assume that you've built the thing you're considering, then ask "so what?" Do customers want your product and can they use it? Customers can be internal or external. Once we were considering making a recommendation strategy involving perfume "notes." In the end we decided against it because we asked a few simple questions. How many people use the perfume filters on the website, and who would consume the recommendations? It turns out most shoppers don't filter search results by perfume note, and there was no enthusiastic consumer waiting on the feature so we didn't build it.

Gutierrez: How do get to know the data once you've decided that the project is worth working on?

Shellman: I spend a lot of time doing basic plotting and visually scanning new data sets. Our director, Jason Gowans, laughs at me because I'll open and full-screen a dataset on my big monitor and just scroll through. I just like seeing it and getting an idea of what the fields are and what it feels like. It's also a good way to find typos and the weird characters that are common in brand names like Lancôme and M•A•C.

Plotting data in scatterplot matrices with R is really helpful because you can quickly start discovering relationships in your data without much work on your end. Heat maps are great for that, too. I use plyr and dplyr a ton to bust data up and look at aggregates. It's not a linear process. I just mess around with it and ask a lot of questions. Usually when I'm looking at new data a lot of thoughts, assumptions, and conclusions come to mind immediately about what things are and how they are related. Once I have those ideas, I'll then try to disprove them. It's a fun way to learn about the structure of data.

Gutierrez: How do you share the knowledge you are building with others?

Shellman: We're obsessed with Confluence from Atlassian, and the Data Lab has a very active Confluence space. Recommendo is fully documented on Confluence, so anyone in the company could learn how it's built, where to find it, and how to use it. We also share exploratory analyses and reports on Confluence so that we can still exchange knowledge even if the work didn't make it into a larger project.

In addition to Confluence, show-and-tell is great for knowledge sharing. We also have brown-bag lunch presentations, and that's a space for more long-form presentations. Sometimes the brown-bags grow from show-and-tell presentations. For example, one of our team members built a REST API, so the show-and-tell presentation expanded into a brown-bag presentation.

Outside the company, our team does a lot of presentations at conferences and meet-ups. Speaking and sharing our technology is not something Nordstrom has done a lot in the past, so I'm really glad we're doing it now because I love talking about my work.

Gutierrez: Whose work is currently inspiring you?

Shellman: I'm kind of obsessed with Hadley Wickham's work. I think in ten years, it will be obvious that Hadley kept R from becoming totally irrelevant for everyone except academics. Or maybe not, but we win either way because we're all better off for libraries like dplyr and ggplot2. I'm also a fan of Wes McKinney, the creator of the pandas library in Python.

Even though I'm out of the biz for now, I still read bioinformatics/systems biology literature, and am really into the work of the Covert Systems Biology Lab at Stanford. They do whole-cell modeling, which is the practice of mathematically modeling cellular systems at the genome scale. I had a failed thesis project in this area, and I love this research. Similarly, the OpenWorm project, which is a whole-organism simulator of a worm, is fascinating.

Gutierrez: When you say you "follow" that project and the literature, what does that mean?

Shellman: Well, besides reading their papers, I download the data and code and tinker with it. The Covert lab has made their code available and I've been messing with it a bit recently, because I've got some ideas left over from grad school incubating in my mind.

Gutierrez: What do you look for in other people's work?

Shellman: Presentation. What can I say? I'm shallow. I don't just mean visual presentation (though it's important), but the ability to convey results both technically and non-technically. The work needs to communicate the point clearly and coherently. I look for whether they did something fancy but without rigor. Sloppy complexity is rampant because so many methods, particularly in machine learning, have been commoditized with external libraries. Presentation is the ability to craft a story, from the reason that you did the research, why I should be interested, what you did, and justifications for your methods.

Unfortunately, I think presentation skills are undervalued, but is actually one of the most important factors contributing to personal success and creating successful projects.

Gutierrez: What do you think the future of data science looks like?

Shellman: If you look at trends in data science-y start-ups, we appear to be moving in the direction of push-button data science. I hear a lot of marketing about "freeing data scientists from having to program" or "freeing data scientists from the technical overhead so that they can get back to the data." I think these products are a response to a lack of supply of people with data science skills. In small companies, who are lucky enough to have one data scientist, these tools promise to make them more efficient. It's not an easy job though, so I'm pretty skeptical of these products and their longevity.

What I think we'll see in the future is an evolution of what a data scientist is. Right now it's typical that a data scientist is an ex-academic with a masters or PhD, but there just aren't enough of those people out there to meet the demand, so I expect we'll see a lot of retraining of software engineers and migration into the data science role. You can see some of that already happening with Coursera and Udacity offering data science courses and certifications.

Gutierrez: What is something a small number of people know about that you think everyone will know about in five years?

Shellman: Ad tech doesn't work, but I think we all know that now … just kidding. Tying it back to the work of the Covert lab I mentioned earlier, I hope that in the next five years we'll see a greater appreciation for predictive modeling in cell science. I think the ability to run experiments computationally and make predictions at the whole-cell system level is immeasurably valuable. The cycle of learning by modeling, testing model predictions with experiments, and updating the model with the results is so obvious to me, but academia as a whole is extremely risk averse and hasn't effectively made use of these models as a tool for experimental design.

Gutierrez: What advice would you give undergrads and grads looking to move into data science?

Shellman: For the person still deciding what to study I would say STEM fields are no-brainers, and in particular the 'TEM ones. Studying a STEM subject will give you tools to test and understand the world. That's how I see math, statistics, and machine learning. I'm not super interested in math *per se*, I'm interested in using math to describe things. These are tool sets after all, so even if you're not stoked on math or statistics, it's still super worth it to invest in them and think about how to apply it in the things you're really passionate about.

For the person who's trying to transition like I did, I would say, for one, it's hard. Be aware that it's difficult to change industries and you are going to have to work hard at it. That's not unique to data science—that's life. Not having any connections in the field is tough but you can work on it through meet-ups and coffee dates with generous people. My number-one rule in life is "follow up." If you talk to somebody who has something you want, follow up.

I heard a story from a director of a data lab similar to ours, about a potential hire that didn't have the skills and experience required to join the team. He suggested the candidate take a machine learning class on Coursera and figure out if data science was more than a fleeting interest. The candidate took the course and every week emailed the director notes on progress and asked follow-up questions. I think this is a really great example of putting in the extra effort and proving to potential employers and to yourself that you can do what it takes to succeed as a data scientist. I'm pretty sure they got the job too.

Gutierrez: What advice would you give to data scientists looking for work?

Shellman: Postings for data scientists can be pretty intimidating because most of them read like a data science glossary. The truth is that the technology changes so quickly that no one possesses experience of everything liable to be written on a posting. When you look at that, it can be overwhelming, and you might feel like, "This isn't for me. I don't have any of these skills and I have nothing to contribute." I would encourage against that mindset as long as you're okay with change and learning new things all the time.

Ultimately, what companies want is a person who can rigorously define problems and design paths to a solution. They also want people who are good at learning. I think those are the core skills.

Daniel Tunkelang

LinkedIn

Daniel Tunkelang is the Head of Search Quality at LinkedIn, the world's largest professional network on the Internet. Started in 2003, LinkedIn has grown to more than 300 million members in over 200 countries—a growth rate that exceeds two new members joining the service per second. LinkedIn's massive and ever-increasing scale makes the ability to find professionally relevant information and other professionals within the LinkedIn ecosystem more critical and more data-intensive. Building and maintaining the search capability to deliver the requisite user experience pose significant challenges given the size of the data involved, the specificity of search entities, the disambiguation of data, and the required relevancy of search.

Tunkelang previously served as LinkedIn's Director of Data Science, leading the Product Data Science team. Prior to joining LinkedIn, he led a local search quality team at Google. Before that, he was a founding employee and Chief Scientist of Endeca, an enterprise search application vendor acquired by Oracle for $1.1 billion. Tunkelang also worked at the IBM Thomas J. Watson Research Center and the AT&T Bell Labs. He holds twenty-two patents, a PhD in Computer Science from Carnegie Mellon University, and BS and MS degrees from MIT.

Gutierrez: Tell me about your journey to becoming head of Search Quality at LinkedIn.

Tunkelang: I was lucky to have an early exposure to both math and computer science. My dad taught me programming, as well as a bit of math and physics when I was still in elementary school. I also had great math teachers at school, which allowed me to take lots of math and computer science classes at Columbia University while I was attending high school a block away.

Then I went to MIT for college, where I double-majored in math and computer science, and also picked up a master's in computer science as part of a program that included internships at the IBM T. J. Watson Research Center. Going to MIT is, as they say, like taking a drink from a fire hose. I was able to take extraordinary classes in probability, combinatorics, and game theory, among many others.

I then went to CMU for my PhD in computer science, where I enrolled in a program called Algorithms, Combinatorics, and Optimization. This interdisciplinary program emphasized the facets of computer science that overlap with math and operations research. I learned a huge amount from my classes, and even more from my peers. I considered doing a dissertation on optimizing shared-ride transportation, but ultimately ended up developing a system for network visualization. Little did I imagine at the time that someday I'd be working at the company that provides the world's largest professional network!

When I finished my dissertation at the end of 1998, I had some soul-searching to do. I realized that I wasn't cut out for a career in academia, but that I really didn't know what industry had to offer. I worked a few months at a consulting firm, getting my first taste of a nonacademic, nonresearch job.

My lucky break was being discovered by the co-founders of Endeca in 1999 and enlisted as chief scientist. My ten years there were an extraordinary adventure. Our initial ambition was to build a better way to find stuff on eBay. Like most startups, we pivoted, and we ultimately developed technology that revolutionized the search experience for online retail, as well as expanding into other domains like manufacturing, business intelligence, and government.

After Endeca, I went to Google, where I worked on improving local search quality. Specifically, I led a team that matched the local search index against the web index to establish the official home pages of local businesses. It was a fun machine learning problem, and there was something very rewarding about improving the search experience on the world's most popular web site.

But when LinkedIn reached out to me in late 2010 with the opportunity to lead and build a data science team, it was an offer I couldn't refuse.

Sebastian Gutierrez: Why was the LinkedIn offer so compelling?

Daniel Tunkelang: Working at LinkedIn is awesome. We work on challenging problems that create massive value—not just online, but offline too. We help people get jobs, companies find talent, and more generally help professionals be more successful every day. To do so, we analyze big data—rich, semistructured, social data. In the process of which we develop and contribute open source tools, cool research results, etc.

Gutierrez: What teams have you worked in at LinkedIn?

Tunkelang: When I joined LinkedIn in late 2010, I led the product data science team, a group of data scientists focused on creating innovative solutions to improve LinkedIn's products and create new ones. It was a great chance to lead and hire people with a rare combination of computer science background, technical skill, creative problem-solving ability, and product sense.

In the spring of 2013, I transferred to engineering to create a team around query understanding. I've been passionate about search—and LinkedIn search in particular—for much of my professional career, and this was an opportunity to focus my passion into product. Working in engineering is great—we have to be a bit more heads down and focused on operations, but it's great to always be working on the live site. And, of course, the people I work with are amazing—brilliant, creative problem solvers who get things done.

Gutierrez: What does being Head of Search Quality entail?

Tunkelang: In the fall of 2014, I took on an expanded role as the head of search quality. I now lead the team that enables LinkedIn's 300M+ members to effectively and efficiently leverage LinkedIn's professional content to satisfy their information needs. The includes query understanding, but also scoring and ranking of results, and everything else we need to do to deliver the right results to the right searchers at the right time.

Gutierrez: How would you compare and contrast leading a team of data scientists to leading a team of engineers?

Tunkelang: The people I hired as data scientists had strong software engineering skills, so the difference wasn't as drastic as it would have been if I were leading a team of data analysts or statisticians. The main change is that I now lead a team that wholly owns a product that serves billions of searches a year. We have to think about everything from coming up with new algorithms to making sure we don't take the site down.

Gutierrez: Why is query understanding an important problem to tackle for LinkedIn?

Tunkelang: Search is what allows LinkedIn's hundreds of millions of members to find each other and be found. LinkedIn members did over 5.7 billion professionally-oriented searches on the platform in 2012, and that number has kept growing since then. Search is one of the most important ways that LinkedIn's members engage with our platform.

It's our job to make sure that our members find what or who they are looking for. And because the search experience on LinkedIn is highly personalized, we face unique challenges in delivering quality results to our members. Our search engine needs to take into account who you are, who you know, and what we know about your network to help you find what you're looking for.

Query understanding has been an important problem for some time. I started thinking about it when I was at Endeca, working with faceted search and semi-structured data sets. Since LinkedIn is a poster child for both, it was natural to see how better query understanding could improve the search experience. I'd been dabbling in this area for a while, and in 2013, I decided to focus on it exclusively.

Gutierrez: Why is query understanding interesting to you?

Tunkelang: Query understanding offers the opportunity to bridge the gap between what the searcher means and what the machine understands. Instead of tackling the squishy, subjective problem of how relevant a piece of content is to the searcher, it focuses on the more objective problem of establishing an unambiguous information need, so we can figure out which content is relevant at all. Also, we're able to improve the language of communication between the searcher and the machine, which is an exciting development in human–computer interaction.

Gutierrez: Why is search relevance an important problem to tackle for LinkedIn?

Tunkelang: Search is a pillar of LinkedIn's platform—it's what enables our 300M+ members to find and be found. But of course the search results have to be relevant. Our members perform billions of searches, and each of those searches is highly personalized based on the searcher's identity and relationships with other professional entities in LinkedIn's economic graph. It's a challenging problem in several dimensions, and solving it delivers enormous values to our members.

Gutierrez: What have you been you working on this year?

Tunkelang: As its name suggests, the Query Understanding team has been working on understanding queries—specifically, the queries our members issue when they search on LinkedIn. We look at understanding queries before deciding which results to retrieve and score. For example, if someone searches for "daniel linkedin", we can figure out that "daniel" is a person's name and "linkedin" is a company, so we only retrieve results corresponding to people named Daniel who work or used to work at LinkedIn. This presearch processing dramatically reduces the set of search results by eliminating irrelevant results, and allows scoring to focus on more subtle differences—such as favoring the people with whom the searcher has stronger professional connections.

Meanwhile, we've been tacking a variety of problems in scoring and ranking. We've been building personalized, machine-learned ranking models that we segment by query type. For example, a networker looking up a new acquaintance by name is different than a recruiter trying to source candidates. We've also started using a new architecture called Galene that we built to address LinkedIn's unique search challenges.

Gutierrez: What types of queries does Galene now allow LinkedIn users to do that they couldn't do before?

Tunkelang: LinkedIn built our early search engines on Lucene, a popular open-source framework. As we grew, we evolved the search stack by adding layers on top of this framework. Our approach to scaling the system was reactive, often narrowly focused, and led to stacking new components to our architecture, each to solve a particular problem without thinking holistically about the overall system needs. This incremental evolution eventually hit a wall requiring us to spend a lot of time keeping systems running, and performing scalability hacks to stretch the limits of the system.

So we decided to completely redesign our platform. The result was Galene, a new search architecture that is now powering a variety of our search products, including the "instant" to find people as you type. Galene has also helped us improve our development culture and processes. For example, the ability to build new indices every week with changes in offline algorithms supports a more agile testing and release process.

Gutierrez: How has the approach evolved since you started the team?

Tunkelang: When we started the team, we were pretty conservative with respect to the search experience. Filtering queries by removing irrelevant results was a pretty radical idea, when conventional wisdom was that you should return everything and rely on ranking.

Our successes have emboldened us since then. Now we're coming up with structured query suggestions as searchers type. Our ultimate goal is a "things-not-strings" experience, where all queries are composed of standardized entities.

We've also embraced more sophisticated ranking approaches. And we're increasingly looking at two-sided relevance approaches that reflect the talent marketplace we're trying to optimize.

Gutierrez: What do you mean by "standardized entities" and why is "things not strings" the ultimate goal?

Tunkelang: At LinkedIn, our entities are people, companies, universities, and similar things. Of course, people enter strings into the search box just like with any other search engine. But they are mostly trying to refer to the entities that participate in LinkedIn's ecosystem—or, as we like to call it, the "economic graph."

The "things-not-strings" idea is simple. Because the strings are intended to refer to entities, let's remove the barrier of uncertainty between the strings and the entities. Doing so helps enable a world of richer queries, where we bring in relationships among entities—for example, acquired companies, similar job titles—and provide a more structured, guided exploration of the information space. Google and Microsoft are starting to do this for the web with their "knowledge graph" projects, but we have an extraordinary ability to provide this experience today because of the rich structure inherent in our data.

Gutierrez: What is a two-sided relevance search approach? And how do you optimize it?

Tunkelang: LinkedIn is a marketplace connecting talent to opportunity at massive scale. And a successful connection requires satisfying both of the parties being connected. It's not enough to help a job-seeker find her dream job or to help a recruiter find his dream candidate— the job or candidate has to be attainable. What we end up with was a multiple-objective optimization problem that incorporates both the searcher's desires and the likelihood of a successful outcome.

Gutierrez: What do LinkedIn members search for?

Tunkelang: LinkedIn members search for people, jobs, companies, and various kinds of content. For certain kinds of professionals, like recruiters and salespeople, finding through LinkedIn is a core part of their livelihood. For others who are less outbound as professionals, being found means having access to unexpected career opportunities. Getting both right allows us to best connect talent to economic opportunity.

Gutierrez: How do you measure if a search was successful or not?

Tunkelang: We have several measures that we use to evaluate search success. We look at whether people click on results, and even more at actions they perform after those clicks—such as inviting a person to connect or applying for a job. We also use human judgments—but that can only go so far when we're evaluating a highly personalized search experience.

Gutierrez: How does LinkedIn personalize the search experience?

Tunkelang: The personalization of your search experience depends mostly on information in your profile, like your company, location, industry, and your network. For some classes of searches, network distance is a particularly strong feature—for example, it helps us disambiguate name searches.

Gutierrez: Search is something you have done in different roles and companies. What about it excited you at the beginning and what excites you now?

Tunkelang: Search is an amazing problem that just keeps giving. It seems so easy when you first encounter it, as it probably seemed to information scientists in the 1950s who had to invent the concept of "relevance" when they realized indexing documents by their words wasn't enough to make them findable. Search is the problem at the heart of the information economy. The information is out there, if only we can find it. What's also great about search is that it's an area full of open problems, many of them pretty fundamental. Maybe search will be boring fifty years from now, but I doubt it.

Gutierrez: If someone wants to get started in search today, what should they do?

Tunkelang: They should start by taking a class on information retrieval or learn from the vast array of resources available offline and online. Given the open source technology for search, they should learn by doing—for instance, implementing a basic search engine for a public data collection. It's not hard to get started with search.

Where things get interesting is in the details. Ranking is still an open area of research, especially for personalized and social search applications. And there are even more opportunities to experiment with new search user interfaces. I'm biased, but that's where I'd direct people interested in pursuing the future of search.

Gutierrez: What did your work on search at Google entail?

Tunkelang: I led at a Google team that mapped businesses in Google's local search index—which is now part of Google+—to their official home pages. In addition to its web search index, Google has a local search index intended to help people find local businesses. Think of it as a 21st-century yellow pages. My team's goal was to improve local search quality. We approached this goal by mapping the local search index to the web search index in order to determine when local business had official home pages, and associating the two.

The results were useful in two ways. First, we could provide a link to the home page on the local business page. Second, we could improve the association as a signal to web search relevance to better determine when the intent of the searcher was to find a local business. When web search determines this intent, it typically shows a map and other information relevant to this class of search queries. This was a fun machine learning problem, and our accuracy not only improved the quality of the local search pages but also helped Google figure out when web searchers were looking for a local business so that it could respond with maps and other appropriate content.

When I arrived at Google, there was already a system in place to map businesses to home pages. It was a machine learning system—specifically, it used logistic regression to assign scores to candidate home pages for businesses. I can't disclose numbers, but there was lots of room to improve its precision and coverage. Moreover, the model was unstable and difficult to interpret, making it difficult to use for work on incremental improvements to it. So we decided to explore other approaches that would not only improve our system's accuracy, but also facilitate ongoing work to improve it.

I can't say too much about our results—the numbers are confidential under my NDA. But what I can say is that we significantly improved accuracy through a series of changes that included switching from a logistic regression model to a decision tree approach. That was surprising, since decision trees are hardly cutting-edge machine learning models. However, they are very interpretable and that interpretability made it much easier for us to gain insight and iterate.

Gutierrez: Do you find that non–cutting-edge models sometimes work better than newer models as they are applied to new domains?

Tunkelang: I'm not saying that non–cutting-edge models work better—indeed, I'd like to think that progress in machine learning ensures the opposite! Rather, it pays to keep things simple when you're trying to understand your data and iteratively develop models for it. In those cases, it's better to optimize for interpretability than accuracy. Once you've learned as much as you can, you can go back to more complex models. When you go back to them, you'll hopefully now have the right training data, objective function, and features to take advantage of the latest and greatest machine learning has to offer.

Gutierrez: How important is it to continue working on models that have already been built?

Tunkelang: There's no preference for replacing versus improving models. We put most of our efforts into collecting better training data and coming up with new features. Those usually require us to train new models. There's some bias towards reusing our existing infrastructure, because that's usually less work and helps us avoid introducing new bugs. But we do our best to evaluate models on their own merits, even if that means doing more work to take advantage of a new approach.

Gutierrez: What were you most proud of for this project?

Tunkelang: I'm most proud of the fact that although there were only three of us doing most of the work for this project, the changes we made improved the quality of a huge fraction of web search queries. It may sound cliché, but it was great to work on something that benefits my mom's day-to-day experience online.

Gutierrez: How did the three of you come together to work on this project—was it one of Google's 20% projects?

Tunkelang: This was our day job and not a 20% project. We were assigned to work as a team, and so we took complete ownership of the project. The previous developers were available for us to consult them, but mostly they were happy to work on new projects while we improved on theirs. I truly have the utmost respect for developers who understand that their products outlive them.

Gutierrez: How was the model for improving local business search built?

Tunkelang: Fortunately, we had a framework in place to compare the performance of different machine learning approaches. We tried a bunch of them, evaluating their accuracy against a golden set, as well as their efficiency, stability, and interpretability. Ultimately, we opted to use decision trees.

We'd expected that switching from regression to decision trees would trade off accuracy for interpretability and stability. But, to our pleasant surprise, we were able to improve all three. And it was a lot easier to work on new model features once we had a decision tree model in place.

Gutierrez: How are decisions made about replacing models that are already in production?

Tunkelang: At both Google and LinkedIn, we make decisions based on metrics. If a change affects only one metric—or if it affects several metrics but all in the same direction—then the decision process is clear: we ship it. The more interesting case is where some metrics go up and others go down. In theory, we use a single utility measure to assess overall impact. In practice, we negotiate whether the tradeoff is net positive to the business. The decisions usually happen at the level of the teams that own the various metrics, but in exceptional cases the tradeoffs get escalated to someone who can arbitrate between competing business goals.

Gutierrez: What lessons did you learn from this project?

Tunkelang: I've always valued interpretability, but this project showed me how crucial it could be in the context of machine learning. I also learned a lot about the challenges of working with unrepresentative training data. While we had large volumes of training data, we also had systematic biases that could trick our machine learning models to overfit for those biases. We had to learn to compensate for those biases and to distrust anything that looked too clever.

Gutierrez: What are your thoughts about systematic biases, overfitting models, and "too-clever" versus "works-way-better"?

Tunkelang: I'm a big fan of Occam's razor, which in modeling translates into a preference for minimum description length. I prefer small, understandable models, where the coefficients use as few bits as possible. If a model is going to be more complex, it has to prove itself to be more accurate in online testing. And an increase in accuracy doesn't always justify an increase in complexity, as more complex models are harder to debug when they break.

As for systematic bias and overfitting, it's always an ongoing concern. We try to anticipate it by carefully reviewing our methods for collecting training data and considering every way we may have introduced bias. Worst case, we discover bias in our training data when models that perform well on our training data fail against withheld data. Then we use these failures to improve our collection process.

Gutierrez: How would you describe your work to someone who is not familiar with it but familiar with data science?

Tunkelang: As data scientists, our job is to extract signal from noise. We do this in many contexts, from performing analyses that drive business strategy to enabling data products like recommender systems. In the context of search quality, that means analyzing the content we index and the way searchers interact with it to deliver relevant results and improve the search experience.

Gutierrez: What in your career are you most proud of so far?

Tunkelang: What my colleagues and I accomplished at Endeca was something extraordinary. We helped change how people think about search, and I see the effects every day that I browse the web.

Gutierrez: When did you realize you wanted to work with data as a career?

Tunkelang: I'm not sure there was any particular moment of realization. I always loved math and computer science. Early on, I was more tempted by theory than practice, obsessed with open problems in combinatorics and computational complexity. But ultimately I couldn't resist working on problems with practical consequences, and that's how I found myself specializing in information retrieval and data science more broadly.

Gutierrez: How did you get interested in working with data?

Tunkelang: One of the problems I worked on at IBM was visualizing semantic networks obtained by applying natural language processing algorithms to large document collections. Even though my focus was on the network visualization algorithms, I couldn't help noticing that the natural language processing algorithms had their good moments and bad moments. And that there was only so much I could do with visualization algorithms if the raw data was noisy.

Several years later, when I was at Endeca, I found myself working on terminology extraction and had to confront the noise problems personally. Ironically, we ended up licensing our terminology extraction algorithms to IBM as part of a search application we built for them.

Gutierrez: What was the first data set you worked with?

Tunkelang: I feel bad that I can't remember my first, as that makes it sound like it wasn't a deep, meaningful experience! I did spend a lot of time working with a Reuters news corpus to test out information retrieval and information extraction algorithms. One of the great things about my time at Endeca was the opportunity to work with our customers' data, especially when we were prototyping new product features.

Gutierrez: How did the fact that the Endeca data was customers' data make you think about the data?

Tunkelang: It was nice to have a diverse set of customers and thus gain exposure to lots of different problems. But the price of working as an enterprise software vendor was that our relationship to the users was always indirect. So we couldn't just decide to run experiments and observe the impact.

Gutierrez: Was there a specific aha! moment when you realized the power of data?

Tunkelang: Not sure there's a single moment, but there were two unforgettable moments in my relationship with data. The first was when I was working with a digital library and realized we could dramatically improve document tagging by algorithmically recycling author-supplied labels. While authors tagged articles with keywords and phrases, the tagging was sparse and inconsistent. As a result of this type of tagging, the use of tags for article retrieval offered high precision but low recall. Unfortunately, the alternative of performing full-text search on the tags provided unacceptably low precision. So we developed a system to bootstrap on author-supplied tags, thus improving tagging across the collection. The result was an order of magnitude increase in recall without sacrificing precision.

The second was using entropy calculations on language models to automatically detect events in a news archive. We started by performing entity extraction on the archive to detect named entities and key phrases. Then, when we performed a search, for example "iraq", we could compute the language model for the search results and track it over the time span of the collection. What we found is that sudden changes in the language model corresponded to events. I only had the opportunity to build prototypes with this system, but I did have the chance to demo them to people at three-letter agencies.

Gutierrez: How would you describe your work to a layperson at a cocktail party?

Tunkelang: If you use a search engine, perhaps you take for granted that the search engine knows what you mean and returns what you're looking for. Except when it doesn't, and you curse our Skynet overlords. My team works on converting what you say—a few words in a search box—into what you mean and then providing the most relevant results. For example, that "ford" is a surname in the context of "john ford", but a company name in the context of "ford engineer". We do this by learning from our data. In particular, how words are used in our search index and how people use words when they search. We also get feedback on when we get it right, based on how people engage or don't engage with the results we show. We use that feedback to improve our algorithms, and hopefully we do better the next time.

Gutierrez: What are the main types of problems being tackled in search in social networking?

Tunkelang: Though there are a lot of problems, you could summarize them with three Rs—relevance, recommendations, and reputation. Relevance obviously comes up in the context of search quality, but also when we're presenting feeds or streams to users and have to select interesting items from an endless array of possibilities. Recommendations are the complement to search. If search relevance is about giving users the information they're looking for, then recommendations are about giving users the information they need but haven't asked for. Finally, reputation encompasses a wide range of problems, from identifying topic experts to quantifying trust.

Gutierrez: Do you tackle all three of these problems on a frequent basis, or do you work on a subset at any given time?

Tunkelang: These days I focus almost exclusively on relevance. LinkedIn has other teams that focus more on recommendations and reputation. Even though I'm focused on relevance, I get excited about challenges in all three areas and I do my best to contribute to our efforts in them as a company. I also try to take part in global conversations about these topics through blogging and conferences.

Gutierrez: Who are the big thought leaders in search and social networking?

Tunkelang: There are too many to list here, especially because working in these fields means being an intellectual omnivore. I read research papers from information retrieval conferences and I also look at how startups are innovating the user experience. One of the upsides of being at the epicenter of the world's largest professional network is that new developments find their way into my inbox through multiple channels.

Gutierrez: How do you decide what to read and what startups to look at?

Tunkelang: These days I rely heavily on my professional network to curate what's out there. I check my LinkedIn feed, naturally. I also see what the people I follow share on Twitter. And I use a few news aggregators—mostly Techmeme and Prismatic. As a result, I get a good mix of front-page technology news and niche information, and I do my best to keep up.

Gutierrez: What types of data are these data scientists working with?

Tunkelang: All kinds—text, numbers, clicks, relationship graphs, geography, time series, and similar data. Part of the challenge of working in data science is that you tend of have lots of different kinds of data, and it's your job to stitch them together.

Gutierrez: Are there data that data scientists are not yet looking at?

Tunkelang: Probably not. But I think there's a lot of work to do on improving how wearable devices collect data in order to truly deliver better living through data. I'd love to have a device that tracked everything from my physical activity to my mood, which then allowed me to benefit from data analysis without risking the exposure of private data that is deeply personal. We're getting there, but these are early days for wearables and sensors in general.

Gutierrez: What resources are helpful to your work?

Tunkelang: Since I focus on search, I keep up with the leading information retrieval conferences, in particular the SIGIR and CIKM conferences sponsored by the ACM. I also keep a foot in the big data world through conferences like O'Reilly Strata. I don't read particular blogs anymore. Rather, I mostly rely on LinkedIn and Twitter to surface relevant content to me. I tend to use books mostly as references. Fortunately, some of the most valuable information is timeless.

Gutierrez: How do you keep track of things when you are out of the office, given the ephemeral nature of feeds?

Tunkelang: Sadly, I'm almost never offline, as I'm a bit of an information addict. I did disconnect for a week recently and I spent most of the following weekend catching up. For all of the effort that has gone into news aggregation, there's certainly more emphasis on providing real-time feeds than in aggregating feeds over several days, let alone several weeks.

Gutierrez: What does a typical day at work look like?

Tunkelang: I'm not sure there is a typical day. Broadly, I lead a great team and then spend the rest of my time on hiring and outreach. Most of my time is spent providing guidance to my very accomplished team—adding value where I can, and staying out of their way where I can't. I also spend a lot of time on hiring and outreach.

Gutierrez: Do you code as well as lead the team?

Tunkelang: It's been a while since I've written production code, but I still look at code and system logs when we're trying to diagnose system behavior. That said, I do miss writing code. When I find the opportunity to code, like most recently hacking up some old-school games for my daughter, I remember how fun it can be to have the instant gratification of making things run. I'm excited that she'll grow up experiencing that same magic with much better tools than I had at her age.

However, I've found that, while I'm okay at software engineering, I'm actually much better at leading software engineers. I've had the opportunity to hire people who are much better developers than I could ever be, and I'm honored that I can help them accomplish great things.

Gutierrez: What does your work process look like and how do you view and measure success?

Tunkelang: Our work proceeds in three stages. The first is hypothesis generation. We come up with hypotheses either reactively by looking at logs or proactively by exercising our intuitions. The second is offline analysis. We use historical data, human judgments, or some other proxy to efficiently test our hypotheses. Our expectation is that most hypotheses won't survive offline testing—that's what the null hypothesis is for. It's important that offline testing be quick and cheap, as that way we only invest in online testing for our most promising hypotheses. The third stage is online testing, where we implement product changes to test our hypotheses and bucket-test those changes against live traffic.

All three of these stages happen in parallel. At any given time, we're engaged in a mix of hypothesis generation, offline testing, and online testing. Think of it as managing a portfolio strategy for data-driven innovation.

Gutierrez: How do you manage the portfolio?

Tunkelang: For portfolio management, we try to be scientific about it, but fall back on intuition when necessary. For example, we'll spend an hour deciding whether something is worth spending a couple of days investigating. Or a person will spend a week on offline analysis deciding whether something is worth a couple of months of engineering effort before we can test it online. The basic principle is fast failure and an exponential increase in effort as we mitigate risk.

It's hard to be completely data-driven about the process, since different hypotheses apply to different problems. But we adapt. If most of our efforts are failures, then we're not being sensitive to risk. If all of them are successes, then we're probably being too risk-averse and leaving big opportunities on the table. There's no easy way to count hypotheses, since we explore them as many different levels of granularity—from whether we should change a relevance-tuning parameter to whether users will be interested in a new product feature.

Gutierrez: What specific tools and techniques do you use?

Tunkelang: We use the usual tricks of the data science trade—machine learning models, A/B testing, crowdsourced evaluation, data collection, and similar techniques. Most importantly, we look at data and logs below the aggregate level. It's easy to be lazy and look at aggregates—for example, favoring one machine-learned model over another because it performs better on average. Drilling down into the differences and looking at specific examples is often what gives us a real understanding of what's going on.

Gutierrez: What nascent tool are you most excited about?

Tunkelang: I'm not sure it still qualifies as nascent, but I'm very excited about human computation. I can't imagine data science today without crowdsourcing for data collection and evaluation. For example, I'm studying Italian using Duolingo, a free language-learning app that doubles as a crowdsourced text translation platform. These are early days for human computation, and I expect we'll see even more powerful applications over the next years.

Gutierrez: You mentioned drilling down to get a real understanding of a model. How do you measure real understanding?

Tunkelang: I don't know of a quantitative metric for understanding. But consequences of understanding are easy to quantify. When we realize that a model improves performance for one user segment, but degrades it for others, we have a starting point to investigate why. And hopefully we end up with a richer model—or perhaps two distinct models—that allow us to perform better for both segments. Ideally, we learn even more as we get a better understanding of what distinguishes our segments and insights that carry over to the rest of our user base beyond those segments.

Gutierrez: How do you communicate your results to other groups in the company?

Tunkelang: How we present and communicate our work to the rest of the company varies. We give presentations to our peers who work on similar relevance and data science problems. But sometimes we work with teams more tightly because our work is highly related. For example, there are relationships between the abusive search engine optimization team and the fraud team. One thing we've learned is that there's no such thing as over-communicating. No one ever complains that they have too much access to information about what their peers are doing.

Gutierrez: You've mentioned logs in previous answers. Do you have a system to help you look at these logs?

Tunkelang: We have a variety of in-house reporting tools that we use for regular log analysis. And when those aren't flexible enough, we use tools like Hive or Pig to perform ad hoc analysis. Of course, a crucial part of this process is that we instrument and track everything. And we've built a variety of open source tools to support our logging needs. I highly recommend a piece that Jay Kreps of LinkedIn wrote on the subject, entitled "The Log: What Every Software Engineer Should Know About Real-Time Data's Unifying Abstraction." He published it as a blog post, but it's more like the definitive book on the subject.

Gutierrez: How does your work evolve through a project's life cycle?

Tunkelang: Early on, our goal is to fail fast. Most crazy ideas are just that: crazy. So, in the earliest stages, it's important to have efficient ways to reject bad ideas based on data—for example, to put an upper bound on the impact of a change by analyzing our logs. But as a hypothesis shows promise through offline testing, we double down on it. Our focus shifts from trying to kill it to make it succeed. We optimize parameter settings and then look for edge cases and related techniques to improve and understand a model. Because this shift in focus is dramatic, it's important that we only make it for ideas that survive a harsh validation filter.

Gutierrez: How do you differentiate between crazy and novel ideas?

Tunkelang: It's tough. If someone believes in an idea, we always give that person the opportunity to try to back it up with data. An important question in this process is how many attempts we allow them to show the model is worth studying before we kill the idea. At some point, we rely on our judgment to decide that we've exhausted the space of possibilities. Or we just lose patience. And sometimes we revive ideas from the morgue when we have new insights.

Gutierrez: How do you keep track of all the ideas in the morgue?

Tunkelang: Frankly, we rely on associative memory. Some of us have ideas that we never really give up on, so it doesn't take much to trigger them again. And if new information comes in that offers the ingredients for a compelling case, it's easy for the original advocate of the idea to justify giving that idea another chance. We may be data-driven, but our ideas come from a place of passion.

Gutierrez: Where do you get ideas for things to study and analyze?

Tunkelang: To a large extent, I draw on my own intuition and experience. I encourage my colleagues to do the same. Though we take a rigorous data-driven approach to experiments, we often rely on our own creativity to figure out which hypotheses to explore. Of course, sometimes our users make it easier for us by giving us feedback or by displaying anomalous behavior in our logs.

Gutierrez: How did you go about developing your own intuition?

Tunkelang: My intuition mostly comes from exposure to lots of different problems. Over time you learn to recognize patterns. Intuition is really a well-trained association network.

Gutierrez: How do you choose which problems to solve?

Tunkelang: I start by thinking about the potential impact using optimistic assumptions. How many people will care if we deliver an optimal solution, and how much better is an optimal solution than what we have today? I try to find the right measure of impact and establish an upper bound. If that's not high enough, then the problem is probably not worth solving.

Then I try to make more conservative assumptions about the expected impact. If we only make it halfway from where we are today to optimal, is it still worth it? What about 10 percent of the way? If more conservative assumptions give me pause, then I try to make our opportunity analysis more rigorous.

Finally, I try to estimate how much work it will take to validate our assumptions. Not how much work it will take to solve the problem, but how much work it will take to remove most of the uncertainty about the project's success. I strongly favor projects that allow for rapid mitigation of uncertainty. Sometimes this means fast failure, and other times it means early promises of success.

Gutierrez: How do you think about whether you're modeling the right thing?

Tunkelang: As George Box said, "All models are wrong, but some are useful." It's tempting to make models as realistic as possible, but it's a temptation I try to resist. I'd rather have a simple model that is easy to explain. Of course, using simple models means having to consider the risks of confusing them for reality.

For example, when I'm working on search quality, it's tempting to model a click as a search success and an abandoned search results page as a search failure. But not all clicks are successful. For example, the searcher may need to click to see more information to determine that the result is irrelevant. On the other side, not all abandoned searches are failures. For example, the searcher may be satisfied with the information presented in the search summaries.

So we work to keep in mind that models aren't supposed to be photorealistic. Rather, they give you something that's close enough to reality to be an adequate proxy, yet simple enough to be measured and optimized. And, as long as we work with models in good faith and keep our eyes open for glaring divergences from reality, our models tend to be kind to us.

Gutierrez: How do you think about whether you have the right data?

Tunkelang: There is no substitute for common sense. You have to look at all of the data you have and validate it against the data you don't have. For example, if your data tells you that the most in-demand skills are all in a particular industry, you should check to make sure your data doesn't overrepresent that industry.

Most of us recognize and laugh at the parable of a drunk looking for his keys under a streetlight because that's where the light is. But we do it all the time. We, as an industry, work with the data we have on hand and optimize what we can measure. That's not an entirely bad thing. It's much better than trying to work without data or trying to improve things we can't measure.

Still, a little bit of humility goes a long way. If our data tells us something that seems incredible, the correct response is skepticism. After all, incredible is Latin for "not to be believed."

Gutierrez: How does technology selection factor into solving problems?

Tunkelang: Technology is obviously important and choosing a technology stack is one of the biggest decisions that you make as a software engineer or data scientist. The wrong technology selection can be a major impediment, as it often leads to kludgey workarounds.

Technology selection by itself is unlikely to solve any problems. Technology is like exercise equipment in that buying the fanciest equipment won't get you in shape unless you take advantage of it. So always put talent before technology. Get the right team of scientists and engineers, and then make sure the technology doesn't get in their way.

Gutierrez: What do you look for when hiring people?

Tunkelang: I look for three things in a candidate. First, they need to be smart, creative problem solvers who not only have analytical skills but also know how and when to apply them. Second, they have to be implementers and show that they have both the ability and passion to build solutions using the appropriate tools. Third, they have to have enough product sense, whether it comes from instinct or experience, to navigate in the problem space they'll be working in and ask the right questions.

Gutierrez: What do you mean by "product sense," and why is it important?

Tunkelang: By "product sense," I mean the ability to see real-world problems from the perspectives of users and other stakeholders. For example, a computer scientist might come up with a system that improves through positive and negative feedback. But someone with product sense would think about what would motivate the users to provide the system with such feedback. On the business side, someone with product sense will use that sense to inform key business metrics—for example, determining when a recommendation system makes suggestions so bad that they incur a cost beyond the user simply not clicking on them.

Product sense is a critical skill for a data scientist. Without product sense, you can be a great software engineer and a great statistician, but it's unlikely you've identified the right problems to solve or picked the right metrics for evaluating your solutions. Finally, product sense can help you find shortcuts, such as getting users to help you solve your problems.

Some people seem to have it naturally, so perhaps it's a form of applied empathy. You can certainly improve it by studying a blend of disciplines, particularly the social sciences, and by working on lots of different real-world problems.

Gutierrez: How do you hire for creativity?

Tunkelang: As far as creativity, we evaluate it by asking candidates about open-ended problems that we've worked on. Good candidates figure out some of the paths we discovered. The best ones surprise us with ideas we haven't thought of. And sometimes they join us and work on these problems and ideas.

Gutierrez: Does being at LinkedIn make it easy to do hiring?

Tunkelang: It's incredibly difficult to find people with the right combinations of skills and attitude. We've done our best to optimize the hiring process to identify them. However, finding them isn't enough, as these rock stars typically have offers from all of the big-name Silicon Valley companies. So when I speak with them, I do my best to figure out their professional aspirations and whether we're in a position to fulfill those aspirations through opportunities at LinkedIn. Hiring is an intensively competitive process, especially here in Silicon Valley, but it's a very exciting one.

Gutierrez: What attitude do you look for in candidates?

Tunkelang: A passion for problem solving, of course, but also a humility that places the value of the work above their personal ego. The best people I've work with take extraordinary pride in their work, but leave their egos at the door, which is especially important when we work together as a team or with other teams.

Gutierrez: How do you figure out a potential employee's professional aspirations?

Tunkelang: I ask them, of course. But most people have a hard time figuring out what they want for dinner, let alone where they see their careers taking them in five years. So I try to paint different pictures of career trajectories and see what resonates with them.

Gutierrez: When looking to hire, how do you evaluate someone's technology experience?

Tunkelang: I ask them about projects they've worked on and what tools they used for those projects. I don't care so much about which technologies they've used, as about whether they made informed choices. I look for people who adapt to new technologies when they need to, whether that means learning a new programming language or building on top of a new computing framework.

Gutierrez: What do you look for in other people's work?

Tunkelang: I love to find people with good taste in problems. For me, worthy problems are more interesting than clever solutions. As data scientists, we are truly in a position to change the world, as we can improve people's health and well-being, optimize allocation of resources, guide better policy decisions, and similarly worthy problems.

I'm inspired by people who work on inspiring problems. Jeff Hammerbacher once said, "The best minds of my generation are thinking about how to make people click ads. That sucks." I wholeheartedly agree, and so that's why I suggest that people should focus their best talent on worthy problems.

Gutierrez: What does it take to do great data science work?

Tunkelang: Hilary Mason and Chris Wiggins said it best: A data scientist is someone who obtains, scrubs, explores, models, and interprets data.[1] Which means, as Drew Conway expressed in his Data Science Venn Diagram, that data scientists need to be armed with hacking skills, math and stats knowledge, and domain knowledge.[2] And, perhaps most importantly, data scientists need to have strong critical-thinking skills and a healthy dose of skepticism.

Gutierrez: When hiring and training people for your group, how do you approach teaching or mentoring people to develop these skills?

Tunkelang: Failure is a great teacher. One of my best learning experiences in college was implementing an algorithm from a paper, only to have it not perform as claimed. I contacted the authors, who told me how they'd tuned their systems for each example in the paper. After overcoming my initial reaction of indignation—after all I'd worked for months on my own competing approach—I realized that I'd learned an important lesson to not believe everything I read in a peer-reviewed publication.

As W. Edwards Deming allegedly said, "In God we trust. All others bring data." In science, the default assumption is the null hypothesis, which puts the burden of evidence on the hypothesis you're trying to prove. These are all variations on the same theme—if it's too good to be true, then don't believe it until you can back up your belief with data. It's easy and important to tell people to be skeptical, but I doubt it's enough to overcome our cognitive biases. This is a case where experience is not only the best teacher, but also perhaps the only teacher.

---

[1]Hilary Mason and Chris Wiggins, "A Taxonomy of Data Science" (September 25, 2010: www.dataists.com/2010/09/a-taxonomy-of-data-science/).
[2]Drew Conway, "The Data Science Venn Diagram" (September 30, 2010: http://drewconway.com/zia/2013/3/26/the-data-science-venn-diagram).

Gutierrez: Whose work is currently inspiring you?

Tunkelang: Perhaps not what you had in mind, but I'm inspired by Bill Gates. Specifically, I'm very inspired by his data-driven approach to philanthropy. Of course, it's humbling to see one of the world's richest people donating almost his entire net worth to make the world a better place. What is truly inspiring is the way he's doing it. He's focusing on measurable improvements and optimizing his philanthropy according to where it does the most measurable good. That makes him more than just a great human being—it makes him a great data scientist.

Gutierrez: What advice would you give advice to someone starting out?

Tunkelang: It depends where they are coming from. To someone coming from math or the physical sciences, I'd suggest investing in learning software skills—especially Hadoop and R, which are the most widely used tools. Someone coming from software engineering should take a class in machine learning and work on a project with real data, lots of which is available for free. As many people have said, the best way to become a data scientist is to do data science. The data is out there and the science isn't that hard to learn, especially for someone trained in math, science, or engineering.

Gutierrez: What is something someone starting out should strive to understand deeply?

Tunkelang: Read "The Unreasonable Effectiveness of Data"—a classic essay by Google researchers Alon Halevy, Peter Norvig, and Fernando Pereira.[3] The essay is usually summarized as "more data beats better algorithms." It is worth reading the whole essay, as it gives a survey of recent successes in using web-scale data to improve speech recognition and machine translation. Then for good measure, listen to what Monica Rogati has to say about how better data beats more data.[4] Understand and internalize these two insights, and you're well on your way to becoming a data scientist.

Gutierrez: In your opinion, what are the necessary critical thinking and analytic skills that educational institutions should be teaching?

Tunkelang: No one should graduate from high school without a solid grounding in the scientific method—basic concepts of hypothesis testing and falsifiability. The same should be said for basic knowledge of probability and statistics. In a world where we're bombarded with data and analyses of data, we should be informed consumers. And, of course, everyone should learn the basics of computation—at least enough to demystify the computers that surround us.

---

[3] Alon Halevy, Peter Norvig, and Fernando Pereira, "The Unreasonable Effectiveness of Data" (March/April 2009, IEEE Intelligent Systems www.computer.org; www.cs.columbia.edu/igert/courses/E6898/Norvig.pdf).
[4] Monica Rogati, "The Model and the Train Wreck: A Training Data How-To" (O'Reilly Strata 2012: www.youtube.com/watch?v=F7iopLnhDik).

For data science in particular, I think it's helpful to study a blend of mathematics, software engineering, and at least some of the social sciences, such as economics or sociology. The best data scientists are well-rounded, able to combine theory, practice, and intuition about how things and especially people work in the real world.

Gutierrez: What does the future of data science look like?

Tunkelang: As Niels Bohr said, "Prediction is very difficult—especially about the future." I see some exciting developments in the present—a growing awareness of the value of data-driven decision making, and recognition of the critical role that data plays in product development. I'm optimistic that this trend will continue, and data will have the primary role it deserves in organizations.

What does the future hold? Certainly we're seeing new sources of data as wearable computing goes mainstream. Only a few years ago, the Quantified Self movement seemed like a futurist fringe. Now, it's well on its way to being a billion-dollar market. Hopefully, we'll see the results in a more data-driven approach to healthcare. More broadly, I hope that anyone working on a hard problem will be in a better position to find the data that can help solve it.

Gutierrez: How do you think the data science workflow will change?

Tunkelang: Today, expertise with big data tools is still fairly specialized. I believe that this is a transient state and what we call "big data" today will simply be "data" tomorrow. We'll find ways to hide the messy details so that the learning curve for doing simple analysis on petabytes of data won't be that different from learning how to use Excel today. However, even as the technology gets more powerful and more efficient, the science itself won't get any easier. It's crucial that our educational institutions teach the necessary critical thinking and analytic skills to the next generation of data scientists.

Gutierrez: What data sets would you love to see be developed?

Tunkelang: Health and well-being are exciting areas that I'm looking forward to data sets being generated for. There's a lot of opportunity to better understand nutrition, exercise, sleep, and similar personal processes. Of course, these are also areas that raise critical concerns about personal privacy. But I'm certain that there's a way to develop data sets that enable the advancement of science, while taking the necessary precautions to protect the individuals whose data they aggregate.

Gutierrez: What is one problem you think the world of data science needs to fix?

Tunkelang: One problem? There are so many! Data should help us make better decisions. As we've learned from the work of Herbert Simon, Daniel Kahneman, and many others, human beings are terrible at making rational decisions. Data scientists should work to improve decision making wherever data has to compete with human irrationality.

Gutierrez: What is something a smallish number of people know about that you think will be huge in the future?

Tunkelang: By now, more than a smallish number of people should know about what Dan Ariely calls our "predictable irrationality." Books by him and Daniel Kahneman have been best sellers. And yet I see little evidence that people are applying the insights they should be deriving from those books. We make most of our decisions without recognizing the cognitive biases that taint our decision-making process. Indeed, by doing so we are exercising a form of overconfidence bias.

As software plays an increasing role in our day-to-day lives, I'm hopeful that it will intercede in some of that decision making. Our computers, mobile devices, and web-based services are witnesses to many of our daily decisions. I look forward to the day that those devices play a more active role in helping us make better decisions.

Gutierrez: What has been the biggest thing you have changed your mind about and how did that change come about?

Tunkelang: When I was a student, I idealized theoretical work. My aspiration was to be a professor contributing to theoretical mathematics and computer science. Perhaps part of my reason was that the problems were so difficult, and I equated difficulty with value.

As I've grown up, my values have been informed by experience. I still have a deep respect for the intellectual acuity of theoreticians, but I'm much more impressed by people who deliver practical impact. In fact, people who find simple solutions to important problems especially impress me. Sometimes it's necessary to work hard, but what matters are the results, not the effort expended.

I've tried to live according to those values myself. My goal is to produce the most valuable results for the least amount of effort. If I have to solve hard problems or make theoretical contributions along the way, then the end justifies the means.

Gutierrez: What personal philosophies and/or theories have you developed from working with data?

Tunkelang: Not sure it's a personal philosophy, but I assume that anything that looks interesting is probably wrong. Even though I assume this, I look at it anyway, because sometimes it really is interesting, and in those cases, it can be extremely interesting.

John Foreman

MailChimp

John Foreman is the Chief Data Scientist at Rocket Science Group, the company behind MailChimp.com, Mandrill.com, TinyLetter.com, and twelve other email-related web and mobile apps. The Rocket Science Group is an email delivery company that focuses on transactional emails, marketing emails, and newsletters for groups and companies ranging from one person all the way up to multinational companies. The three main services send over ten billion emails per month for their seven million customers. Ten thousand new members sign up each day. With this growth comes challenges, as the increasing numbers of members, recipients, email addresses, and emails sent make it harder and harder to fight spam intelligently, model email sender and receiver behavior, and understand the overall email ecosystem.

Before becoming a data scientist, Foreman's career spanned positions at the National Security Agency, at MIT in dynamic inventory management, at Booz Allen Hamilton as an analytics and management consultant, and at Revenue Analytics, Inc., where he helped Fortune 500 companies with revenue management, price optimization, and sales forecasting. In these various roles, Foreman interacted with a wide range of stakeholders—from the US Department of Defense and Internal Revenue Service to companies such as Coca-Cola, Royal Caribbean International, and Intercontinental Hotels Group. He holds an MS in Operations Research from MIT and a BS in Mathematics from the University of Georgia.

Now a "recovering consultant" who has retained his client-centered perspective, Foreman has embraced the world of data science. He is the author of *Data Smart: Using Data Science to Transform Information into Insight*, which demystifies the field by explaining key data science techniques using powerful examples in simple spreadsheets.

Foreman typifies the data scientist who culls the latest data science techniques to best serve his end users' needs. His prioritization of the end user informs his conception of his role as a translator between customers and the data science team and his counterintuitive observation that overreliance on KPIs can harm the end-user experience. Foreman's thoughtfulness in selecting tools, his desire to help other data scientists be successful, and his constant end-user orientation characterize his interview.

Sebastian Gutierrez: Tell me about where you work.

John Foreman: I work for MailChimp.com, which is a web application that allows businesses to have email conversations with their customers. We help about seven million customers send over ten billion emails a month, which to me is very exciting. Whether large or small, nonprofit or for profit, businesses choose us for the excellent user experience we provide for creating and sending marketing email. Our application is exceptionally powerful yet simple while still being playful and fun.

From a business point of view, it's great to help people connect with their audience. From a data science point of view, it's exciting because we get back all sorts of data about who was emailed, which people opened the email, which people clicked on links, which links were clicked, which people went to which websites, and then what they did on your own website. All of that engagement creates this really fascinating set of data that we can then turn around and use to create even more value for our customers.

Gutierrez: Why is this data interesting?

Foreman: If you think about a Facebook "like" in terms of following a page, it's not entirely clear why you "liked" that page or company, as that's a very public-facing action. The things you engage with on Facebook are very public. They're partly for you, but they're also partly aspirational. They're actions for the people watching you. Same with Twitter. Your actions on Twitter are also aspirational. They're for the people looking at you.

However, it's the complete opposite with your email subscriptions, because you're doing it for you. No one sees your email subscriptions except for you. These are the newsletters you want to receive because you're actually interested in getting the content in your inbox, which is very different from the notion of having the information be in a stream that you check in on every now and then. So it's really fascinating to have this data because it represents all of the subscriptions that you really wanted, as well as how you actually interacted with them. For me personally, I get newsletters from places like my church, my local pool, nonprofits I've engaged with in the past, and organizations like my local craft brewery store. So this data is a representation of a wide range of interests that are more personal and a lot more interesting than a "like," "favorite," or "follow."

Furthermore, email data is powerful, because as a communications channel it generates more revenue per recipient than social channels. That's why companies are always asking you for your email address, not your Twitter handle. A business can have a more personal conversation with you and generate more revenue from a customer via than inbox than a waterfall of ads, gifs, and the latest fear-mongering news.

Not only is the depth of this data powerful, but MailChimp is so large, that its breadth and network effects become an asset. 60% of our business is international. We send to three billion unique email addresses, making the network of recipients of MailChimp's email about 10 times larger than Twitter's user base. So it's really cool to think about and see in aggregate what people are interested in and how they engage with the content. And not only that, this is a dynamic data set, as some newsletters are daily, weekly, biweekly, or even monthly. This means it's an evolving and growing data set. It's fun and really cool to work with it.

Gutierrez: What is it like working the MailChimp team?

Foreman: I like the people at MailChimp because we're a very diverse group of people, and we're very choosy about who gets to come in. MailChimp is known not just as a solid, technically advanced product, which it is, but it's also known as a really beautiful product. People have really paid attention to the user experience and design within the application. So it looks beautiful, it feels intuitive, and it's a very pleasant experience for folks to use. Because of all of that, we've really spent a lot of time as a company bringing in the people that can continue to make this happen.

Whether it's customer support, or knowledge-base, or the ops team, or the dev team, or the creatives, or UX—the people at MailChimp are excellent at what they do and very collaborative to boot. That makes MailChimp an excellent place to be challenged. One day I might have an internal client who's a graphic designer and the next day I'll be working with an anti-abuse officer. All located in the same office in Atlanta. So I have to stay on my toes and keep the internal customer in mind; a designer is going to understand and interact with data products and research in a fundamentally different way than a developer, so I have to keep my ears open and my communication skills—verbal and written—sharp.

When you talk to a designer, the way that they think about a problem and the tools they employ to solve that problem are going to be very different from, say, a data scientist or very different from a dev ops person. Those differing perspectives are healthy. There's a—perhaps apocryphal—operations research story about an operations research [OR] person who was hired to fix an elevator-scheduling problem. The OR person initially thinks that they should build a model to solve the rush-hour traffic problem at this elevator bank. Instead, they end up solving it by putting mirrors in the elevator so that people feel like it's more spacious and so they pack in more appropriately. That's not an analytical, data-driven solution, rather it's more of a design solution. And that's kind of the perspective you get when working with all these other kinds of people.

Gutierrez: Was math combined with beautiful design already present in your work or is it a new thing since you joined MailChimp?

Foreman: No, it's something that has developed since I joined MailChimp. I worked at Booz Allen for many years working as a consultant for the government. Then after that I was at a boutique consulting firm called Revenue Analytics, which builds large-scale pricing models for Fortune 500 companies. These consulting engagements offered fascinating opportunities for data scientists. The problems were complex, affected top line revenue, and had excellent data sets. You can make the argument that companies like Intercontinental Hotels were doing big data before there was big data. But given the customer on the receiving end of these models, and given their already ugly, difficult-to-use set of enterprise BI tools—I won't name names!—what I provided was rigorous but often less-than-beautiful.

When a customer books a hotel room or a seat on an airline, they don't interface directly with a pricing model, so who cares if it's ugly or not? I worked with one Fortune 500 company on a very complex production optimization model that involved a lot of decisions, a lot of money, and a lot of raw materials. The model was a huge cost-saver for their business and even garnered some press coverage. It was all true. None the less, the "user interface" was an Excel sheet hot-glued to an Oracle database, IBM CPLEX, and some other systems. Don't let the words "Fortune 500" fool you; the big guys can do work to match anything janky the start-up world might produce.

Getting back to my current job, one of the cool things about working at MailChimp is that I get to build analytics products that may be used by the customer directly. One thing I've worked on at MailChimp is Discover Similar Subscribers. For this project, I built a clustering algorithm that helps users detect segments within their email list. But just as much thought went into displaying the tool to the user as went into the math.

I had to work with others to figure out how to communicate data mining concepts to the layman in a way that they could use and understand the product effectively. So now I became not only a data scientist/analyst, but I also became a communicator/translator to a broad group of people. And that's where this interplay of math, beauty, and design comes in. How do you think of data science in a user experience context to make your data products really work for businesses? It's new to me at this job compared to previous analytics jobs. It's been refreshing.

Gutierrez: Why is Discover Similar Subscribers important to MailChimp?

Foreman: As an email company, MailChimp's reason for existence is to help people engage with their audience. Whether it's a nonprofit that wants to engage with their donors, a small business that runs a newsletter, or a larger business reaching out to previous clients. What we have found working with groups of various sizes is that almost all of them want to effectively understand how people in their list differ from each other so that they can converse with them appropriately. They understand that different people respond well to different kinds of communication. MailChimp considers itself very much a tech company, so there's an element of self-service to our tools. So it was important for us to build a tool to help our customers be able to segment their lists easily and effectively on their own. Discover Similar Subscribers facilitates that self-service audience research aspect of the application.

At MailChimp, because we have such a large user population and recipient address population, we really understand readers at a global level. So we can provide intelligence to our users about their own lists that competitors cannot. We can do this because we're pulling from a global set of email data that includes things like fantasy football league lists and quilting-bee newsletters, to the much larger company newsletter lists. We have the data, and so we can to find ways to make it available to folks. So the Discover Similar Subscribers product is one way we bring our understanding of emails to our customers—by saying, "Based on the criteria you've given us, we think these people on your list are in this segment and these other people are in this other segment."

Gutierrez: Is this a tool that you use internally as well?

Foreman: Yes. One of the ways we've used this ourselves was when we released a new product called Mandrill about two years ago. Mandrill is a transactional email product that is operated through an API. So this would be something that apps would use to send email. It's very different from MailChimp. With MailChimp, you log in, you upload your email list, and put together your content in a drag-and-drop editor. Mandrill is different in that you hook it into your app. So you can use it to send receipts, password-reset emails, and similar transactional emails. It's an amazing product.

When we went to announce Mandrill, we used the Discover Similar Subscribers to mine which of our customers would want to know about our new product. At the time, we had an email list of about three million email addresses for our various customers. So the question was, do we let all three million people know that if they're building an app, they can also use this other product we've developed? The answer, not surprisingly, is no. After all, my local church, which sends a regular newsletter, doesn't need to know about Mandrill. They have no reason to know, and we don't want to bother people. So what we did was use the Discover Similar Subscribers tool to figure out whom we should email about the new product.

Gutierrez: How does the Discover Similar Subscribers tool work?

Foreman: When you think about the billions of email addresses that we have, you can think of them as a globally connected graph where they're all connected to each other based on mutual subscriptions or interests. With all of these connections, I can then figure out who "lives" (in an interest sense) next or near to each other based on proximity calculations. Basically, you can think of it using the concept of neighborhoods—that is, which emails are neighbors to each other. So a user with a list of people, who have already subscribed to their content, can pull different segments based on the globally connected graph. They can then create and name these specific segments of their list.

Going back to Mandrill, that's what we did—we created a segment, based on software development as an interest. We then sent our product announcement email specifically to that segment saying, "Hey, just so you know, Mandrill now exists. It's really cool if you want to use it." We got amazing engagement from that email. And not only did we get great engagement from it, we also didn't bother all of the other people who wouldn't have wanted to know about the new product. That's where I see the power in having this kind of data and providing this kind of segmenting capability to users.

Gutierrez: How do you describe your work to someone who's not familiar with the math behind it?

Foreman: For those who aren't familiar with the math, I talk about how I serve three roles. First is that I build data products, second is that I am a translator between customers and our data science team, and third is that I am an ambassador for MailChimp. In regards to building data products, I do this for two types of customers. The first type of customer is the MailChimp user. The data products for this customer tend to be external-facing products or artifacts like reports or blog posts. We do research on our global data set and provide artifacts or products, like Discover Similar Subscribers or Send Time Optimization, to external customers.

The other set of customers are actually internal customers. There's a huge need at MailChimp across teams to understand the data we have, especially around understanding our customers and their unique characteristics. Those are needs the data science team can facilitate, so we also build tools for these internal customers. These tools range from things like anti-abuse models to support scheduling tools to likelihood to pay models.

The second piece is being a translator between internal/external customers and the data science team. Being a translator helps me to figure out what to build. I want other people's input; I don't just make our workload up. That's actually scary when some analytics teams do that. They just kind of show up and say, "Well, we should build this," and they haven't talked to anybody. They just thought it would be cool to use some tools. So a large piece of what I do is talking to people, understanding their needs and their problems, and translating that back to technical folks within MailChimp.

The third piece of what I do is having conversations with my peers to understand where the data science world is moving and communicating what MailChimp is doing to the rest of the world. It's about engaging in the emerging, global data science conversation. I think MailChimp is doing some things that are cutting edge, and I want to tell people about them. Furthermore, I want find out what everybody else is doing. This falls more under questions like: Where are the tools moving? What types of data are becoming available? What kinds of practices are being done? I want to understand the metaphors and analogies other people are using to think about their work. For instance, at the Strata Conference, I attend talks on healthcare, oceanography, defense, etc. Where the speaker's world is not my world, but so many of the problems are the same; these metaphors really help me.

Gutierrez: How do you describe your work to someone who is familiar with the math behind it?

Foreman: Other than perhaps geeking out on the implementation of some of our data science products, I wouldn't change that description much. Technical folks (perhaps even more than non-technical folks) need to hear the emphasis on communication as part of a data scientist's daily routine.

Gutierrez: How does MailChimp think about spam?

Foreman: First, let's talk about email. Email is a messaging standard that exists that no one really controls. Other messaging systems, like Facebook Messenger, which is operated by Facebook, are under the control of an entity that gets to decide what happens on its system. Email, on the other hand, is powerful because anyone anywhere can send you an email at any time about anything—without having a larger entity dictate the terms of the exchange. Anyone can set up a mail server and suddenly have the capability to send and receive email. When someone receives emails that they do not want, we call it spam. This has always been an issue with email and so people have dealt with spam for a long, long time.

Fighting spam is especially important to MailChimp for two big reasons. One is that we don't want our users to send spam and two is that big email companies like Gmail, AOL, Hotmail, etc. will block our IP addresses if their users report spam coming from our IP addresses. Potentially having our IP addresses blocked is a big issue. We send emails for seven million users, but we don't have seven million IP addresses. Outside of a few users who have dedicated IP addresses, most of our users send with a pool of other users over one IP address. If we allow a bad user to send really evil stuff, then the email receivers will block that IP address, which is bad for the whole pool of users using that particular IP address. So we do a great deal of work combating spam.

Gutierrez: How is spam typically combated?

Foreman: Over a decade ago, Paul Graham wrote an essay called "A Plan for Spam." Since then, there's been a whole lot of work done around spam detection. If you go back and read those early essays, the way it was dealt with was looking at the actual content of the email. People looked at the words on the page and put them through a model to get a sense of whether the words in the email were about something bad or something spammy. So of course, this started an escalation between spammers and people trying to stop them. All of a sudden Viagra has an @ symbol in it, a one numeral instead of an "I", or similar things like that. The spammers were trying to get around matching tokens, naive Bayes filters, and similar spam detection techniques. This kept escalating and escalating. Then, rather than using words, spammers started using images with words, which meant that the people trying to stop spammers had to start doing OCR [Optical Character Recognition] on images to get out the tokens to put them through their models in order to identify spam.

This continued until it got to the point where there needed to be a new approach. It became very clear to us that it's not good enough to just look at the content spammers are sending. We're an international company, so the content we are analyzing could be in many, many languages. It could also be all images. Or, most worrying—and perhaps the biggest problem—is that the content could look perfectly fine. And this is something I don't think people realize about spam. The modern assumption that classical spam filters operate under is that spam embodies a platonic ideal such that spam is about Nigerian princes, Viagra, or similar sorts of things.

Gutierrez: How is MailChimp combatting spam?

Foreman: We think about spam using the postmodern definition that "spam is in the eye of the reader." We know this to be true, because email clients have spam buttons. It's you—the person who receives the email—that deems an email to be spam. It not words on the page. It's *how you interpret* those words and your relationship with the sender. This is fascinating from a data science perspective.

How we combat spam is by going after the relationship between the sender and the receiver. The content of the email that is sent out could be very innocuous. It could be local real estate ads, let's say. So in this example it could be a real estate agent sending out ads for houses that are for sale in a specific market. What we ask and try to figure out is how this real estate agent procured the list of people they are contacting via email. Are these people clients they've worked with in the past, and they told the agent, "Yeah, I'd love to see your content in the future!"? Or did the agent go to a local Chamber of Commerce meeting and take the email list? Did the agent do a public information request to the state and get public university employee email addresses—which we've seen people do? Did the agent scrape email addresses off of the Internet? Did the agent go to Facebook profiles that were public and take them? All of these different ways to acquire emails will show up as different relationships between sender and receiver.

So what we need to do is determine what the relationship is between sender and receiver to figure out if it looks like a spammer or non-spammer relationship. Once again, because MailChimp, unlike other competitors, has a massive global set of information about how emails are connected, what people are reading, and how users have interacted with email addresses in the past, we can bring all of that information to bear on the problem. So we can calculate if the relationship between a user and the list they've constructed feels right or not. We do that by building a lot of AI models around transactional data, metadata, and email list data. So when a new user comes in, we can run them and their list through the AI models and either permit them to go through or shut them down immediately. The great thing about this approach is that we can actually determine, before a user has even created content for an email, whether we should shut them down because the relationship looks really fishy. Of course, we communicate with them that we think the relationship is complete bunk and that they're going to have to go elsewhere because we can tell immediately that this looks like a spammer account.

Gutierrez: How did you get involved with programming?

Foreman: Back in college, I was a pure math major, and I really enjoyed it. I thought abstract algebra was awesome. I also really enjoyed things I learned in number theory. I just thought all of it was a blast.

But my advisor, for better or for worse, sat me down one day and said, "You know what? You could go all the way and become a pretty average mathematician in academia. You're never going to be great, but you could be one of those guys that toils along as an average mathematician if you wanted to do that." I left that meeting thinking well, you know, he shot me straight. He told me more or less where my future lies in academia, which is being average, and that is a perfectly noble pursuit.

As I was mulling over that meeting and my future, I got a job working for another professor in the math department modeling knot tying. Knot tying focuses on how knots or curves that loop back on themselves interact, what happens when you tighten them up, and how compact can they get. This has applications in things like physical cosmology, protein folding, and other important subjects. So we did a lot of 3D modeling of these sorts of problems, as well as determining if you have a knot, how do you even understand what type it is? There's a huge typing system in knots, so there was a lot of coding necessary for this work.

This was my first experience coding in C—finding memory leaks—and I kind of enjoyed it. I had a huge bank of Mac Pro towers, and I got to parallelize code across all the towers. It was just a real blast. During this experience I started thinking that there were some aspects to the "real world" that I could be interested in. However, I still didn't really know where to go with it.

I ended up briefly at the NSA for a summer internship program called the Director's Summer Program. They basically get a bunch of math nerds, put them in a room, and throw food at them while they do math. I did that and realized I didn't really want to be in government. Not because they don't tackle important problems—they do, but because I encountered a lot of folks who were just waiting to retire. For example, I worked with one person who had a picture of a golf course up above their desk who would say, "This is me in two years." I was hoping for more excitement than this in my work life.

Gutierrez: Having ruled out government and academic math, what did you do upon graduation?

Foreman: I went into consulting, in part, because I wanted to be able to attack problems at a fast pace. So I joined Booz Allen and started doing a lot of different analytics projects for the government. The team I worked with mainly dealt with quick-hit analytics projects. For example, one of the projects was around BRAC [Base Realignment and Closure] for the Army. Since the Army can't shut down training when it moves bases, we modeled the move of the Armor School, with all of their tanks and heavy vehicles, from Fort Knox to Fort Benning with respect to training demands. Our mathematical models allowed the military to train personnel to operate heavy artillery all while shutting part of one base down and relocating it. Don't ever let someone tell you that you can't change the tires on a moving car!

The great thing about doing analytics for the government is that there are lots of opportunities to improve things. They've got a lot of very complex problems that have very interesting constraints. For example, how do you predict tax return fraud when you're not allowed to use discriminatory features, such as zip code, which are quite correlated with the fraud itself? There's a challenge!

From government, I moved into consulting for large enterprises doing pricing models, and some blending models for juice products over in China, which was a blast. So at this point, I had had some large enterprise experiences, but I wanted to get a sense of data science in the startup world. One of the things that I wasn't given a chance to do a whole lot of in the enterprise world was supervised artificial intelligence. I did a lot of forecasting, which is a type of predictive modeling. You're using time-series data, along with a lot of optimization modeling when that's around pricing or supply chain, but not a whole lot of true machine learning.

Gutierrez: How did you end up at MailChimp?

Foreman: I met with Neil Bainton, MailChimp's COO at the time, and spoke with him about the problems at MailChimp and how data could potentially solve some of them. Because it's a successful online company with two domains in the Alexa 500, the data set being produced is just massive. So it was an amazing opportunity to use data to solve problems. I initially thought in my meeting with him that we were talking because he was looking for talent, so I started suggesting other people he should speak with. After some conversations, he asked, "Well, what about you?" As we talked more about it, the scope of the opportunity became clear to me, and I ended up here.

It's been an absolute blast, because MailChimp operates so differently from the government or large enterprise companies. In the government, you have layers and layers of management, and there are really well-defined chains of communication. You find the same thing in the enterprise world, too. MailChimp is a small organization, so it's very flat. It's not the kind of place where you go and tell other people what to do or go talk to their manager and their manager tells them what to do. No, it's all done through close communication and collaboration, and so it was a new challenge to do analytics in this type of environment. I have found that I really liked the independence and the chaos that came with it.

Gutierrez: When did you realize the power of data?

Foreman: When you look at operations research problems or optimization problems, the data's important, but what's really important is the formulation. You have a really small amount of input data—and really complex decisions need to be made from that data. When I went and started doing price optimization models, that's where the power of the data really hit me and opened my eyes. This is because you're forecasting demand for some product, whether it's a hotel room, or a room on a cruise ship, or a seat on an airplane, or something else using historical data, and then updating the pricing of this product as more data comes in.

What's amazing about these pricing models—I think they're some of the coolest data science models around—is that their decisions directly affect revenue. If I choose to lower prices on my hotel because I think that's going to maximize revenue, that's a pretty audacious decision for a model to make. The fact that it can say, "You need to take the price down $15," and have that actually affect your bottom line in a positive way is amazing. The model can actually say that, and then we can go back later and prove it was right. We saw some Fortune 500s get 2 percent revenue uplift per year, which is amazing at that scale.

That really opened my eyes to the fact that if you can appropriately gather and keep track of lots and lots of transactional data, there's a competitive edge there.

Gutierrez: How are you learning about and keeping up-to-date with the data science industry?

Foreman: Though I love reading blogs and enjoy engaging online to keep up to date, for stuff like learning, there is no replacement for books. In the data science world, there are lots of great books coming out. There are just so many great ones that come with real-world examples and accompanying data sets. Obviously, there's the Hastie, Tibshirani, Friedman book, *Elements of Statistical Learning.*[1] That's kind of the data science bible. Just recently, Khun and Johnson's new book on predictive modeling, *Applied Predictive Modeling,* came out and I've been reading it.[2] It's excellent. So books are one place I go for all my learning.

The interesting thing about a lot of data science books is that they are written with a specific tool already picked, whether it's R or Python or something else. If you're going to learn how to really do this stuff, you've got to do it with the available tools, so the books make tool decisions. Learning to code examples out of these books can help you get your feet wet. But it can also become a distraction when you're trying to truly understand a technique.

This is, in part, why I wrote my book, *Data Smart.*[3] I felt like a lot of the examples in these books were simply, "Let's load the support vector machine package, train our model, and then look at the results." It was like wait, wait, wait, wait, and wait. You need to explain in detail how that support vector machine just got built. You can't just build it. That's cheating, which is totally what you do in a real job—you trust that the packages work—but for learning purposes, you don't. So I wrote my book to break down all these things in detail and not use really complex formulas, maybe the way Hastie's book would.

Gutierrez: Where else do you engage with the community?

Foreman: The great thing about data science right now is that there's a very active, engaged community both in the physical world—at conferences like O'Reilly's Strata Conference—and online at websites like Cross Validated and Twitter. In person and online are both great places to have conversations with other practitioners. Amazingly enough, Twitter is probably the best place to *start* conversations about data science, although I find myself often turning to email to finish them. You can find the experts who know this stuff and then further that conversation in a longer form.

---

[1]Springer, 2nd ed., 2009.
[2]Springer, 2013.
[3]Wiley, 2013.

Gutierrez: What does a typical day look like for you at MailChimp?

Foreman: I lead the data science team at MailChimp, and I like to get my hands dirty too. Some of the big pieces of my day involve working with my team to take stock of current projects and figure out where to go next, doing my own work and prototyping things—I do projects just like my peers do—and then also my talking with other teams, talking with management, and planning for the future.

On our team, we've got different folks facing different kinds of projects. We've got one person who really owns compliance and looks at our compliance processes. We've got another data scientist who focuses on more of the user experience side of the house and understanding our customers. I help them and others as is needed while I do the three things mentioned earlier—build data products, be a translator, and have conversations with the data science community.

We also have developers on the team who can take prototypes and turn them into production tools.

And on top of that, we have a group of qualitative researchers on the team who work alongside us to source ideas and projects—and to verify findings and product directions—through surveys and a heck of a lot of customer visits.

That's one thing that makes the MailChimp data science unique—we've placed quantitative and qualitative research in a single place right next to some great engineers. It's a powerful mix.

Getting back to how my day is laid out, first thing when I get into work I take stock of where we're at across the various projects that we work on. It's important to understand where we're at on projects—whether it's surveying users, mining data, or building models. Once our progress is clear to me, I carve out time to prototype before my brain gets fried. I am not a developer, and I don't claim to be one. However, I can write terrible spaghetti code that will run, though no one will probably want to read it.

And so one of the things I try to do is get my hands dirty and actually get into the data and build prototype models. An example prototype I put together recently focused on helping people segment their list based on predicted email address demographics—for example, age range.

Once the day is going, a big component is communicating with other teams to figure out what we can do to help them. For instance, we may be working with the integrations and partnerships team, who might have a big user who's interested in doing a spherical K-means clustering of their list to investigate different segments. So we'll work together to figure out if we can run the clustering, and then talk through the results with them to figure out what came out of it and who the readers are on this list. This type of work then benefits this particular team, because they can provide this as a report to the customer. This benefits me because I learn a little bit more about the power of our data, as well as what our customers want and need.

Another part of the day is spent very carefully planning and thinking about the type of infrastructure we build. Now that we have a data science team at MailChimp, it's no longer just about building out the infrastructure that just keeps the application running. That's obviously priority number one. But now we have all of these people who want to access the data for analysis. This type of access involves very different patterns than our users accessing their own data, so we need to think carefully about that. People run for the hills whenever I execute an SQL query, because the queries that are run from the application only hit indices, whereas the ones that I run involve a whole lot of joining or subqueries.

So when I come along and I've got something that's some crazy nested thing with a million joins, and then God help us, there's a cross-join in there and I'm blowing some data set out, and I'm doing window functions on it in SQL and things like that, I make all sorts of stoplight charts turn red on our NOC [Network Operations Center] wall. And that's not fun for them and that's not fun for me. So there are definitely discussions that have to occur during the day around what systems we're building and how they're going to be used.

We have to make space for analysts to hit these boxes as hard as they want or hit different data sets as hard as they want. It's a back and forth of what data do you need? What data can we provide you? What are your needs? Are you building something for production or building something for analysis? Are you okay with a query running for a really long time? Does it need to run super quick? So there's just a piece of my day that involves technical requirements.

Another part of the day is spent speaking to executives and other customer-facing teams to ask them what they are hearing from users in terms of features that they want. Then thinking about whether data can be brought to bear to make any of those requested features possible. One requested feature we recently did was send-time optimization. We spent a lot of time discussing how to use our data to appropriately predict and optimize when a customer should send their newsletter to best engage their audience. One of the things that drives that engagement might be the time zone of the email receiver. Everyone's different and everyone's on a different schedule, but they might cluster around a few different specific times of the day, especially based on whether they're on one side of the globe or another, as most people work day shifts and most people sleep at night. Those are obvious things. But then it gets way more complex from there. So we talked to executives and our qualitative research team to really understand if people wanted this and then figured out what data was going to be needed. And if we did build it, how we thought it should be presented to the user. We spent a good amount of time making sure that if we invested the time and energy into the project, that it would be a good investment.

Another part of my day, once we've spent time figuring what customers want, is mapping out where we're headed next in terms of the products we're going to build, and then taking that back to the teams we are working with to explain how we're going to build certain products. For instance, we do a lot of things internally via internal API calls, so we need to make sure we build those APIs in a way that'll work for our internal customers.

Gutierrez: How do you view and measure success?

Foreman: We are very gun-shy, perhaps to the point of being hostile, about using what someone might call "KPIs" to measure success. We see a lot of competitors measuring success in really stodgy ways. Internet companies look at things like ARPU [Average Revenue Per User], retention, or conversion to paid plans. And we could do that too. But we don't. The organization measures none of these. Specifically, because we feel as an organization—not just the data science team, but the organization as a whole—that when you start looking at those quantitative measures of success, things can get perverse.

I'll give you an example of this perversion caused by KPIs. I'm a huge Popeye's Fried Chicken fan. I love their chicken and their red beans and rice. However, I went to one of their drive-thrus recently and placed my order. I pulled up to the window, and they've got a timer running. That timer is essentially there to evaluate their performance for how quickly they can serve each customer. If that's the metric that has been incentivized, if that's the metric that everyone looks at, then in an ideal world, people would strive to improve that metric in the appropriate way.

What they do instead is improve the metric in a perverse way. Every customer that pulls up immediately gets told, "Hey, can you go park in a space? I'll walk your food out to you later." So immediately, they're actually degrading performance and not necessarily speeding anything up. They've improved the time to serve each customer metric that's being measured because they've now reset the clock—because you just drove into a space. After all, you can't really tell them no, as they're making you your food and you don't want to mess with that. But now they've degraded performance because they've got to walk it out there. So this is time they've spent going around the counter, going out the door, trying to figure out which car is the right one for this particular order, bringing the food to my window, checking it with me to make sure they've got the right car, then walking back inside, and going back around the counter and serving the next customer, which, as you can imagine, takes a long time. So because this organization has focused on this time metric KPI, they've actually degraded performance globally.

Going back to our business, we've seen that type of behavior with our competitors, as well. Where they've focused on certain metrics that have actually led them to not serve their users appropriately—maybe they've generated more revenue in the short term and driven up their stock price, but have actually created a lot of animosity with their users. We don't have to worry about that because we're a private company that's not owned by someone else. Instead what we focus on—and this is going to sound goofy, especially for a data scientist—is the happiness of our users. So we stay in very close communication with our users, not just by being very engaged online through social media, but also flying all around the globe to talk with our users. We use these conversations to gauge what our customers are saying, what they like, and what they don't like. And we focus on their happiness exclusively in order to provide the best products possible.

Metrics aside, if you're doing what you can to solve all of those things that you're learning about via conversation, and focusing on the things that you're doing well, then that's really the best you can do. And that doesn't create any weird, perverse measures, where instead of looking at your users and looking at your product, you avoid your users and products to instead look at a KPI number. In some cases, you're even competing internally with other teams over this number and asking questions like, "How much of this number can I attribute to the data science team?" That's a terrible way to look at things. How much revenue can be attributed to data science versus marketing versus user experience? The moment you do that, then all of a sudden, you've put teams at odds, you've put leaders at odds, and now there's less of a desire to collaborate.

So by not focusing in on those measures, we've actually done a great service to the organization, I think, by just encouraging everyone to focus on making our users happy. It's much easier to say here's the problem that needs to be solved and ask who can bring what to bear on it. Maybe there's not a number behind the problem, so the way I measure success is the feedback I get from customers—whether they are internal or external customers. Focusing on getting very positive feedback ensures that my goal is to see folks using my features in an appropriate way and getting good stuff out of them.

Now, of course, we also measure certain really down-to-earth metrics that help to ensure our customers stay happy. These low-level metrics are more around server uptime, whether send-time optimization is working, if people are using it, what segments look like, if it is helping, whether bad users are getting through, what the spam reporting looks like, whether emails are bouncing, and similar local metrics. So there are operational metrics like that that we look at—just none of those crazy global stupid metrics.

Gutierrez: How do you know you're solving the right problems?

Foreman: That is a tough one because the interesting thing about figuring out what problems you should solve is that it's going to be a combination of what your users think they want, what your users actually want, and where the space is going. Talking to users is crucial because they point you in the right direction. Often though, the way a problem is characterized by an internal or external user is the result of the thought process of someone already having thought through and processed the problem in a particular way, which means we have to figure out if the way they're posing the problem is the *actual* problem.

Going back to the elevator bank example mentioned earlier, maybe the problem is posed as, "I need you to build an optimization to speed up my elevators," when, in fact, the problem is, "I want more people to get on the elevators in any way possible." The solution would be very different depending on which problem you solved. Once it's understood that people just need to get through, the right solution is to put mirrors in there so that people pack in tighter. So you have to be very careful that, when people request features or tools, you actually solve their underlying problem.

Furthermore, part of being a company in a very competitive space is coming up with stuff maybe people aren't asking for yet, but as soon as they see it, they're going be like, "Oh, yeah, that's much better." So there's also this notion of creativity. We have to make sure to understand where the space is going, not only in our industry, but also watching what's happening elsewhere. This way we can learn those metaphors and think about how to apply them to our world. So that's where a little bit of the black arts for data science comes in—with this creative piece. It's not just about understanding the techniques, the data, and the technologies available to me—that's the base line. It's about taking the time to dream and get creative, instead of waiting for people to tell me what to do.

So there's the notion of really actively listening to figure out what people are actually saying, and there's also the notion of hearing what's going on in the world and understanding how it affects my world. The French word for it is *bricolage*, which is taking these disparate things and combining them into something new. In my case, it's about thinking how people have solved problems in other industries and how that applies to what we are working on. Sometimes you go down some dead ends, where you realize that for our customers and what we're trying to do, this has no applications. Other times you strike gold and you realize, oh man, if we did what they're doing, if we could actually create this, it would be awesome for our customers.

Gutierrez: How do you think about the toolset, and what technologies you are currently using?

Foreman: I'm very conservative in the way I think about tools. What has struck me in the data science world is that the tools are wagging the analytics, which is to say that a lot of organizations get attracted to new, bright, shiny tools rather than thinking of what problem they are trying to solve. They say, "Ooh, we really want to use this tool," and then they come up with a reason to use it. Vendors probably drive a part of that. Vendors are there to sell you a tool for a problem you may or may not have yet, and they're very good at convincing you that you need it whether you actually need it or not. I frequently encounter people running Hadoop and they are excited to tell me that they now have all of their data in HDFS. I ask them how much data they have and if it is structured. I'm always amazed when they tell me that it's a few gigs of structured data. That size and type of data could fit into a tiny free SQLite database. This tells me that they encountered a very good salesperson and they haven't actually thought through the problem they are solving.

If you do it this way, which is backward, it's a lot like most people's New Year's resolution for getting healthy and losing weight. It's January 1st, and I go get a gym membership and buy a bunch of workout gear and new clothes. What have I done? Nothing. I'm just as fat as I've always been, but I feel like I'm making progress because I've spent money and bought things. That's how I see the businesses that go out and procure tools. They say to themselves, "We've got to do big data and we've got to do data science, so let's go get tools and get consultants, and then we'll be ready to go." And before they know it, all they have to show for it is a bunch of money spent, a bunch of tooling, and maybe an infographic, because they never took the time to do the one thing that's very hard to show progress on, which is thinking. They never sat down and thought through: What problems should we be attacking? What data do we have, and how should we attack these problems given the data that we have? Instead, they went out and spent their budget, because that's a great way to show you're doing something. You're spending money. Something must be happening. Everyone's waiting for someone else to make something happen while they spend the money.

We're different and very conservative in the sense that the way I think about tools is problem-focused. We start with the problem we want to solve or a general understanding of several problems we want to solve. Then we take stock of the data that's available to us. We think about the techniques that are available to us. We think about the technologies that are available to us. And then, and only then, do we select the technologies that are going to solve those problems.

For instance, on some of the AI models we built for compliance, there are some really sexy tools that we could have used. However, what we realized is that all the data we cared about for these models was already structured. And because it was already structured, it already worked well within an SQL context. Furthermore, a lot of the queries we needed to run for the training sets were queries that were best accomplished via SQL window functions, as we were looking at a lot of lagged time-series data. So once we hit that point, we realized that it would fit fine in a sharded PostgreSQL database, as the data was probably smaller than 30 terabytes. Having realized this, we asked ourselves why would we need something else? This is a tool that's very robust and stable. It's a tool that the devs know how to work with really well. They can spin it up fast. Our compliance team needs our models yesterday. Why would I choose to go after tools that are a little bit less stable but sexier and that our devs and our ops people don't quite understand yet? Why would I risk it when I can use this other stuff that we already understand? So I'm very conservative with the tools that are selected.

Gutierrez: What do you look for when hiring people?

Foreman: Obviously, if we know the technologies we're using and are going to be using, we do want to hire people who are familiar with those technologies. We want to go after the folks that really get the technologies we already use, are going to use in the future, or are in the realm of tools that we might use in the future even if we don't know now.

However, I don't think that's the only thing you should look for. I look for a couple of things that I think are very important, which is the ability to learn and the ability to communicate. We want to understand that though maybe you haven't used exactly what we're going to use in the future and maybe you've used something else, that you can still learn these things. That's a tough thing to grapple with and understand about a person, so you have to ask for examples. You've got to hear them articulate problems they've encountered in the past. And if they can't articulate situations in which they've encountered problems, and how they used various approaches and showed resourcefulness by saying something along the lines of, "I didn't know how to use it, but I grabbed it and we used it this way," if they can't articulate something like that, then that could be a red flag. So we always make sure to look for that learning capability.

Part of being able to articulate a story like that is that you've got to be able to communicate. I think it's essential for a data science team to hire people who can really speak about the technical things they've done in a way that nontechnical people can understand. This is because as a data science team, you end up working with a lot of very nontechnical teams. I work with the marketing team just as much as I work with the dev team, so if I have a bunch of folks on my team who can't communicate well, who just wait for work to be thrown to them like some wild animal waiting for raw meat to be thrown over the fence for them, then that is unsustainable and unproductive.

At certain organizations that have one key analytics project that just needs to get better and better and better and better, it can make sense to hire people who don't need to focus on communicating with others. For instance, maybe it's ad placement in a giant search company, where you just need to hire people who wrote their PhD thesis on that particular area, and you don't need them to know how to communicate. You just need them to do what they do and you'll get huge marginal benefits out of having them improve your AI models.

But at an organization like MailChimp, where gains are made through collaboration and solving new problems, you've got to know how to communicate. So that is a key piece I really go after. It's not just technological familiarity, but also the ability to learn and the ability to communicate with other folks.

Gutierrez: Do you find it easy or hard to find and hire the right people?

Foreman: I find it tough to find and hire the right people. It's actually a really hard thing to do, because when we think about the university system as it is, whether undergrad or grad school, you focus in on only one thing. You specialize. But data scientists are kind of like the new Renaissance folks, because data science is inherently multidisciplinary.

This is what leads to the big joke of how a data scientist is someone who knows more stats than a computer programmer and can program better than a statistician. What is this joke saying? It's saying that a data scientist is someone who knows a little bit about two things. But I'd say they know about more than just two things. They also have to know to communicate. They also need to know more than just basic statistics; they've got to know probability, combinatorics, calculus, etc. Some visualization chops wouldn't hurt. They also need to know how to push around data, use databases, and maybe even a little OR. There are a lot of things they need to know. And so it becomes really hard to find these people because they have to have touched a lot of disciplines and they have to be able to speak about their experience intelligently. It's a tall order for any applicant.

It takes a long time to hire somebody, which is why I think people keep talking about how there is not enough talent out there for data science right now. I think that's true to a degree. I think that some of the degree programs that are starting up are going to help. But even still, coming out of those degree programs, for MailChimp we would look at how you articulate and communicate to us how you've used the data science chops across many disciplines that this particular program taught you. That's something that's going to weed out so many people. I wish more programs would focus on the communication and collaboration aspect of being a data scientist in the workplace.

Gutierrez: How do you deal with privacy and user expectations concerning their data at MailChimp as you build data science products?

Foreman: To start, MailChimp doesn't make its money through any means that's at odds with our users or readers. We make our revenue via monthly subscriptions from users with permission-based lists. That means that I don't have to use our data in ways that trick, manipulate, or in any way suck more money out of our customers (think ad placement). When I use data, it's in line with what our customers want, which is fantastic.

Obviously, the in-house lawyer's a huge help. But the law isn't what it's really about, is it? It's about user expectations. So it's great to have such an active user base that stays in touch with us because they always make their preferences clear, more than any law would. Our users say, "This is what I think this is acceptable and this is what I don't think is acceptable." Which is another great reason that we look to maximize user happiness rather than some cold inhuman KPI. So we're constantly listening to folks to understand how to properly use data to add value to our service in a way that our users and ourselves feel comfortable with.

Gutierrez: Can you elaborate further why you thought it was necessary to write a book on data science, even though there were already so many great books out there?

Foreman: It goes back to this thing that I'm passionate about that perhaps a lot of people aren't, which is that it is really easy to get obsessed with tools—and you have to watch yourself. It is really easy to get the tools to do all the work for you. But I think it's important that data science practitioners know what these tools are doing. You don't have to know everything, but you should have a general idea.

This was really driven into me in grad school. In one class, my professor made us build an optimization model by hand. We had to do all of the pivots of the simplex algorithm by hand on paper. It was awful. Anytime you have to do lots of operations to matrices by hand, it's just a nightmare. He made us do it once, and after that we never had to do it by hand again. Though it was terrible, now when I run the simplex method, I know exactly what's going on because I've done it by hand.

That's very helpful in terms of intuitively understanding what you are doing. When you formulate a model, you will now intuitively know what you're doing and what the method is going to do when you call it. If you don't do it once, if you don't really learn it and internalize it, then I feel like there's always going to be this secret doubt you're going to harbor. And you're not going to be able to fully justify what you've done or believe in because there's this magical incantation you make at some point like, "Call this AI model." And you don't really know what it's doing. You might expect too little out of it or you might expect too much, but you won't really fully comprehend it.

I think a lot of AI models are very dumb. Naive Bayes is the dumbest thing you've ever seen. So now that I know how it's built, I've actually done one by hand, I kind of know what to expect out of it because I know what it's doing internally.

And so the purpose of the book that I wrote, *Data Smart*, is that it takes readers through a bunch of different types of modeling. It takes people through unsupervised artificial intelligence to data mining. It takes people through supervised artificial intelligence modeling several times. It takes people through forecasting, outlier detection, optimization, and simulation as well. And the way that they go through the modeling is that they do every single model by hand in a spreadsheet.

They do it in a spreadsheet because there's nowhere to run or hide. You're doing every single step yourself and you can see how it works. I did it this way for a couple of reasons. One is to show people that it's not black magic. Data science and machine learning have a lot of really amazing words around them that make them sound like *Terminator 2,* where robots are going to come and destroy the Earth. When you look at something like a boosted trees model, however, it's relatively cutting-edge but not that hard. I think someone with a year of freshman college math could build that model and understand everything about it.

So I wrote the book because I wanted to take people through these things in great detail to help build their confidence and so that they understand they can prototype in them and not be afraid of them. I want them to see how even though these models are easy or silly, they still get the job done amazingly well. I want to broaden the conversation. I feel like the way a lot of people talk about this stuff is just so mystical. They don't really want to tell you what they're doing because their job security is wrapped up around being some sort of shaman-like persona. But that's not what your job security should really be based on—it should be based on solving problems. If you're solving problems appropriately and you can explain yourself well, you're not going to lose your job. You don't have to hide behind the fact that no one else knows what this model does.

So that was the purpose of this book. It's for the people who are gluttons for pain and who really want to understand how these models work. You can work the spreadsheets with me. And by the end, if you survive, then you're really going to know this stuff. You'll know how to talk about it intelligently. So I think it's very fun for a particular kind of reader. It's not bedtime reading. You're not going to read it in bed. It's much more about sitting at your desk, leaning forward, and doing it step-by-step as I walk you through it. And so I thought it was a pretty cool addition to a lot of the other books already out there.

Gutierrez: Why did you choose Excel as the tool to use to learn the techniques?

Foreman: There are already a lot of books on R, Python, and similar programming languages. Unfortunately, there are a lot of people in the enterprise space that don't do R and don't do Python. If you look at accounting, or finance, or the government, what you'll see is that for a lot of these places, the analytics system of record is freaking Excel spreadsheets and VBA macros. This means that books using the aforementioned programming languages are leaving these people behind.

How do you start a conversation with them about these techniques? You do it by working in a tool that's familiar and comfortable with them, and then you can slowly usher them into more modern tools. In the last chapter of my book, I actually do introduce R and say, "Remember that forecast model that we built that took us 50 pages in the book and 10 tabs in Excel? It turns out that you can call it in R in three lines." I take people through that and show them that you can just call the forecast function, and it just does it, which is amazing. Could I have done that on page 1? Absolutely not. You can't do that at the start. You've got to teach people first.

So that's the book. I think it just goes back to my general philosophy, which is that if you're going to do data science, you can't just buy the tools and sit there. You have to know what tools are available to you. You have to know the techniques intimately because you've worked through a book like mine. Then when a problem comes along—wow, you'll actually have a toolbox that doesn't just have a hammer in it. You'll have a toolbox that's got all of the tools inside of it.

A lot of data scientists only do AI, whereas there are a lot of problems that are solved and could be solved with optimization. There's even a whole range of problems that can be solved with simulation. You've got to have all of these tools in your toolbox. That way, when you know all these things, you can then select the appropriate tool to solve the problem in a way that matters.

Roger Ehrenberg

IA Ventures

Roger Ehrenberg *is the founder and Managing Partner of IA Ventures, a seed-stage venture capital firm investing in companies creating competitive advantage through data. As a result of his investments, Ehrenberg sits on the boards of DataSift, Metamarkets, PlaceIQ, Sight Machine, The Trade Desk, and Twice, and serves as a board observer for MemSQL and SavingStar. An early investor in the "Big Data" theme and more recently in "data-centric companies", Ehrenberg commands a comprehensive view of the opportunities and challenges posed by data-driven competitive advantage. Since IA Ventures' inception in 2010, he and his team have invested in more than 30 companies and helped them to implement strategies that leverage data to drive differentiation.*

From 2004 to 2009, Ehrenberg seeded forty companies as an angel investor through IA Capital Partners—including bit.ly, Buddy Media (acquired by Salesforce.com), Global Bay Mobile Technologies (acquired by VeriFone), Invite Media (acquired by Google), Magnetic, TicketFly, TubeMogul (IPO), and TweetDeck (acquired by Twitter).

Before becoming a startup investor, Ehrenberg worked for over seventeen years on Wall Street in M&A, capital structuring, derivatives, and quantitative trading. He served as the Global Co-head of Deutsche Bank's Strategic Equity Transactions Group and finally as the President and CEO OF DB Advisors, Deutsche Bank's internal hedge fund trading platform. He holds an MBA in Finance, Accounting and Management from Columbia Business School and a BBA in Finance, Economics and Organizational Psychology from the University of Michigan.

Ehrenberg is a paramount example of someone who foresaw and championed the transformational power of Big Data and data-centric companies and, by investing in and mentoring a select subset, has been able to help a range of organizations and the entrepreneurs behind them achieve significant growth and impact. Ehrenberg's unique background—helping to drive the high-speed data evolution on Wall Street and then investing in data-driven approaches to transforming global industries— lends authority to his discussion of the value of having a data scientist in residence at IA Ventures, why he looks for entrepreneurs with data-centric DNA, and what he thinks data-driven companies of the future will be working on. While he is clearly passionate about the power of data, what really stands out in Ehrenberg's interview is how much he helps and cares for the people behind the companies he invests in.

Sebastian Gutierrez: How do you think about IA Ventures?

Roger Ehrenberg: IA is the culmination of everything I've done previously, from my eighteen years on Wall Street and running data-intensive businesses there—where I was both one who utilized and also purchased what's now commonly referred to as "big data technologies"—to my years as an angel investor. As a principal with IA, being able to invest in leading-edge technologies is something that's very, very exciting.

Gutierrez: What was the transition from DB Advisors on Wall Street to being an angel investor like?

Ehrenberg: The transition was definitely a challenge. Culturally, it was a shock going from having a corner office on Park Avenue managing a global team of 130 quants to working in my home office and having a clean slate. On one hand, it was very exciting. On the other hand, it was terrifying. I needed to build a suite of relationships to insert myself into the early-stage tech community. It took five meetings a day, five days a week for five years to lay the right foundation.

Gutierrez: Were there similarities in the two cultures that made the transition easier?

Ehrenberg: Certainly. The psychology and the makeup of the quantitative traders whom I recruited and managed at DB felt somewhat similar to those of the technical startup founders I work with at IA. In many cases, their personal and academic backgrounds were quite similar, even though their chosen vocations were quite different. In terms of their thought processes, I felt a lot of commonality between the two.

Gutierrez: Were there differences that made the transition interesting?

Ehrenberg: I felt like the quantitative traders were much more in their own heads, because they're largely focused on building models to leverage data and generate returns, which is a somewhat abstract concept. A technical startup founder is generally focused on producing a thing—not just a model, but a model that does something that touches and affects people. The DNA of wanting to build something for many people, as opposed to building something for oneself to make money, reads very, very differently.

Gutierrez: What in your career have you been the most proud of?

Ehrenberg: The ability to recruit and retain great people. I've had fantastic mentors from whom I've learned a lot about how to build great teams which has augmented my own experiences. Over time, I've learned not only how to recruit people but to retain them, and how to build a culture of growth not just for the sake of the firm's success, but also personal success. This has required taking a longitudinal view. I'm probably most proud of that.

Gutierrez: What have you learned from managing teams and helping others learn to manage teams?

Ehrenberg: The first thing—and I tell this to my children as well—is that it has to be about passion. The people I've worked with, whether the quant traders or the technical startup founders, have myriad things they could be doing. They are all incredibly capable, very creative, deeply interested, and unquestionably interesting. However, they need to feel deeply about what they're doing or they're not going to be the best at it.

The other thing, notwithstanding their technical brilliance, is that ultimately what's going to make them successful at scale is people skills. Hence they shouldn't simply fall back on the things that make them feel comfortable—namely, coding and product development. They should instead push themselves to learn how to manage, how to lead, and how to recruit. Learning these skills is an essential element to becoming a fully formed professional—not to mention an evolved human being.

Gutierrez: How did you think about the process of building the right mix of people for your team after leaving Wall Street to start investing in data-centric startups?

Ehrenberg: Not jumping right into it and thinking honestly about my strengths, weaknesses, and areas of competitive advantage. I went through a somewhat similar process before going to graduate school. I didn't go straight from undergraduate school to graduate school. I actually worked for four years in between. During this time, I spent a lot of time thinking about where the gaps were in my knowledge, what were the things I cared deeply about, and what were things I wanted to augment in my knowledge base.

I did very much the same thing in trying to develop profiles of people that I felt would be great long-term partners who would be complementary to me. So the first thing I thought about was that, even though I'm technical, I'm not a PhD. And a lot of the things that I was interested in considering for investment required a higher level of scientific knowledge to do the fundamental due diligence on the technology itself. That's what led me to develop a profile for the so-called "unicorn." This person would be someone who both is high in technical skills and has a strong scientific base but at the same time has built stuff—building not just models but teams and maybe even businesses.

The search for a person who fit this profile led me to my partner, Brad Gillespie, who is one of the few unicorns who has the combined knowledge and experience I felt the team needed. We were then fortunate enough to subsequently find our colleague, Jesse Beyroutey, who embodies a similar profile. He had gone through the Fisher Program at Penn, obtained his master's in CS, worked at Microsoft, and then worked at Insight Ventures. So again, we went looking for and found somebody who had exceptionally strong technical skills but also the perspective and passion to be a builder. He's grown tremendously over the last three years. We also have Drew Conway as a scientist-in-residence who is a globally-known data hacker and data visualization expert. Again, he is someone who has incredible horizontal skills.

Gutierrez: How is Drew, as a scientist-in-residence, a competitive advantage for your team?

Ehrenberg: We felt that having Drew work and live with us would seep into our DNA. He would also be able to work with some of our portfolio companies on very prescribed projects. The way that we've been able to best apply his skill sets, give him the best learning experience, and be the best resource for our companies is when projects are very clearly scoped out. That is, when there's a particular data set or a particular product initiative that would benefit from Drew's experience in hacking data. Drew's worked with some of our portfolio companies like PlaceIQ, Next Big Sound, and Recorded Future that tend to have these very interesting, rich data sets that, from Drew's perspective, can be mined to extract even greater value.

Drew sits in on our deal meetings and that has been worthwhile as well. His greatest value has really been post-investment, because when we're first investing, there generally isn't a rich data set for somebody like Drew to play with. In the deal meetings, because we're investing so early, it's much more about the technical due diligence and our assessment of the market opportunity. That's where Brad, Jesse, and I really excel, and that's what our job entails. Then, as the companies execute their plans and achieve scale, rich data assets are developed. That's when Drew can come in and provide a valuable lens on data assets.

Gutierrez: What are some lessons you've learned from investing in data-centric companies?

Ehrenberg: It's interesting that the most valuable companies look a lot like platforms, but they don't start out that way. They tend to look like companies with very specific vertical applications. Over time, as success in a particular use case or vertical is demonstrated, the companies expand into other verticals. I think that the notion of having a very crisp early use case is vital because it's very seductive, when you're backing these brilliant technologists with a platform vision, to spread yourself too thin and "boil the ocean."

Building a generalized platform from the start is often not successful because you aren't solving a known pain point for a specific set of customers who have said they'll pay for it. So I guess the biggest lesson is to have a very clear set of customers that you're going to serve, notwithstanding the fact you may be building something that can ultimately help many different types of customers. That laser focus early on is very important to demonstrate the power of the technology and to prove product/market fit.

Gutierrez: When you're looking at teams of brilliant technologists, do you expect them to have a fully formed idea, or is it more about working with and learning about them during the preinvestment relationship?

Ehrenberg: It's almost always the latter. It's a special thing that happens. We're a fairly young firm, and we are building our reputation and our experience based on a network of relationships with founders. We're not somebody like Sequoia Capital, which has been around for and excelled across generations, where startup founders are on their second, third, fourth, or fifth companies. For us, we're in large part backing first-time founders, so not only do they not have all the answers, they often lack business-building experience.

What we're looking for are great people who have a compelling vision, a demonstrated ability to write and ship code, and with whom we're excited to work for the next 8-10 years. Given the kind of high degree of engagement that we have with our companies, it's essential that the chemistry be right. We're spending a tremendous amount of our time with these younger, less-experienced teams on helping them focus, grow as founders and as builders, and execute against their vision.

Gutierrez: What does the early engagement with companies look like?

Ehrenberg: We work very early on with our companies to identify what the KPIs are and should be as they execute against their plan. KPIs aren't always knowable Day One. There are certainly generalizable KPIs that everybody needs to track but, depending upon the business, there may be other elements of data that you want to be collecting whose importance doesn't become manifest until customers start interacting with the product. So we work with the companies to try and build in a measurement culture. We have a bias toward founders that intuitively embrace data collection and analysis. We work with these teams to identify the right data to track given their particular business and how this data can be used to improve the product and/or marketing strategies.

Gutierrez: In interviews, you talk about investing in companies that transform passive data into active data assets. How do you find these companies?

Ehrenberg: There are two principal approaches to opportunity identification. One is curating our significant inbound deal flow that either comes to us "warm" through trusted connections or "cold" through direct outreach from someone not in our network. The other results from hypotheses we have about a particular market opportunity or emerging space. In this case, we're actively looking for people trying to solve problems where we see a market gap and where we're passionate about finding solutions.

The issue of helping to transform passive data into active data happens later. We invest in a team that is building a business. We do not invest in a team building a data set. The data set is exhaust that emerges from interacting with customers in many cases. Once a company is interacting with customers, we can then think about the exhaust and how the company can improve the product or customer experience by gaining insights from the data.

Gutierrez: Who or what helps shape your views on the data space?

Ehrenberg: Aside from reading current literature, I would say just watching how society grapples with rapidly changing conditions and the wave of new technologies. As our portfolio grows and as we have greater longitudinal experience with our companies, we are able to learn more and more about what's working, what's not working, and what are some big problems that simply haven't been solved yet. We're always collecting data, whether it's internally generated data from our own companies or things we're observing in the outside world.

Gutierrez: How did you form your first views on data-centric companies?

Ehrenberg: A great deal of it was through osmosis from my experience on Wall Street. The last five years of my career entailed living in the world of high-speed data and feeds. This gave me an understanding that while the infrastructure that existed in the late '90s and early 2000s was good, there was a huge gap relative to what I saw as the inexorable increase in the velocity and volume of data. I understood that new technologies needed to emerge to handle the changing world.

The other experience that helped shape my views was how predictive analytics and semantic intelligence could be practically applied. Again, these were things we were dabbling with in some of our strategies at DB Advisors. So, even though I had left Wall Street, I found both areas very compelling. What was interesting was that most of the conventional quantitative approaches were already reasonably well-known. The literature was out there for everybody to see, so there was going to need to be innovation either in the kind of data that was being parsed to generate insight or in the technologies to parse existing data to generate better, faster insights.

So those were some of the early influences that helped to inform my thinking. Then when I saw what was occurring in the realm of advertising technologies, it was intuitively obvious that the structure of advertising markets and financial markets would converge over time. Now, when that was going to happen, I wasn't sure, but that's what informed a lot of my early ad tech investing in companies such as Invite Media and TubeMogul. These early insights and the way they have subsequently played out have carried through to IA Ventures.

Gutierrez: Having had these views and insights, did you start a company to capitalize on them?

Ehrenberg: I did start a company myself. Unfortunately, we were way early and made a ton of mistakes and we failed. While helping build this company I continued angel investing, so collected a lot of data about how successful founders build companies and how that differed from my first-hand experience. After my failure as a company founder I decided there was a way I could continue to be a builder—by becoming a full-time investor. This is what prompted me to start a venture firm and to better leverage my core competencies. So far it has gone pretty well, but venture investing and company building is a very long time-scale business.

Gutierrez: How do you guide first-time founders through the thought process of finding their core competencies?

Ehrenberg: When we go into an investment, we already have a sense of what a founder's core competencies are. Just because a founder has an incomplete set of skills doesn't mean that they can't grow and develop, especially given that most of the founders we're backing are in their 20s and 30s—these are young people with lots and lots of time for growth and development. However, some people simply don't want to change. Let's say a founder is super-tech-focused and they only want to code. So they don't really want to be the CTO and they don't want to be the VP of engineering. What they really want is to be the Chief Architect. What we try and do is back great small teams with lots of talent. Then, over time, they and we reassess roles and responsibilities as the company executes their plan and staff accordingly.

There are some technical founders that absolutely, positively want to be the CEO, yet they know they've got knowledge and experience gaps relative to those who have successfully scaled big businesses. What many of these technical founders do is to actively seek out mentoring. They're open-minded and eager about expanding their skill sets. However, sometimes a founder thinks that they want to be CEO, but then when they're tasked with recruiting and finance and general oversight, they realize (a) they suck at it and (b) they hate it. And that's fine. Part of our job is not only putting our founders in the position to succeed but also helping them to be honest about their strengths, weaknesses, and personal objectives. Once they make an honest assessment, we help to build a support system around them to help them do what they want to do at the same time as helping the company be as successful as it can be.

Gutierrez: How have you learned these lessons?

Ehrenberg: I would say it's a combination of my successes and failures. Over the five years after Wall Street but before starting IA Ventures, I seeded forty companies as an angel investor. This experience gave me a good amount of data, both good and bad, with respect to what makes for a successful company in terms of founder personas, the common threads separating success and failure, and what is most productive in my own interactions with founders and their companies.

Through my reflections I came to appreciate the fact that the human element—psychology—was as important as—if not more important than—the quantitative and technical assessment of a team. This is why—and I've written about this many times—venture investing is an artisanal business. It doesn't matter how bright somebody is—I've known phenomenally brilliant people who are *abysmal* failures when it comes to starting companies because they just don't have the empathy, they don't have the people skills, and they don't have the perspective. So it's not just about building in a vacuum. It's about colliding with the market and with other human beings. I think that's a lesson that deeply resonated with me.

Gutierrez: What's an example of founders who displayed great psychology?

Ehrenberg: My biggest success as an angel investor is Buddy Media. If you know Mike and Kass Lazerow—these are very smart people, but their raw processing power, which is great, is *dwarfed* by their team-building and leadership skills. They created a culture at that company that is second to none. Whether the team was three people or three hundred people—and I saw it from both ends—they built a team of A+ players with an esprit de corps that even as it became hard for them to deeply know every single person, there was still that sense of belonging to Buddy, and that meant something. The job Mike and Kass did in building culture and communicating mission was absolutely stunning, and it led to an equally stunning result for employees and investors alike.

Now, of course, Mike had the vision when Facebook was 20 million people that it was going to change the world. That was the big thing he got right. There were actually three different attempts at figuring out the right path for Buddy. There was virtual currency on Facebook, there was app development, and, finally, there was instrumenting Facebook pages and becoming a dashboard for big brands and marketers to measure, monetize, and reward engagement on the social nets. However, that didn't happen overnight. Mike, Kass and the team were flexible enough and humble enough to know what was working, what wasn't working, staying cool, testing, adjusting, and keeping the faith of their investors and their team all the way along. Then once they hit it they stepped on the gas, which lead to one of the most successful outcomes in New York start-up history.

Gutierrez: What did you see when they first came in to talk to you that made you think they were the right team to back?

Ehrenberg: So I hadn't met Kass at first. I met Mike. Mike was very personable and was already a three-time serial entrepreneur. He had just sold GOLF.com to Time Inc. and he had previously run a student loan marketing company during the go-go days of the dot-com era. When we spoke he had a vision for why this thing called Facebook was really going to change the way brands thought about touching their customers. Obviously, I knew what Facebook was, but it certainly wasn't a phenomenon. This was back in the middle of '07. He talked about his knowledge of brands and marketers, and how he knew that they didn't understand how to communicate in this new medium. They didn't understand social media. So Buddy's raison d'être was going to be helping them understand and speak to people on social media, initially Facebook. That was the idea.

Now, they had this initial idea for how they were going to do this, but to me that almost didn't matter. One of the most important things I've learned, which I still carry to this day, is that as a seed stage investor you can get a lot wrong, but you need to get a couple of things right. The most important is the right people. Another is that there needs to be enough "white space"—opportunities in the market to reach customers and be successful. It doesn't mean there are no competitors. In fact, competition is validating. If people validate a space before you arrive, you can do a better job and if there's enough white space, then it's fantastic. What Mike did was identify an area that had so much white space, and he himself was so great, that he could have lots of false starts. The market was so nascent that as long as he got the big thing right—that yes, Facebook was going to be a phenomenon, and, yes, brands and marketers were going to want to speak to people on the social nets, Mike was going to figure out how to make money—before running out of money!

What's hard is when you see the fifty-eighth NoSQL database company, where the amount of white space is rapidly shrinking. You could put in an A+++ team, but if they're operating in a space that's now insanely competitive, super crowded, and the ability to differentiate is actually quite small, it's almost a waste of time. Conversely, let's say you have the Buddy situation again, but you don't have Mike and Kass. You have a B team that identifies the big idea correctly, but as executors, they're B-players, they'll fail. So it's that intersection of an A team with enough white space in the market that makes for a compelling investment opportunity.

Gutierrez: When you invested in Buddy Media, were you thinking in terms of a data play?

Ehrenberg: There was no doubt in my mind that Buddy would create a supervaluable data asset by virtue of people interacting with brands and Buddy-powered applications on Facebook. This data asset would then inform better targeting and better campaigns. It would be very much in line with the way DSPs [Demand-Side Platforms] and DMPs [Data Management Platforms] work in ad tech. "Demand-side platforms" allow buyers of digital advertising to manage multiple ad exchange and data exchange accounts through one interface, while "data management platforms" serve as data warehouses. While it's not a perfect metaphor it's pretty close. There was always that sense, from my perspective, that there was this data play, but it's very rare—again, at least in the way we invest—that there's a large data opportunity up front. Data value emerges from successful businesses at scale, so that was the hypothesis I had about Buddy.

Gutierrez: What does your workday look like and how do you measure success?

Ehrenberg: I think about this almost every day, as it is an ever-present challenge of how to best optimize my time. My day is generally comprised of three elements. There is information ingestion and sharing. There is sitting on boards and helping portfolio companies. And finally, there is looking at new deal flow. For the information ingestion and sharing, I do all my reading relatively early in the day. I'm very active on Twitter and I share a lot of information. Part of that is simply that I like to share. I know some people use my stream like a newsreader because of the curation that I do with Twitter. My feed and what I share also reinforces to the world the things that I'm most interested in and passionate about. I think it's important to share, as entrepreneurs are looking for the just-right investor to really understand who they are. I put it all out there so people have a pretty good understanding of my interests and my vibe. And I'm pretty clear about the things that interest me—whether it's data, entrepreneurship, financial markets, youth baseball, or the University of Michigan. It's all out there. So that's the first part of my day and it starts very, very early in the morning.

Then there's the biggest part, which is I sit on my boards and work to support our portfolio companies. I'm deeply involved with our companies, especially the least mature that want the most help, so there is almost always existing portfolio company that I spend a chunk of my day on. I frequently speak with potential recruits, assist with business development, discuss and help with financing strategy, and sometimes just provide a sounding board to our founders. That's the portfolio part of my day.

The third part of my day is looking at new things, whether it's companies that are coming into our deal funnel or digging deeper into hypotheses about different markets and emerging spaces.

Gutierrez: How does the time you spend with your companies evolve as you go through the investment life cycle?

Ehrenberg: It changes dramatically as a company goes through the investment lifecycle. This is something that I'm experiencing now as several of our Fund I companies have recently raised Series C rounds that have taken in 30 or 40 million dollars, where, honestly, the kinds of things that I can do and the value I bring is much less than it was earlier in a company's life. At these companies, we have brought on new board members—both independents with tremendous amounts of domain experience and later-stage growth investors that understand how to help companies go from 20 million to 200 million to 2 billion in revenue. I've never done that, so it's great for me to have a seat at the table and soak up the learning. While I still help with recruiting and financing strategy, the more detailed discussions around business model and achieving product/market fit fall away when the emphasis shifts from "figuring it out" to rapid scaling.

Gutierrez: How do you stay so grounded?

Ehrenberg: This is easily the most humbling business I've ever been a part of. There's not a day that goes by that I don't feel stupid at one time or another. This business is so difficult.

Though Fred Wilson is obviously one of the most successful venture investors—certainly of this era, if not of all time—he is also unbelievably humble. Fred continues to be one of my principal mentors and was one of my early supporters. He said to me very early on that he always feels like he's learning and regularly feels like he's making mistakes. That really stayed with me. I also think that as I've gotten older and I've had successes and failures that I have much greater empathy now than I did when I was young, cocky, and "taking on the world." At this point, I feel very much in my zone and it's not about proving anything to anybody but myself and trying to help however I can, so I think that helps keep me grounded.

Gutierrez: How do you evaluate new opportunities when facing this level of business difficulty?

Ehrenberg: Generally, when opportunities come in, there's going to be three different types of deals. A cold deal comes in with no introduction from a trusted source. Then there are more warm introductions from other investors. Finally, there are warm introductions from entrepreneurs in our ecosystem. Every piece of inbound is implicitly weighted by virtue of its provenance. Things that are from trusted and respected sources will be prioritized. We have a pretty clear philosophy, as well as a very defined approach at which point we are interested in investing. We are early-stage investors. So it's got to be somewhere between incubation and pre-Series A. The area between Seed and "Seed prime" [or seed extension], is probably the fat part of the distribution of when we initiate an investment. So there's a lot of stuff that has to line up for us to even want to look at a company in depth.

Let's say there's a company at the right stage in an area we're passionate about with a founding team that in first interactions seems compelling. One of us is the deal lead and generally first meets with the founding team. If there's continued interest at that point, then we'll really start to figure out the hypotheses we have about the space and the company. We work to figure out the gap analysis between where the company is and what we think this next check should enable them to accomplish. We also start to do homework with the team around the company and around the space to quantify what those gaps are and whether or not the amount of money they're looking to raise is appropriate. A lot of times we'll find out—when we're digging into a company and helping out—that their financial model shows that they are generally underestimating the amount of resources they'll need to hit key milestones to prove what they need to prove to raise to the next round. Around this time the company will come in to meet all three of us, and get to know us as people and as a partnership. We then figure out if we want to work together as partners to build a great business.

One of the things I've learned is that the worst thing you can do is undercapitalize a business, especially in an environment where it's very hard to raise follow-up financing. Again, we're talking about the very early stages, so this isn't the issue of writing a $30 million check and overfunding their company. The issue at this stage is whether you raise 1.5 million or 2.5 million dollars. That's the order of magnitude we're talking about. Sometimes that extra half million or million, which provides an extra six months of runway, can mean the difference between having a super-successful Series A or "Oh my God, we need to do a bridge because we haven't proven enough." The latter type of conversation is never a fun conversation. This happens quite frequently at the seed stage, so we work hard to try and avoid this by properly financing the plan upfront..

Gutierrez: How do you manage the data you're generating about these investment hypotheses?

Ehrenberg: We have a very, very structured file-sharing system. We're very careful about categorization. This makes it very easy for us to access our data. Then, given the fact that we're running a fairly concentrated investment portfolio, we aren't burdened by the fact that we don't have a hundred portfolio companies or more. We're not a micro VC. We are a classic, conventional, old-style VC, which makes our data very concentrated. So we are pretty good at retrieving that data and leveraging it again and again and again. Even though we leverage this data frequently, we are constantly refining how we analyze companies and the manner in which we engage with companies. At this point, I don't feel like we've reached steady state.

Gutierrez: You've previously talked about finding people who are data-centric and have data-centric DNA. How can you tell?

Ehrenberg: We can tell by going through the process with them. Let's say there's a team operating in a space that we find intriguing, and we have two to three meetings with the founders. It's super easy to tell at that point whether or not they are metrics-driven, where they are thoughtful about architecting the business to best leverage the data that they are collecting. It's something in the natural course of due diligence that just becomes apparent.

There's no way to short-circuit the process of getting to know someone. That's why I constantly refer to my friend Mark Suster's post on "Lines, not Dots,"[1] and why it's so important to build relationships over time to really get to know founders—and vice versa, for them to get to know us. We are extremely firm about the fact that we don't chase and we don't get into bidding wars. If somebody has five term sheets and they are selecting on the basis of price—that's not a founder for us. We don't begrudge it. By all means, especially if it's a seasoned founder and they really don't see that much value coming from who the specific investor is, then we're not the right investor.

I believe—and I believe if you were to speak to our founders, they would corroborate this —that people who end up working with us specifically want IA Ventures. People want me, Brad, and Jesse because they want us as individuals and as a team, like the way we interact with companies, and value our knowledge and domain experience. Those are people we want to partner with for years. It's got to be where they want us as much as we want them. If it's heavily skewed where we want them, but they don't really want us, then that's just a bad fit.

Gutierrez: How soon do you know if it's going to be a good or bad fit?

Ehrenberg: Remember, most of the people we're backing are first-time founders. If somebody is optimizing for price as their objective function and price is it, then that's not the right company for us. Somebody who generally wants money and doesn't want engagement or support is either a seasoned start-up veteran or someone who believes investors aren't partners but simply sources of capital. While we have backed serial founders who are very experienced and neither need nor want much support, this is not the usual profile. If a first-time founder who lacks experience is also a poor communicator, this type of founder generally doesn't work for us.

The Holy Grail for us is a great founding team who is interested in and highly capable of running and building a business. If we can assist in ways they find valuable, great, but we generally just let them do their thing. However, at this early stage of our own firm development, we don't have many repeat founders. So we're getting a lot of new entrepreneurs who value our help on an array of topics. This is not about "imparting our wisdom," but simply trying to help a young company in concrete ways that they value and that are important to the business. This what I mean by chemistry—where the combination of brilliant, visionary, and self-aware fits with our passions and interests.

Gutierrez: In a competitive hiring environment where it seems like you could throw a stone and hit a hundred new startups, how do you coach a team on finding the right people with the right combination of talents?

---

[1]Mark Suster, "Invest in Lines, Not Dots" (Bothsid.es, November 15, 2010: www.bothsidesofthetable.com/2010/11/15/invest-in-lines-not-dots/).

Ehrenberg: It's comes right back to the people issue. You could say, "Well, we've got this super-great technology. We're better, faster. We've got this algorithm." Honestly, no one cares. What you're selling is an experience, whether you're selling to a customer or you're selling to a potential employee. You have to sell the experience. Yes, economics such as salary and stock options are part of that experience. However, just because you pay a little more doesn't mean you're going to get somebody.

What's going to get somebody is your ability to sell them on a vision, sell them on the team, and demonstrate your passion. That's what attracts and closes top talent. And when I try to talk to some of my more technical founders about this it's hard, because to them, why they're doing what they're doing is so obvious. Why wouldn't you want a hundred grand and a point in the company to work on our awesome technical challenges?

This is when startup founders need to learn how to sell. And I don't care how technical they are—they need to sell. So coaching them on how to sell, how to communicate the vision, how to have their passion come across, how to make somebody feel wanted, not like a commodity—moving beyond the transaction is very important. You've got to make it personal, and that's hard. But that's what you have to do to win. That's how you hire great teams.

Gutierrez: When you raised your initial and subsequent LP [Limited Partner] funds, how did you sell yourself and what was the big vision?

Ehrenberg: Back in '09, when I started in the midst of the "nuclear winter" in the LP market in the wake of the financial crisis, I really didn't spend time with traditional venture LPs. My initial money was raised from a small group of top venture general partners who knew me, as well as a bunch of strategics, in addition to my own money. I had this extremely clear thesis around big data being an investable theme. I was the first person to talk about this theme. I also sold the idea of having this traditional venture mindset, but in a small fund that was initiating investments at the seed stage, but that was also going to make follow-on investments. The thesis, structure, and strategy combination was something that resonated with a certain set of sophisticated investors.

After we held our initial closing of $17 million back at the very beginning of 2010, I said to our investors:

> We're going to be heads-down for six months. We're not going to think about raising a dime. We're going to focus on building the IA Ventures brand as being synonymous with big data. We're going to build strong relationships with the data science communities on both coasts. We're going to become a recognized thought leader in the area, and we're going to hopefully work with some really great entrepreneurs on some really interesting businesses that are going to demonstrate our thesis.

Fast-forward to the middle of 2010 and the LP market had thawed. As the market became more liquid, emerging managers were hot. Big data was scorching hot and there we were. So we had significant institutional demand for IA Ventures at that point. Initially, when we closed the first fund, it was largely around selling the big data thesis, but also a higher level of investment rigor. My Wall Street background and strong emphasis on risk management resonated with a lot of LPs because it was so unusual. And the way I talked about portfolio management was also appealing as it was highly structured and analytical and different than many other venture managers.

When we raised our second fund, we tapped a group of tremendous institutional LPs through a small set of my VC mentors. We had spent time actively cultivating those relationships and keeping them abreast of our progress with Fund 1. We said to them, "Here's how we're going to build our first fund. And we'll come back to you when we're ready to raise our second fund and show you how we executed our plan." We did, and then we were able to quietly, quickly, and from whom I believe to be the best partners in the world, raise $105 million for our second fund. It's been a wonderful partnership thus far.

Gutierrez: How do impart to your technical founders your knowledge of relationship and mentorship building?

Ehrenberg: I try to introduce them to people who could be good mentors for them. I've found that there are a lot of tremendously successful people who are more than willing to "pay it forward" by helping young founders. Just like there were people that helped me, I try and help others, and then those who have been successful as technical founders are happy to help others as well. There's nothing like learning by analogy, so when you're able to sit in front of somebody that's done it and you can draw those parallels to yourself, it makes it real. It's not just words. So what I do is to help cultivate a set of relationships around these super-smart young founders so they can build their own coaching network.

Gutierrez: You were one of the first to talk about the big data investment thesis, what was going to happen, and how it was going to evolve. As it continues to evolve, what do you think the future looks like?

Ehrenberg: The term "big data" was something that I used very early on. Back then it was very powerful because (a) nobody was using it and (b) it had some meaning. The term has now become so distorted and overused that we never use it. I don't even know what it even means anymore. That is why I've moved on to this notion of data-centricity. In fact, I would say if you look at our latest investments, it's not clear that everybody would label them "big data" investments. For us, it's much more that we have a distinct set of skills and interests and passions with a strong technical bent. We're looking for companies that we feel are prosecuting very interesting missions where we are the best investors for these companies. So the definition of what we're doing, I'm going to say, has evolved.

I think that when most people think about big data, they think infrastructure. They think of enabling technologies. We've done a bunch of investments in that area. We may or may not do a lot more of that, at least in this next wave of opportunities. If you look at what we've done lately, they're much more applications. Whether it's reshaping how quality inspection is done in manufacturing processes, or infrastructure-as-a-service for the developer community. We've also made three investments in the healthcare space.

As the market has evolved, we've evolved. We've gotten very clear about what we're really good at and how we can help the most. That's naturally caused us to gravitate toward certain kinds of founders and certain use cases. If we were to sit down and have this discussion in three years, I'd be fascinated to hear what I was saying, since we've evolved and sharpened our focus and investment methodology since the early days of 2010.

Gutierrez: A term that has come into prevalence is "data scientist." Do you think it's a similar thing where right now in 2014 it means something, but it's generally going to lose its meaning, or do you think that it eventually becomes a much more well-defined role?

Ehrenberg: I think it's something now that's distinct because we need to make more of them: there aren't enough skilled people to manipulate and extract meaning from data. One way the market is trying to solve this shortage is through new technologies and applications that give the power of data science to non–data scientists. We back a company called Data Robot that essentially places the power of a data scientist in the hands of a non–data scientist. So I think we're going to see many more tools and technologies to democratize the power of the data scientist.

At the same time, I think we'll see the value of people who can go beyond this continue to rise, because while these tools are powerful and are sufficient for a wide range of use cases, they are not a panacea. It's kind of like machine learning and AI. These are areas where tremendous progress has been made, and yet you still need human input to get the most value out of these systems. I think data science is much the same.

Gutierrez: IA Ventures is one of the few investment funds with a data scientist serving as a scientist-in-residence. How will that role evolve?

Ehrenberg: Having somebody like Drew, our scientist-in-residence, on the team is valuable. Whether or not it's going to be somebody who's more on the investment team that has that set of knowledge or somebody fulfilling a similar role to Drew today—I don't know. It may be that one of our investment people will look a lot like Drew in makeup but still be an investor and not a data scientist. I think it will evolve. As data science becomes built into everything, it's not clear to me that we will need someone of his stature.

Gutierrez: What are some of the areas where you think the data-centric investment thesis has room to grow?

Ehrenberg: I think with healthcare IT—I don't even know if we're at the tip of the iceberg yet. I think the areas where the biggest opportunities are also have the most challenges. Healthcare data obviously has some of the biggest issues with PII and privacy concerns. Added to that, you've also got sclerotic bureaucracies, fossilized infrastructures, and data silos that make it very hard to solve hard problems requiring integration across multiple data sets. It will happen, and I think a lot the technologies we've talked about here are directly relevant to making health care better, more affordable, and more distributed. I see this representing a generational opportunity.

Another huge area in its early days is risk management—whether it's in finance, trading, or insurance. It's a really hard problem when you're talking about incorporating new data sets into risk assessment—especially when applying these technologies to an industry like insurance, which, like health care, has lots of privacy issues and data trapped within large bureaucracies. At the same time, these old fossilized companies are just now starting to open up and figure out how to best interact with the startup community in order to leverage new technologies. This is another area that I find incredibly exciting.

The third area I'm passionate about is reshaping manufacturing and making it more efficient. There has been a trend towards manufacturing moving back onshore. A stronger manufacturing sector could be a bridge to recreating a vibrant middle class in the US. I think technology can help hasten this beneficial trend.

Gutierrez: How do you and your companies communicate to middle managers in manufacturing companies the benefits of data, data techniques, and cutting-edge technology?

Ehrenberg: The most important thing—we're dealing with this in real time with our companies—is that you've got to find a person that has the greatest incentive to change. As an example, our company Sight Machine is investing huge amounts of time in communicating with plant managers and heads of quality at plants. You know—the people with goggles, hard hats, and steel-toe boots. It is very important culturally to be able to speak to these people. These managers and heads of quality are heavily incentivized to reduce the costs of quality, so if you can show them a way of substantially improving their results, which directly affects how their plant is assessed and how they are compensated, then that's pretty powerful.

It's a little bit like the way successful businesses crack Wall Street. They don't start selling at the top; they go to employees on the front lines feeling the greatest pain. So they avoid the IT department and sell directly to the bankers and traders. That's how BlackBerry became ubiquitous on Wall Street: they sold direct and got users hooked on the product. And once the device had infiltrated the trading floors and bankers' offices, these producers closest to the customer went to IT and said, "Support this." IT staffers first responded by saying, "We don't support this." Then the bankers and traders said, "You do now." This worked because the people closest to the money ultimately get what they want.

Gutierrez: How do you think about the move to being more open with data given fear-inducing stories of job loss and data privacy?

Ehrenberg: The way that I operate, as somebody who knows the good and the bad, is that I am quite open with my data. Why is that? Because I want to live in a world where I see things and have opportunities that are meaningful to me. As I've made my data available, I have found that the offers I get and the opportunities that I see are increasingly relevant and better tailored to me. Overall, this had led to very positive experiences.

Clearly there is a difference between PII and other types of data. For example, the difference between your Social Security number versus being cookied or tagged where somebody keeps track of your clickstream. I'm extremely careful with my PII. As it relates to my clickstream and preferences, I don't care, because my own experience gets better and better. So I would say to people who are concerned, "Look, if you're spending time on the computer and you like to check the news or you want to shop, if you are willing to let people follow you and follow your clicks, then you will read better and more relevant stuff, you'll get offers that you care about, and the whole experience will benefit you." That said, I do know people who are sensitive about these issues and opt-out, but they represent a very small segment of the population.

Conversely, when it comes to your Social Security number and your highly personal data, you're right to be concerned and careful about what you share. You should always be extremely careful what sites you put that information on because there are bad actors, as in the Target case, where you can be doing something that seems innocuous and yet have your personal data be exposed. So you can never have zero risk if you're on the grid. If you're willing to engage on the Internet, you are going to encounter risk.

Gutierrez: In terms of education, as more and more people come into contact with data and come into contact with companies that are doing interesting things with data, what do you view as the future of getting educated in data literacy?

Ehrenberg: I'm actually putting my money where my mouth is. I'm very active with the University of Michigan School of Information [UMSI]. UMSI focuses on the intersection of programming, data science, user interaction, and entrepreneurship. I've invested a lot in the program, and believe that their broad-based curriculum is the wave of the future. It is geared towards asking hard questions and acquiring practical tools for solving hard problems. This has application in both for-profit and not-for-profit settings, and UMSI has a strong social mission in addition to supplying talent to startups and companies such as Facebook and Google.

I believe we'll see these topics get pushed earlier and earlier in a students' education. What was once the province of PhDs is now being taught to Masters students, Masters curriculum is being offered to undergrads, then to high schoolers and so on. So instead of just math, it will be math and stats. It will be not just taking Spanish or French, but taking Spanish or French and scripting. Whether it's CodeNow or any one of the fantastic initiatives that have emerged, we are seeing these topics brought into schools earlier and earlier in the lifecycle of students, which is something I find super exciting. A greater emphasis on data literacy will happen. It just depends on how fast and how soon. In New York City, you're seeing it happen on an experimental basis in many schools, both public and private. I think it will rapidly become more prevalent. The question is when will it be everywhere, not just in the intense tech centers of New York, San Francisco, and Boston.

Gutierrez: What is something you believe is already here but just not evenly distributed in regards to data?

Ehrenberg: I think its education and the knowledge of how to teach these topics. The foundations of programming and data analysis are well-known and understood. There's a relatively small group of people that are benefiting from this knowledge. How this becomes a core part of the curriculum across the country is a challenge. We need more dynamic and fluid measures of achievement that take into account abstract problem solving, statistics, and interpretation of data—the stuff you actually need in the real world. I think that's what's going to raise the entire country's standard of living and recreate the middle that's been kind of squeezed out and pushed to the ends, which isn't at all healthy for our country or society. The rise of data literacy will happen, and it will be tremendous for society.

Claudia Perlich

Dstillery

Claudia Perlich is the Chief Scientist at Dstillery (formerly Media6Degrees) and teaches data mining for business intelligence in the MBA program of the Stern School of Business, New York University. In the top 50 of Forbes' Top 100 Most Promising Companies in America, Dstillery is an advertising technology company that uses scientific methods to capture the full customer journey by collecting massive quantities of data from both the digital and the physical worlds to identify patterns that are unique to the brand and indicate consumer intent. Their models predict outcomes prescribed by the marketer—right person, right time, right platform, right channel. The scope of these models, the data necessary to build them, and the ever-changing tastes and behaviors of customers bring challenges in terms of data infrastructure (decisions and responses to ad bid requests have to occur in less than 30 milliseconds over 30 billion times a day), privacy, fraud, model accuracy, and knowing how to think about the correctness of models.

Perlich grew up in East Germany and did not interact with a computer until she was 15 years old. After taking her MA in Computer Science from the University of Colorado and her PhD in Information Systems from NYU, Perlich worked in data analytics research at IBM's Watson Research Center, where she won the KDD CUP—a data science competition sponsored by the Association for Computing Machinery (ACM)—three years in a row (2007-2009). She holds multiple patents in machine learning and is the author more than 50 scientific papers. She has taught at NYU, MIT, Wharton, and Columbia. Her contributions to data science have been recognized by many organizations, including the Association for Computing Machinery (ACM), Crain's New York Business, Fast Company, the Advertising Research Foundation, and the American Marketing Association.

Perlich exemplifies the data scientist who makes her own rules and succeeds in multiple areas of business, marketing, computing, and academics. Her multifarious talents are critical for her unique role in Dstillery as a scientist, mentor, and ambassador for the company. Perlich's enthusiasm for learning, sharing, and mentoring both at work and in her academic role at NYU invigorates her interview.

Sebastian Gutierrez: Tell me about how you got started with working with data.

Claudia Perlich: I started out in computer science. I was sort of a math geek, but I was more interested in applications rather than just theory. Through an accident, I ended up in a university exchange program and came to The University of Colorado at Boulder for a year. While selecting courses, I found a course on artificial neural networks taught by a German professor and, on the spur of a moment, I decided, "Yeah, why not?"

The German professor was Andreas Weigend. He later became chief scientist of Amazon. He is one of my mentors and a great friend.

Gutierrez: What happened after the exchange program?

Perlich: I went back to my university really excited about data. So I dove into it and really started working more with data. I finished my degree in Germany and started looking for a PhD program in computer science. Meanwhile, Andreas had gotten a job at NYU Stern Business School in the Information Systems Department. When I asked him for recommendation letters for those computer science programs, he suggested I apply to NYU Stern. I told him, "Look, I don't know anything about business." He said, "Well, neither do I." Through that connection, I ended up at NYU Stern where I obtained a PhD in Information Systems, working with Foster Provost on machine learning and data mining. My dissertation was on predictive modeling.

Gutierrez: Did you have any exposure to the industry side of research?

Perlich: I did a summer internship at IBM's Watson Research Center in one of the summers during my PhD. I really enjoyed the atmosphere and the group there, so as I contemplated my life after finishing the PhD, the decision was to either go into academia or to go to IBM. In the end, I decided to go to IBM and ended up working there for six years in the predictive modeling group. I really liked the diversity of projects, because they have all kinds of internal and external consulting projects coming in. All these projects centered around data and building data-driven solutions. We also had time to publish and participate in data mining competitions, so it was a great atmosphere for me.

Gutierrez: What drove you to leave IBM for a startup?

Perlich: My good friend and PhD advisor, Foster Provost, met with me and told me that he had been working with a startup that was doing targeted display advertising. He had been their academic advisor for a while and thought the group was really amazing and super smart. He suggested I have lunch with them. I went for lunch and the rest is history. So now I am the chief scientist for what is called Dstillery, though it used to be called Media6Degrees. I have been doing this for the past four years and have never regretted the move.

Gutierrez: Tell me about Dstillery's history and focus.

Perlich: Media6Degrees/Dstillery started out six years ago to focus on display advertising. Initially, the premise was around social advertising—that is, being able to target people's friends. The idea was that we could work with the marketer to observe who is buying their product online. We would then use the purchaser's browsing history to see what sites they had visited and compare it to the browsing history of people they were connected to on a social network. This co-visitation of sites was used to establish connectivity between people. We would then reach out to those people who were connected to the original seed of people purchasing and show them the marketer's advertising.

We initially started out with MySpace. What we realized pretty quickly was that while social data was okay as a predictor for certain types of products, it just did not work for other types of products. Once you got to products or services that did not have a social connection, looking at people's interests and the actions they had taken was more useful. Reading a blog on the New York marathon is actually a much better predictor of the fact that they are probably a runner, than the fact that they had, at some point, looked at the MySpace page of a person who was a runner. Being able to find that signal at scale, that is really what we are doing right now.

Over time, Media6Degrees/Dstillery grew into a very different concept, where the view was that it actually does not matter whether you are friends with the person who purchased something. What actually really mattered—if you are really technical about it—is being able to predict the probability that somebody is going to buy a product. That is the conceptual view, which is great because it is what I love to do—solve large-scale machine learning problems.

We do it for financial, travel, and any other type of consumer products/services you can buy online across all the world's borders. All that is necessary is that you have some online touchpoint with your customer. Recently, we have started focusing more on mobile and videos, so there is a bit of a broadening. That said, I am not able to advertise milk. So few people buy milk online that it is basically hopeless. So other than the things not found for sale online, when we find some kind of a signal that we can track online, then we use it in the predictive models.

Gutierrez: What data goes into your predictive models now?

Perlich: What we have in terms of data is partial URL history of actual visitations that we receive from data providers and bid requests from advertising exchanges where we actually buy the impression. Unless a person has cookies disabled, we can observe some of the sites the person has been to. Before we use the URLs, we encode and hash them, as we are not interested in the web page's content or what the actual URL was. What we care about is whether a person's browsing history shows that they have or have not visited any one of the millions URLs that the data stream contains.

From this, we now basically have a binary indicator for a millions of URLs for any one person—actually cookie, as a person can use multiple computers and more than one person could be using the same computer. Based on this data, we can then predict whether the person will buy a product based on having seen a couple thousand other people on the website of the product/brand we are working with. This works very generically on any type of URL data, because we do not have to rely on the data being meaningful for the hashing. It could be a photo-sharing URL, it could be a video URL, it could be a blog URL, it could be a retail site URL, or some other type of completely different website URL. Because we use binary indicators, we are able to use this very generic representation of the data that lends itself to all kinds of URL data for the models. This is now the core value proposition and in some sense it supports privacy as we are not interested in extracting meaningful behavior patterns or link it back to a particular person.

Gutierrez: How do you and other data scientists fit into Dstillery?

Perlich: Right now, our team is about six-and-a-half data scientists out of approximately a hundred people. At this point, the group is fairly large given what we do and given the fact that we are in a startup. Over time the group has grown into something that has a bit of a hierarchy, but it is still very flat. We have a VP of data science who is formally in charge and has to deal with managerial responsibilities. I, as chief scientist, do not have managerial responsibilities, though I might, arguably, have some notional and intellectual leadership when people have problems. They value my opinion and come to me.

This separation of responsibilities means I do not head a team per se—I just get to pick and choose what I want to do. From time to time I might ask people to do something for me, but I have never really enjoyed telling people what to do.

One of the benefits of not having managerial responsibilities is that exchanges with other data scientists are easier. I really like the eye-to-eye exchange, where you can just bounce ideas off of one another and discuss various things. It does not work very well if people feel like subordinates to you. I much prefer to just reach out to someone and say, "Hey! Can we talk about this? I need a second pair of eyes or a second brain to look at this." Not having them be a subordinate makes this type of conversation better. So we try to keep it pretty level here.

Gutierrez: What are the responsibilities of the Data Science Group?

Perlich: We have three main responsibilities—models, performance monitoring, and fraud detection in addition to communicating with people outside of our group. On the model side, we now build on the order of 10,000 predictive models a week, each of which lives in this very high-dimensional space. These models are based on the URL histories that we prune down to maybe 2 million URLs from the data set of 10 million URLs or more. This process is completely automated. Even with a team of six people, we are not going to look at 10,000 models. It is not happening.

Sometimes the modeling work means building very specific models and prototypes as well. For instance, one thing we did recently is build a bidding model that evaluates not just the history of what a person has done before, but specifically estimates what is the correct bid price for this person in a real-time advertising auction based on what the person is doing right now or reconditioning the bid based on how likely we think that person is a runner. So we build a prototype, we run it on a small scale of production to see if it works, and then we supervise the automation. Then it is built by our engineering team with a full-strength and fully automated process that contains a quality assurance part that sends warnings if things go wrong.

On the monitoring side, we supervise the performance of what is going on with our models and how they are performing. Some of this is watching the performance, and other parts are dealing with the QA process if/when it sends out warning that things are going wrong. A final part is actually doing the exploring if something is wrong.

On the fraud detection side, this is always going on. We have to deal with a great deal of advertising fraud. We receive about 30 billion bid requests a day. We have about 30 milliseconds to decide whether or not we want to bid on a specific request when it comes in. If I bid and win, then our system shows an ad in this specific real-time auction. The problem is that a good chunk of those bid requests are bots, artificial traffic, or nonintentional web page visits that are unlikely to actually ever be seen by anybody. This causes the fraud problem on our side to actually be two problems: one—deciding whether or not the traffic is fraudulent, and so whether or not to show ads, and two—understanding how the traffic that is deemed fraudulent affects our models.

As traffic data on ads we have shown is part of our models, fraudulent traffic is fed into the models, which means we have to think very hard about how to counteract how fraud data affects our models. Interestingly, models are much better at finding out who is a bot and who is not because bots display deterministic behavior. This stands out because our models are predicting the very noisy human decision process of going and buying a pair of shoes. So the reason we took upon ourselves the fraud detection is because we need to clean out the data streams as they come in. So that has been a big focus over the last year or so.

On the communication side, we have to communicate with people outside of our group. This is where we spend time on data visualization. When a marketer ask you how you actually find your audiences, it is really, really hard to tell them, "Oh trust me, it is a black box that builds models in 10 million dimensions." That does not really fly. So being able to communicate some of what the model is doing is a big part of our work. Lately, one of the things we have been doing is embedding the signal from the model into a geoloca-tion, kind of marrying the desktop with the mobile world, and actually seeing where people have the highest probability for purchase and projecting this into a geographical region and making a graphic out of it that people can inter-act with. This really helps with communication.

Finally, our team and I serve as ambassadors for our company and our work. I teach a high-level overview course on data mining for the NYU Stern MBA program to give people a good understanding of what the opportunities are and how to manage them instead of really teaching them how to do it. So that is a slightly different perspective than what you get in a computer sci-ence department. We also publish and we write papers. Increasingly, my time has been taken up by helping to organize the KDD 2014 Conference in New York City.

Gutierrez: What about this work is interesting and exciting for you?

Perlich: I have always been fascinated by math puzzles and puzzles in general. The work that I do is a real-world version of puzzles that life just presents. Data is the footprint of real life in some form, and so it is always interesting. It is like a detective game to figure out what is really going on. Most of my time I am debugging data with a sense of finding out what is wrong with it or where it disagrees with my assumption of what it was supposed to have meant. So these are games that I am just inherently getting really excited about.

People laugh at me because I am not a GUI [Graphical Usert Interface] person. Because of privacy, firewalls, and other data security measures, we cannot open any windows to look at the raw data, so it is a moot point. This means I liter-ally have ASCII files running on my screen. I look at them and look for things like "these numbers either are ordered when they should not be" or "this is supposed to be a continuous number and it has too many zeros." Things like that get me really, really excited to figure out what is going on there. And hey, I'm getting paid for solving puzzles, so that is absolutely perfect.

I also like the freedom that I have here in terms of giving talks and being out there. I tend to get bored doing the same thing over and over, so I need a variety of things to do at work. There are always different problems with data, so there is no need to worry about ever being stuck with the same problem for too long. I never really had the patience and the control for being a formal, good coder, so I can just hack my way around in Perl or shell or whatever the hell I want, and that suits my computer science skills the best from what I can tell.

Gutierrez: What exciting nonwork data puzzles have you played with?

Perlich: A great example comes from a data set that we used to predict breast cancer. Siemens Medical provided an fMRI [Functional Magnetic Resonance Imaging] data set that came coded with 117 numeric variables. The puzzle was to use this data set to predict whether a data sample showed a malignant cancer or not. It turned out that the most predictive feature in that data set was the patient ID, the random number that they had assigned to the patient. The reason for this is that they had to pool different data sets together because they did not get enough positive data from any one particular source. So they had to source data sets from different places, and as it turned out, some of the treatment centers had very high breast cancer prevalence rates. Patients had random numbers that were assigned by each location. So the model was able to figure out that a patient being in a specific treatment center was a great indicator of whether the sample was malignant or not.

The bigger picture of this work is asking the question: "Is this data set suitable if we really wanted to build a model that identifies breast cancer?" The answer is, "It depends." You cannot ignore it, because even if you do not use the patient identifier, which, of course, you do not really want to have in your model, the model still finds a kind of the calibration of the grayscale. So the model still implicitly learns from the location. If you want to use that model on a different set of locations, it is obviously not going to work at all. If you want to use it on the same set of locations, you should just basically put an identifier for location in there. That is the best model that you can build.

The interesting observation from this is that you really had to change the data set or augment it if you want to make it useful. That was just one of these accidents where you are looking at it and you think to yourself, "This is strange. This seems like a weird story." That is what was really fun. These are the hidden stories in the data collection that I want to get to the bottom of when I work with data. I find that type of thinking and work makes me very happy. I get really excited by the somewhat abstract intellectual challenge of it. What amazes me is how much a data set contradicts my expectations. If the data is just what I expected it to be—it is surprisingly clean maybe, but it does not have the puzzle about it, then it does not really get me excited.

Gutierrez: Do you remember the first data set you worked with?

Perlich: The first data set I ever worked with was based on information regarding educational tests and whether people had passed some finals. This was in Boulder during the exchange year. We were trying to predict how well they would do on these tests based on what the student had in their college applications. The problem with this data set was that it was boring because it was so aggregated. It did not contain people's actual actions. Instead, it was the average GPA that they had over some number of years. There were very few interesting things going on in the data set.

Maybe this is my answer to the big data movement and what makes it actually much more exciting than previous data work. Increasingly, we are not having to look at aggregated, prepared, preset, meddled-with, and whatever else data sets. What I love is that now we can go to the original recording. It is like our conversation right now. The conversation is not what is in the transcript. It is the nuances of the voice that get the excitement, more so than the language, and that is hard to catch. That is exactly the same as what happens with data.

Gutierrez: How does this insight translate to your work?

Perlich: Rather than looking at aggregated URL histories, we can now spot something like "this browser went to this web site, and then two seconds later this browser went to this other web site." For instance, we might see that eighty percent of the people who go to a women's health page, then go to wrestling news, and then go to try Netflix on TV. This makes no sense whatsoever and that is exactly when it gets exciting. This looks like fraud. I found it!

If you look at aggregates, this type of pattern is very hard to find. In the old days, without the infrastructure that big data has given us, you just could not keep these data sets accessible and explorable. You could store them, maybe, but you could not go back and analyze them because you could not get the bandwidth and access. Now you can go back to the raw material and work with it, and that is what really makes it fascinating to me. So, actually, it only gets better and better. Right now, as far as I am concerned, it just keeps getting better.

Gutierrez: What are some challenges the advertising targeting industry is facing?

Perlich: Let me try to give you the lay of the land of the three main challenges—attribution, metrics, and fraud. The first important thing is actually figuring out whom the right person is to show an ad to. The shift there is that historically this was all about crude audiences, like the middle-aged wealthy soccer moms or NASCAR dads, or whatever else you want to call them. That is just such a crude and broad—and honestly, meaningless—characterization that I do not think it does any justice to the actual complexity of people's lives and who they are. Just looking at myself, I have many different instantiations of me, of who I am at work, of who I am at home, and what I do with my hobbies. So labeling people in this kind of categorical way is meaningless. But that is the way the industry has worked for the last hundred years.

And now we find that the industry is shifting to this very fine-grained, specific data that people can see about what consumers actually do, how they express themselves in their actions, which means they can stop using some arbitrary label that is typically wrong. I think this shift that the industry is facing right now—and we are part of that —will take a while before that really gets embraced. With it come notions of communicating the value proposition of advertising.

From a consumer perspective, most of them are not terribly happy about seeing a whole bunch of banner ads. The fact that roughly ninety-five percent of the Internet infrastructure is paid for by advertising is very far in the back, back, background of people's thoughts. You will not find a free blog-hosting system anymore once you ban all advertising. So there is a discussion that needs to occur about the tradeoff of finding the right balance and level of advertising. The discussion should center around giving people the large-scale voice and choice as to where they want to stand on the continuum of paying for things and not being bothered and tracked, or having systems that are not paid-for content, and therefore having to deal with the fact that you will see advertising. So that is one of the current big issues.

Another challenge facing the industry right now is about how to integrate mobile, because more and more people are spending time on mobile. The conversation is based around how to properly deal with even more sensitive information about where exactly people spend their lives. Dealing with sensitive information is a very interesting challenge, as well as how to, on the other side, actually connect it to metrics so that we can make sure we are not wasting anybody's time. Advertising should be able to be relevant to the person. In order to be able to deliver that, I need to tie your identity back to when you take an action. I can only optimize it if I can relate the same identity back to the person purchasing.

However, there are a lot of issues around attribution and last click. These issues really create a lot of incentive mismatch, which is a big problem for the industry. If you say the credit goes to the last impression just before purchase, then my incentive is to show you 50,000 ads all over the board at minimal cost—I am not going to be very specific about this—whereas if you try to do a more causal analysis to see which impressions really seem to have in aggregate—obviously, you cannot do this for one person—tipped the needle and changed that person's behavior, it is a much harder problem. There are analytical solutions available to do it, but they are really difficult to make reliable. So that is the attribution issue of which metrics is another big part of it.

And then the third challenge is the immense amount of fraud in ad exchanges. When we talk about ad exchanges, we are basically talking about a marketplace, which is a place for programmatic buying and selling of display advertising. Like any marketplace, whether it is eBay or the stock exchange or display advertising, there is, of course, a certain part of the market that is almost constantly attempting to take advantage of the system. In display advertising, it is harder to keep track of what is going wrong because the customer is not complaining as loudly. If I am on eBay and I buy something, which then does not show up, am I going to complain? Yes, and eBay will take care of it. But if the customer is a marketer that shows an ad and people never see it, how would they ever know? The public outcry is not quite there, so it is a more difficult problem in that respect.

It is also technically more challenging to very clearly say, "This is an instance of fraud, whereas this other one is probably okay." It is a challenge that as an industry we will have to face together. We will probably have to collaborate and raise the perception of it because most metrics that marketers look at are very easy to fake.

"Clicks" used to be one of those. A lot of the "clicks" we see are clearly not natural or are coming from different countries than from where the ad was shown in. I think it is easier to observe click fraud. When it comes to the question of whether there is really a person on the other side, that is when it gets harder. We are now moving to visibility and other ways to verify a person is really there. My guess is that this can be faked, too. Which makes it very much like the old adversarial problem of spam filtering—every time you build a better model, your opponent is trying to figure out how to beat your better model. Which turns the game into a game of escalation. It will keep the data scientists in business, I guess.

Gutierrez: What really drives this problem of fraud?

Perlich: One of the main reasons for the fraud problem is the highly fragmented nature of the industry. There are the marketers themselves. They have historically been relying on agencies to help them execute campaigns. Then there is the creative part—trying to package the message of the brand for some audience. Then you have all the different types of media including not just print, TV, online, and social, but even out of home—basically billboards—all siloed out as well because it is a completely different form of measurement and interaction. And now we are adding to the already existing layer of industry participants, data providers, and real-time exchanges. On top of which live firms like Dstillery, who are then asked to do the targeting. As you can imagine, every participant has a slightly different perspective on what the problem is, where it is occurring, and how to solve it. And, not only that, they also have very different incentives as to what they should really be achieving.

Depending on where you stand, the fraud issue is not a problem, or it is a small one, or it is a very large one. For instance, there are the real-time exchanges. They are paid by volume. They are basically saying, "Figuring out whether this is a good opportunity or bad, or whether this is fraud or not, should not really be my problem." Now, you can argue, gee, if you are eBay, then it is your problem and you should clean it up, but these real-time exchanges are paid by bulk, so they do not necessarily have that incentive. Furthermore, they say, "Look, I even have advertisers who want it because it looks great along certain metrics. You say I should not sell it, but you know what? I have customers who consider this absolutely what they want to have," so it puts them in a very tough position.

Now, the ad agencies say, "If I am going to tell my client that it is all fraud, guess what happens? I will lose their business. That is not in my real interest either. I am going to see if I can get around that problem in some way."

Then a similar problem happens with measuring the ad effectiveness of the creative. We have developed some internal tools where you can actually see whether showing the ad had any effect on purchasing behavior. And we have seen ad campaigns where the creative was very much click-oriented and we demonstrated that it had zero impact on purchasing behavior. We offered these tools as a service for a while, but the issue was that whenever the answer was not positive, nobody wanted to know the answer. It is kind of bad if you go back to a large advertiser and say, "Oh, by the way, your creatives do not do anything."

Gutierrez: No one wants to hear that.

Perlich: No one wants to hear that. That is what I mean by incentive issues. So a large part of how things are presented, communicated, and represented carry very different messages from very different angles, depending on what you are reading, so you probably need a very broad depth to understand the issues. There is not really one place to do that. A lot of the research-oriented entities—Journal of Advertising Research and American Marketing Association—that sit a lot more on the scientific side of it, are a good place to start. I think they are in the process of catching up with where the reality has gone. There has been so much change in the industry in the last five years that there really has been a revolution. This means research has been catching up as to establishing good standards of how to analyze this stuff. Nobody really knows what to do with it.

Gutierrez: What does a typical day look like for you?

Perlich: A typical day for me is very much circumstantial to the niche I have built for myself in here. Formally, after about two years of my being here, my CMO, CEO, CTO, and I came to the formal understanding of how my position is defined. Let me explain my very specific setup here. I spend twenty percent of my time on management, which is intellectual leadership and related activities. I spend forty percent of my time contributing to our core business, meaning I am building models and doing analysis. Lastly, I spend forty percent of my time being an ambassador for the company.

Gutierrez: What does a typical brand ambassador week look like?

Perlich: This setup means that I am able to spend quite a notable amount on giving talks. I would say on the order of maybe one per week. For example, I will be going to Washington, DC, for an AAAI [Association for the Advancement of Artificial Intelligence] event on Saturday. Next week, on Thursday, I am going to speak at a panel on lawmaking in Georgetown in Washington, DC. And then, the following day, I fly out to San Francisco, California, to give a small presentation at Berkeley. That is kind of what my schedule looks like. Further, I gave two talks at the O'Reilly Strata Conference, and then I went to give a keynote at Data IO conference. I do a lot of speaking, primarily because I enjoy it and am happy doing it, and because it is something that the company appreciates me doing. I also go to many local meetups and get-togethers.

Gutierrez: What does a typical intellectual leadership day look like?

Perlich: On the actual in-house work, we have very informal meetings here, so I will just connect to two or three of my colleagues and have conversations about what it is they are working on right now and what makes sense in that context and just bounce a few ideas around. We are also currently talking about what we want to publish, so we had a brainstorming session the other day in terms of what are the nice things that we feel comfortable writing about and maybe submitting to KDD or some other event. This type of teamwork is constantly going on here.

Gutierrez: What does a typical modeling and analysis day look like?

Perlich: My day-to-day is divided by routine tasks, special projects, and building new prototypes. As an example of what types of special projects I personally work on, recently I spent a couple of hours deep-diving on a specific issue that our CTO brought to my attention. The issue was that something was not working with the way we were crosswalking physical locations to desktop identities. An example of this process is that if we see people show up at airports, we then crosswalk them into "here is their desktop identity." Once we have done that, we can then try to target them for travel advertisement to see how much we can deliver, and then measure whether that has some value, meaning higher predictive power or getting higher conversion rates. Our CTO asked me to make sure that the process of the translation from some mobile advertisement bid request location was getting correctly identified and translated through the IP into a cookie identity. And then to check whether those cookie identities were correctly running on the back end through the bidding system. And finally, when we bid on them, whether were we actually getting some impressions.

This means a detective game or puzzle, right? You expect to see about a hundred thousand impressions. You find that the system only sees three impressions, and you start wondering what the hell happened to them on the way through. To figure this out, you need to pull very specific data sets out of a large Hadoop cluster and trace the data, like specific events, as it goes through the system, and see whether they come out on the other side. And if they do not come out, you have to figure out where they got lost. That is the detective work in action, just figuring out what is wrong with the data.

Gutierrez: What about your routine tasks?

Perlich: In regards to my more routine tasks, almost every week I rebuild and study our inventory models. This is a basic process done with a couple of semi-cron jobs that I run. The inventory models are run in the back end every time there is a bid request. Remember, we get as many as 30 billion bid requests per day. So these models are predictive models that estimate how good the current URL for the current bid request is for a particular campaign. If it is good, then we bid on the request. Otherwise, we let it pass.

As an example, with our travel campaigns, we know that if we get a bid request from Kayak, travel campaigns convert with a factor of 4. So what happens is that for Kayak URLs and the travel campaigns, there is a factor in the system that multiplies the bid price by 4, or by 3.5, or by 4.7. So depending on what the model is estimating, these models will decide whether or not to bid and at what level.

So my routine is to basically start up and run them. When they are done, I quickly eyeball the results. This process is not fully automated because the technology it relies on is a little bit sensitive to the data stream and bid request. Everything that has a bid request—supply and demand in the auctions—is a little bit tricky. I prefer to look at the results in order to make sure it passes my personal smell test. I will probably spend some time just looking at the results and making sure they are within the ballpark of where we wanted them. Then I hand them over to Gabriel, the head of account management, so that we can discuss them. The way we interface with brands, like Nike and others, is that there is a human that manages that relationship and also the campaigns. So they pick how many ads are delivered and for what goals.

To some extent, we give recommendations on which of the many models is the best, but they have some override because models optimize to one thing. Sometimes the campaign is measuring something else or has a different attribution metric, so there is a little bit of a translation. There is a human step involved, which is why we revisit the results at least once a week. Often we have a deep-dive on the models, results, and performance. We might discuss something like: "We feel that this campaign isn't performing as well as it could. Can you tell us whether we should change the frequencies or whether we have to change the bid strategy in a certain way?" And so I spend some of my time working with people upstairs, some account management time, just translating what I know about how these things work into strategies on how to best run a certain campaign.

Gutierrez: What about your prototype building tasks?

Perlich: I have lots of favorite little toy problems. Sometimes these are developed into new prototypes. These toy problems encompass things like feedback saying we were actually performing too well on certain campaigns, or feedback that we are not necessarily realizing the margin that we could on other campaigns. Much of the prototype work is looking and asking if there are there some smart ways of going into our system, looking at the data, and seeing if there are strategies we should employ to help our customers and our margins.

Once I decide to build a prototype, I have some idea of what to do. I implement it and then I test it out on some small data set. We call these small data sets our "sandbox," as they do not use the production system of data. So I run the prototype and I see how that would work in the sandbox. When that comes back as good, then we will try it out with bigger data sets.

That was the case recently, where we felt something was wrong with the relationship between our bid price and the price we end up paying. The auctions work a bit like eBay where the winner is supposedly only paying the second highest bid price. For these campaigns we wanted a black list of publishers that we should not bid on the bid list because, for some reason, the bids that we were winning were really expensive and then they were performing really badly. In this context we discovered that it seemed like we could reduce the bid price for these types of requests and we would save margin, and for some of the campaigns, the performance would even go up. I do not know why this was happening or what was causing it, so this was a prototype that we ended up building and it was used to select campaigns to be put on different bid strategies. But it started out as my little pet project with a toy model that I worked on for three days until we had a prototype. It does not add value consistently on all campaigns so we have not automated it but keep it rather as a tool to use as needed.

Gutierrez: What kind of tools do you use?

Perlich: We are very much a self-made shop, so we build the things that we need. I am not really hardcore on top of what the production side looks like, so I will talk about our side. On our side we have two main technology areas.

The first area is for the data collection and storage for analytics. The other area is responsible for being able to do a real-time response in 30 milliseconds to bid requests. So there you have technology with NoSQL—very high-speed lookup tables, like Cassandra and other things.

For my day-to-day, we have a Hadoop cluster. All of the incoming events are put into a standard format and then stored. We have event logs for everything—bid requests, impressions, clicks, conversions, all the visitation data, and so on. I want everything logged. We record them as event logs, with certain lookback times and fields.

They are housed in a Hadoop cluster, on top of which we have Apache Hive. Hive is a tool that basically lets you query this data with more or less standard SQL. It is not necessarily a real-time response. It is a little bit slow because of the whole interaction with Hadoop, but I do not need real time. I just need to get the data that I want. So I use Hive to get data out of Hadoop.

The key to working with this data is to figure out what exact data sample you need, so it is about figuring out which Hive query will give it to you. Typically, I try to avoid going beyond 10 GB of data. For most things I need to do, I can downsample significantly, as I do not need to process all the data. Once I have figured out which downsampled data set I need, then I just need to write the correct query to get the piece that I need.

Gutierrez: What kind of tools do you use to preprocess the data?

Perlich: I use a lot of UNIX tools—sed, awk, sort, grep, and others. You name it, I probably use it. I also write a lot of my own code in Perl. I do a lot of scripting that runs over that data. The scripting is done not so much for analytics. Rather it is done to preprocess the data into a state where I can then run it through some special-purpose tool.

Gutierrez: What types of special-purpose tools have you built?

Perlich: For the hard-core modeling that we do, we have our own implementation of a stochastic gradient descent logistic regression. That thing takes something along the lines of 10 million examples with 10 million features and within 5 to 10 minutes you get an answer. It is not parallelized, but it is really to the point and implemented very well for our specific use case of dealing with this kind of sparse data.

We are very much a self-made shop, so we are not using any kind of commercial tooling. So we build our own specialized solutions. When we need something, I typically start digging around in the academic literature and say something like, "Okay, let's see what SVMlight (some specific implementation of a Support Vector Machine Algorithm out of Cornell) is doing." I first check on performance, and even if it takes three hours, that is fine. Maybe we try random forest. I can get examples of this code, and then we see what works well for a certain problem. When we come to the conclusion that seems to be the right technical solution, we then typically reimplement it and tune it toward the exact setting we need. Then we have this as the in-house solution.

Gutierrez: What types of non-special-purpose tools do you use for data analysis?

Perlich: We still do a lot with R, but that requires more kinds of downsampling. You cannot run stochastic gradient descent on the data set sizes that we want. Or at least I do not know how to do it, put it this way, so I will leave that to others. Occasionally we do data visualizations, though mostly for communication purposes. There is nothing really to look at in our world of very high dimensional models. For the data visualization, I have played around a little bit with KML [Keyhole Markup Language] files for making maps. We also use D3.js for our customer-facing side, where we actually show graphs of the stats on the campaigns that we run. The consumer-facing side is more the analytics team. So it is not so much the data science team that is involved in that part.

Gutierrez: What lessons have you learned from using these tools to transform a toy project into a production system?

Perlich: I am not sure it falls under lessons for these tools specifically, but as usual, when you start looking into a data set that nobody has paid that close attention to, you end up finding things that you did not expect to see. For instance, speaking about the instance of bids to performance issues, we realized that on some inventories, meaning URLs, we were always paying what we were bidding. Now, according to the rules, this is a second-prize auction. This is not supposed to happen. So we found a couple of cases where we felt that the way the billing system was set up was not necessarily correct. We also found instances where we had just hard-coded the minimum bid price in the wrong way. I guess the overarching lesson is that if nobody looks at a data set for more than a month, it becomes useless pretty quickly because it is actually almost totally wrong somewhere, so only regularly looked-at and worked-with data sets are reliable.

Even if the project would have been a complete failure for all other reasons, I think it found enough issues in our setup that it was very well worth having me spend three days on it. This is something we realize again and again—side observations and insights almost always add value beyond the primary purpose. This kind of extra value happens very consistently whenever you look at data.

Another stunning example of learning lessons from really looking at the data is what happened when exploring a fraud case recently. My CTO came in and said, "Look guys. You managed to double performance in the last two weeks across all our campaigns. Do you have anything to tell me?" And we scratched our heads, because we had not really done anything. The only thing we had done is that we had added some new data from the exchanges. But once you know how hard it is to predict human behavior, the fact that you doubled performance—that is kind of scary. What ended up happening is that, yeah, it had doubled the performance because there were bots committing fraud. We were able to figure this out from the data because the bots were behaving very deterministically.

The whole thing had started out as a pet project to figure out why we had doubled performance, which is in some ways is a great thing. But we were just skeptical enough not to believe it. It turned out that a whole bot network was fooling us. Okay, fine, I am glad we talked about this and figured it out. From that we then implemented a whole overhead system to watch for this. That is very typical for these things. We start looking into some things and saying, "Hey, there's something really surprising going on," and typically, the answer's completely different from where you started. That is the lesson learned: you always have an open mind to the side things. I think it is a good skill set that you are not just narrowing down, looking at the particular problem you are looking at. Just keep your mind open and see what else is going on here. You will typically find a lot more going on in that process.

Gutierrez: What makes a good data scientist?

Perlich: I think this is really the marksmanship of a good data scientist—you have to have some amount of intuition about what should be happening. You do not have to be a medical specialist to realize that the patient ID being predictive is a problem. It just takes some amount of common sense to observe that. I think what this intuition develops with a lot of experience. You cannot just make a data scientist out of a computer scientist or a mathematician necessarily.

What I have observed is that there is a group of people who can embrace uncertainty and noise and what it means. There is another group of people who love to live in a deterministic black-and-white world. In a sense, they believe that when you sort the list, it is sorted. And once the algorithm sorts the list, it will always sort things right, because that is what it was made to do. The algorithm is either correct or it is not, but you have a very clear metric for correct.

Once you move to the side of data, the whole world develops a lot more gray areas. It is actually very interesting for me to see the interactions with some of our engineering team. Some get that, some actually figure this out, and some just feel that they are done when they implemented the steps on the list. This last group does not get the part that, once you have implemented the steps, you have to start looking at the output to check whether the output makes sense. "Makes sense" or "this should not really be happening" are not part of these programmers' informal checklist.

So data scientists need this intuitive part that revolves around data. It is something that is part of their personality, kind of like skepticism, as well as having expectations. I had a long talk with my friend Andreas Weigend, who was the first guy who ever got me excited about data. His belief was even stronger. He felt that, "You have to have an emotional reaction." I said, "You just have to have expectations. You need to know what you would expect to see, and then you can see whether it deviates." He felt that this is good, but he felt much stronger about having to have the emotional response.

Gutierrez: Is this intrinsic to a person or is it something that can be taught and/or learned?

Perlich: Something I have consistently seen in myself and other data scientists is an ability to say, "You know, something does not look right," even though it may take a while to translate that feeling into something you can communicate and make a formal case for it being wrong. It typically starts with "Hmmm, I am not sure about this." To develop this, I think takes good apprenticeship. You need to be shown that process. You need to make peace with it. I have had the blessing of having really good mentors, starting with Andreas, and then Foster, who has a very similar pedigree and attitude toward these things. The interesting thing is that the process is not terribly formal, but having seen the process a few times, you start to get the hang of it.

That said, you cannot force it down everybody's throat. It takes experience and apprenticeship, but that is a necessary condition, not a sufficient condition. In addition to the experience and apprenticeship, you still need the intuition about what is happening.

Gutierrez: How do you think about whether you are solving the right problem or modeling the right thing? How do you even know you have the right data?

Perlich: This is one of the biggest generic problems to arise in data science. It takes more than technical skill to be able to answer these questions. I may know how to solve a problem, but the ability to provide feedback on whether or not the question I am being posed with is meaningful in the first place—that is a very difficult problem.

One of the reasons I love working here at Dstillery is that there is a lot of appreciation for what data science does and how we can help. We are part of almost all of the decisioning from a business perspective in the first place. What I have seen in some of the consulting engagements of IBM was that if a conversation does not involve somebody who actually knows what can be done with data, you often end up solving the wrong problem or no problem at all.

I like to explain it like matchmaking. First, there is the problem you are trying to solve. Second, there is the data that you may or may not have or want to get. And lastly, there is the algorithm. My primary challenge as a data scientist is to use the right algorithm to connect the right data to the problem you actually want solved. However, by trying to match these three things up, it may also mean that the problem cannot be addressed with any algorithm I am aware of. It may also mean that we might have the wrong data. And finally, there is still the question of whether that problem we are solving is relevant, well specified, and is the right problem to work on in the first place. And the best way to perform this iteration between problem, data, and algorithm, is that you need to have a team of business people and data scientists working together. The data and algorithm knowledge resides with the data scientists, but to be able to really connect it to what business problems you want to solve, it is great if you can bring the data scientists into the room and have them be part of the discussion from the start.

An example comes to my mind. Recently somebody came to me and asked, "What is the average age of the cookies that we are seeing?" which on the surface sounds like a meaningful question, except that it is not actually a very meaningful question. To answer this question, I can come up with any number between one hour and three months. Not only that, each time period answer would be justified. The reason there is so much spread is that if somebody has third-party cookies disabled on their computer, it looks as if the cookie lives for zero seconds, so I write the cookie, it just never comes back.

Now, the question is, do I count them or not? If they are part of the average, we are now talking of a really long-tailed distribution with a huge spike at zero. Averages are meaningless for a long-tailed distribution with a spike somewhere. If I leave the zero in, the answer is an hour. If you ask me what is the average age of stable cookies that we see at least twice, then my answer will jump from one hour to three months. So you can see that averages are meaningless without more specific information.

So when someone asks me questions like this, my response is to ask a series of questions: Why do you want to know? What are you going to do with that thing that I am telling you? What are you going to use it to do? What decision or decisions are you going to make based on my answer? I am not going to make a statement or answer the question until I understand what you are doing.

Gutierrez: Is this part of the reason why it is helpful to have data scientists part of the conversation from the beginning?

Perlich: Yes, because—depending on what the goal and context are—the answer could be completely different. The problem is that many people think they know enough about data to talk about averages and other statistical measures, which makes them feel like they can ask meaningful questions. The problem is that in the real world, data is not like data they saw in classrooms and in books. They do not realize that they do not know enough about how distributions and other statistical measures really work. If they knew, they would realize that single-numbered aggregates are almost always completely useless. And that is the conversation part I need to have, because I am not going to give you a number until I know what you want to do. You may be solving the right problem, you may just not know how to ask the right question. It is not even that the problem is the wrong one, you might just be unable to formulate it in a meaningful way when you hand it over.

Gutierrez: How do you approach the issue of how non–data scientists ask questions of you and the data?

Perlich: You cannot expect everybody to become data scientists. I do not really need to have my CEO or other management people understand the intricacies of long-tailed distributions. I think there are other core competencies that are more important for the management to have and develop. The most important thing is getting out of the silo mode you saw historically in many institutions that had statisticians. In these institutions, you would have the statisticians sitting in the basement or somewhere far away from the business units. Nobody knew them and they were kept completely separate. You would just order a report from them and it would come back to you with an answer. You would then make whatever decisions you wanted based on those reports. The only times I have seen things work out well when working with statisticians and data scientists is if the problem-solving was done in teams, where you actually had enough face time to have a conversation with the person who executes it in the end.

What helps with education and with data literacy is that you do not spend three days agreeing on the vocabulary every time you discuss something. At least we have to know how to talk to each other. When you ask me for the average, having a common language of what that means before we start the conversation makes a huge difference. Otherwise, people walk away after a few minutes if I do not get to the point or understand what they are asking.

This means that the business side should probably have some high-level analytical understanding. They do not need to be able to do it, but they need to know enough of the vocabulary to understand what their techie or data scientist is telling them. In the same way, forcing techies or data scientists to actually talk to the practitioners is great. It is not the worst thing in the world to have them actually learn what matters on the business side and the common language of that world. It goes both ways. It is not about making everybody well rounded, it is more of agreeing and learning a vocabulary to communicate with. Once you do that, it is a huge step forward.

Gutierrez: When you are looking to hire someone, how do you think about the process and how does the communication aspect play into it?

Perlich: Internally, we have been having an interesting conversation around the idea that being a data scientist is one of the hardest jobs ever, and so how do we properly hire one. This discussion has been ongoing as we recently hired two people. The interview happens in two main stages. First, we give them half an hour to make a presentation on their previous work. This presentation can be one or two things that they have done before. Then we talk a bit about some of the problems we are working on and ask them for their thoughts. Throughout the interview we look for four main things: smell test, critical thinking, communication, and how they drive the process to understanding the data.

I look at the smell test first. I need you to have the sense of when something in the data feels wrong. That is the thing I cannot teach you if you do not at least have some sense. I can teach you language. What I do not need at all is domain knowledge. When we first talked to our HR people about needing data scientists, they asked if we needed them to know about digital advertising. We told them, "No, absolutely not. We could not care less." On the contrary, I prefer somebody who has done ten different things in ten different domains because they will have hopefully learned something new about data from each of different places and domains. I would rather have the breadth than the depth. So forget domain knowledge. I assume they are smart enough to learn they need to know about the domain in two weeks to three months.

After the smell test, we look for the rest of the criteria. I do not think communication is explicitly stated in any requirement set, but it is something that we pick up as we talk with the person. In terms of critical thinking, we may show them plots of data and say, "Is there something that strikes you odd here?" We want to see how they process that information and come up with questions that they feel are important to really get to the bottom of what is going on.

Then we move on to saying, "Okay, here's a problem." For us, even more important than talking about the problem, is that they are able and comfortable to start asking the right questions. When I present a scenario, I want them to ask, "What do you mean by that?" They need to know when they really understand something and when they do not, so being open-minded and being able to drive that process is very important. After all, that is what they will have to do as data scientists here.

Gutierrez: What problem did you explore with one of the recent hires?

Perlich: When we interviewed Melinda, we looked at a project we had recently worked on. We said, "We have to optimize Nielsen reports." Nielsen is one of the companies that provide feedback on advertising campaigns. For instance, they may tell you that of all the ads that you showed, females saw 73 percent of the ads. The interesting part about this is Nielsen has some internal panel. That panel does not cover all the people you showed ads to but just some subset. Part of this panel is then matched against Facebook. Then they figure out from this percentage that was on Facebook which ones self-identified as being female. Whether or not they are female is a separate question. But this is the basis of the report that tells you that females saw 73 percent of the ads. So the data is some subset of some subset and it is hard to tell whether ultimately it is a representative sample of my ads. And now the problem is that I am then supposed to optimize this without having any access to any of the underlying data. But because it is not a predictive model, for no instance/person do I get the answer. I only get aggregate feedback on sets of a hundred thousand impressions.

This is a problem we had been working and thinking about recently. Internally, we had brainstormed about it and had basically developed a methodology. So when we interviewed Melinda, we asked her questions like: "How can you optimize it?" and "How can you build a model to optimize for females, if this is what you want." This is not something we typically want, but we wanted to hear her thought process. We said, "Tell us what to do about it. You have an hour. Ask questions if you want to. This is a problem we are working on right now." It was quite interesting to have this conversation.

Gutierrez: How did Melinda approach the problem?

Perlich: Melinda went into probability theory, saying, "You have one group that is 80 percent female. This other group is 70 percent female. The intersection: Should it be higher than 70 percent or should it be lower? Is the fact that you show up in both of them increasing my belief that you are female, or decreasing it?"

So we discussed how to go at this problem with the Bayesian theory of probabilities—in particular, where it was possible to assume independence versus overlapping, and so on. Ultimately this idea was not what we implemented since the overlap was not sufficient. But we did take some of the ideas forward and made it into a predictive modeling task: "Well, let's use it to randomly label examples. Let's get a whole bunch of those that Nielsen thinks are female. If they say it's 80 percent, then we will label these things as female with 80 percent probability." We did this for all kinds of segments and then actually built a model on it. So we faked the outcome, and built the model based on probabilities, which is, in fact, what we ended up building.

Gutierrez: The interview is basically a working session then.

Perlich: Correct. In our daily work, we have this type conversation often: "We have a tricky problem. Let's get somebody to think about how we should solve it." This work and their presentation let us get a sense of their smell test, critical thinking, communication, and how they drive the process to understanding the data. It also lets us see how they ask questions. In particular we are interested in what questions they ask about what else they need to know about the problem, what are the constraints, what is the environment in which the problem happens, and what is relevant about the industry or the specific setting. Of course, we are also very interested in how they come up with ideas for solving the problem. We think it is a good process, and so far the people we have interviewed seemed to have liked it.

Gutierrez: Is it hard for a company without data scientists to hire data scientists?

Perlich: Yes and I am not even talking about convincing one to join you and the cost associated with it. Companies, with the help of some vendors, may have started collecting data. Now that the companies have data or will soon be looking at data, they have to go out and hire someone to do things with data. But what then happens when looking for data scientists is that "data scientist" is a completely undefined job description. The biggest issue for companies is that if they were to try to hire a data scientist, they would not even know how to tell if they were interviewing one, because they do not really know what "data scientist" means. As long as they do not have one data scientist recognizing a second one, it is actually quite a daunting task. Perhaps a little too negative, but it seems to me that many data scientists basically just changed the label on their résumé, for some of them it makes sense but for some it does not. It is a major issue that data scientists have no agreed-upon skill set as of today. If you hire a database administrator, you know what you are getting. Today, if you hire a data scientist, you do not know what you are getting.

Gutierrez: Is it hard for a company with data scientists to hire data scientists?

Perlich: It is easier because the second is more likely to accept the job if there is a first and because you have somebody who can evaluate the candidate to some extend, but it is still far from easy. One of the big concerns with data science is what I call "quality control." If I build a model, I have a good gut guess of the overall quality of the model. However, I cannot tell you how good it is in the sense that I do not know if my working on it for another week will increase the performance by 5 percent or by 50 percent. I have a gut feeling about this, but the reality is that there is a noise part that comes from the data that has nothing to do with the algorithm. This makes it extremely hard to really know where you stand on your own model's quality. I am pretty okay predicting model quality for my own models based on knowing how much time I have spent, how much I have explored, and how much is left on the table that I have not looked at.

So now a company with data scientists wants to hire a data scientist. The data scientist gives the company one of their projects and now the company is supposed to judge how good the data scientist's model is? The data scientist does not even know it. The company does not know what the data scientist really did or could have done. And, unless it is in the same industry, the company may not even really know the data either. So, looking at a project from the semi-outside, even from a managerial perspective or hiring perspective, and saying how good it really is in terms of what it could be or should be is almost impossible. And that directly translates into also measuring the skill set of the data scientist because that is just how good a project can be produced. So quality control is a problem of the project and it translates into a real difficulty on evaluating the skill and the proposal of somebody you are hiring or talking to. It is easier if you already have a crew that can look at it and have a smell test on the data scientist as well.

Gutierrez: What do you think of the present and future of data science?

Perlich: In the present, I am not quite sure where we stand as a group. Part of this uncertainty is that the narrative of data science and big data has really been shaped a lot by people who have a stake in the game, whether it is vendors or consultants. They focus on telling you all the cool things you can do with big data. This has generated expectations that a lot of people struggle with fulfilling, which is a negative.

At the same time, I think it is very good to have much more awareness of the opportunities that come with data. I think that is a great thing, because we have been arguing about data and the future of it for the last twenty years. I am glad somebody is finally listening. I think it is all for the good. There are many people who have been doing cool stuff with data that may have been overlooked for a long time. They are now getting some notice.

In the future, I think we will have a much larger universe of skilled people. I also think that the education, language, appreciation, and the ability to perceive opportunities of data is going to move up as well. Particularly in management side, we will see this increase in knowledge.

The future will also have a clearer understanding where data science is not always useful. There are plenty of areas where you do not need a data scientist, and do not let anybody tell you otherwise. There is only so much you can do with data at the end of the day. A lot of what I see people pushing today are things that cause me to roll my eyes and say, "Okay, fine—whatever." Just because something can be done with data does not make it valuable even if it looks cool. At the end of the day the question always has to be: can you and are you willing to act on it. If not, you are wasting money and resources on data porn.

I think we need another five to ten years to get a better sense of best practices and expectations on when things really work and are useful, and when they are not. I do not think that anything will change fundamentally about how data science is done. With the hype going down, I think some of the more hard-core concepts that have been going around, like optimization, will gain a little bit in appreciation, and some of the hype-y things will go down a little bit when people realize that ultimately they cannot really make any impact.

For instance, I am very hesitant to embrace every time somebody asks me for actionable insights. It scares me. It really scares me because how can you expect me to tell you the actionable insights if you do not tell me first what your actions are? Unless you communicate very clearly and think about what exactly it is you can do and are willing to do based on data, sending me off on a wild goose chase with data to come back with actionable insights is a fool's errand. The more data you give me, the worse it gets because I will find more stuff that is probably meaningless. I think we need to change that process and that comes back to communication. I think that will improve, but I do not think anything very fundamentally will change.

I do not believe data science will be automated. I do not believe that your secretary will do data science for you, as much as that is kind of the way it is positioned—that anybody can do data science. That is not true, because more often than not, it is a problem with the data that you have to find first, and if you do not know data, then you cannot do that. Just because you can run an algorithm on data does not mean that you get something meaningful. You do need to look at the data. And that skill has to be in there. There has to be a human there, somewhere. You cannot hide it all away and abstract it in data layers and cool tooling. It is not going to work.

Gutierrez: Are there any areas you think data science should focus on?

Perlich: I think there is a tremendous opportunity for data science to have a huge impact in the medical field, in particular general well-being. To make this impact occur, we have to go back and really figure out how we deal with privacy and data sharing in order to get a good stake in the ground. I have personally walked away from medical applications repeatedly because I just got so hung up in HIPPA regulations. It was depressing because I felt like I was getting my hands tied behind my back. It was very depressing to know that I could have otherwise been making a large impact.

In the medical field, I think it is crucial that we understand a lot better what the tradeoffs are with these regulations. We need to have the medical experts communicate exactly what can be done and who the people doing the work with the data would be. As with most things, part of the problem is a lack of information sharing. Every doctor basically has his cases to read and knows what he has done over the last twenty years of his career. So when doctors see a new patient, they are harnessing their experience and what cases they have read—maybe even some from a medical journal. What would be better is to be able to combine the particular instance of a person, their condition, medication, and symptoms in a privacy-protecting way, to then be fed into a system. Then we could have a system tell the doctor, "We have seen this about 15 times. If you do treatment X, the guy will do really poorly, so that is not really what you want to do. However, if you do treatment Y, the guy will get better."

So being able to have the collective experience of medical conditions, symptoms, diagnosis, medications, treatments, and results would be tremendous. After all, that is really what data is—a collective experience of what happened being brought together. This is important because clinical trials are problematic for two reasons: (a) they never have real-life environmental settings where people take all of the stuff, because people are screened very clearly; and (b) the other part is that I do have my concerns, and they have been voiced repeatedly, about how reliable some clinical trials are. I want to leave it at that.

I think that from observational data, we could learn so much more as to what are effective solutions and what are ineffective solutions. Along with more observational data, we also need better forms of data regulation. Regulating the data and privacy implications have to be dealt with. I think this is true for advertising and privacy in general. Not just in medical applications and social data people share on Facebook, which are used for all kinds of purposes. I think the medical application is the best case where you can make a statement of, "Here is what should really be done." We need to start being able to compare the benefits to the costs without sensationalism.

I get upset when I hear of singleton sensational examples like the Target case where the dad found out his daughter was pregnant. Dads get to know that their precious daughters are pregnant every single day, through all kinds of means. Whether it is the daughter's friend who tells them, or a nosy neighbor who lets it slip out, or something else—these things happen all the time. That is part of life. I do not want to necessarily push this as a call for more targeting. Rather I think we have to confront the fact that once in a while, privacy might be violated. After all, it happens in real life every single day.

So I get upset at the boulevard newspaper atmosphere that exploits this situation to instigate a huge outcry. The fact is that you cannot make privacy airtight. That cannot be the goal. We need to have a really meaningful conversation of what the tradeoffs are. Sadly, I do not really see that occurring in the near future because often privacy advocates sit on one side with a lot of complaints and no proposal as to how to do it better, while on the other side, the general attitude is more along the lines: "The sky is the limit. Let's just do whatever we want."

So neither side comes to the middle to further the discussion. I am looking for an honest middle ground there, and I do not see it. I am not sure how to get there, but I would love to have a more constructive debate about it. Once we get there, I think there is a tremendous opportunity for data science to have a huge impact in the medical field.

Gutierrez: For someone starting out, what should they strive to understand deeply?

Perlich: I think, ultimately, learning how to do data science is like learning to ski. You have to do it. You can only listen to so many videos and watch it happen. At the end of the day, you have to get on your damn skis and go down that hill. You will crash a few times on the way and that is fine. That is the learning experience you need. I actually much prefer to ask interviewees about things that did not go well rather than what did work, because that tells me what they learned in the process.

Whenever people come to me and ask, "What should I do?" I say, "Yeah, sure, take online courses on machine learning techniques. There is no doubt that this is useful. You clearly have to be able to program, at least somewhat. You do not have to be a Java programmer, but you must get something done somehow. I do not care how."

Ultimately, whether it is volunteering at DataKind to spend your time at NGOs to help them, or going to the Kaggle website and participating in some of their data mining competitions—just get your hands and feet wet. Especially on Kaggle, read the discussion forums of what other people tell you about the problem, because that is where you learn what people do, what worked for them, and what did not work for them. So anything that gets you actually involved in doing something with data, even if you are not paid being for it, is a great thing.

Remember, you have to ski down that hill. There is no way around it. You cannot learn any other way. So volunteer your time, get your hands dirty in any which way you can think, and if you have a chance to do internships—perfect. Otherwise, there are many opportunities where you can just get started. So just do it.

Jonathan Lenaghan

PlaceIQ

Jonathan Lenaghan is the Head of Data Science at PlaceIQ, a mobile geoloca-
tion intelligence company aggregating and analyzing spatial data for marketers. In
early 2014, mobile Internet users eclipsed desktop Internet users for the first time.
Projections show that this trend will only accelerate as we move into the future.
PlaceIQ is at the forefront of the intersection of mobile ads and location intelligence. Rapid growth at this dynamic intersection brings with it challenges in terms of
privacy, data infrastructure, size of data, and the ability to process the data intelligently and put what has been learned from the data to good use.

Lenaghan's precursory career before venturing into data science spanned theoretical
physics research, editing prestigious science journals, and being a quant researcher
for algorithmic quantitative equity trading on Wall Street. After taking his PhD in
physics from Yale, he conducted research on the statistical properties of strongly
interacting quark-gluon plasma at Brookhaven National Laboratory, the Niels Bohr
Institute, and the University of Virginia. He served as an editor of two journals of the
American Physical Society: Physical Review C (nuclear physics) and Physical Review D
(particle physics and cosmology).

Lenaghan stands out as a prime example of a data scientist who has migrated
from physical science to data science via quantitative finance. This richly varied
background informs Lenaghan's nuanced appreciation of the risk dimensions of
data science, his optimistic pragmatism, and his conviction that useful data science
depends critically on a sound engineering foundation.

Sebastian Gutierrez: Tell me about your journey to becoming a data scientist at PlaceIQ.

Jonathan Lenaghan: Prior to joining PlaceIQ as a data scientist in March of 2012, I worked in the financial services industry doing algorithmic trading. Before that, I worked for eight years in academic physics. So I've always worked with a great deal of data—although, compared to my algorithmic trading work, my physics work was a bit more weighted toward the analytical than the computational.

Gutierrez: Algorithmic trading sounds like an interesting job with interesting data sets. What drove the transition to PlaceIQ?

Lenaghan: I liked the style of work I was doing in the financial services industry, solving quantitative problems, but it began to feel like I was solving the same problem year after year with slight variations. After five years, I was ready for a new challenge.

The data startup community in New York was growing very rapidly, and I started going to the New York tech meetups. At a meetup in February of 2012, I found myself sitting next to a VC who told me about the sort of companies he was investing in. I told him about the work I was doing with data in algorithmic trading. He told me that a couple of his companies needed a data scientist and suggested I speak with them. The next day I spoke with PlaceIQ's CTO, Steve Milton, and then a couple of days later with the CEO, Duncan McCall. A week and a half later I started work as PlaceIQ's head of Data Science.

Gutierrez: Wow, that was a pretty fast transition!

Lenaghan: Two weeks.

Gutierrez: Something must have really excited you about PlaceIQ. What was it?

Lenaghan: The size and variety of the data sets. PlaceIQ is a location intelligence platform, and so we ingest all kinds of data—social data, GIS [Geographic Information System] data, POI and AOI [Point/Area of Interest] data, and some consumer behavior data

Gutierrez: What do you do with these data sets?

Lenaghan: We've divided up the world into 100-meter-by-100-meter tiles. We ingest all this data, cleanse it, normalize it, and project it onto our 100-meter-square tiles integrated with what we call our "PlaceIQ time periods." Using this unified data layer, we then apply machine learning to contextualize locations and movement data.

Gutierrez: The data sets you ingest seem pretty straightforward other than the movement data. Where does this data come from and how do you use it?

Lenaghan: The sources of our movement data are ad-request logs. So even though our geospatial analytics platform is very horizontal, the main vertical we've been working in is mobile advertising. We're now moving into consumer insights and building up our geospatial analytics platform vertical-by-vertical.

Gutierrez: So the idea is that, as I'm walking around with my mobile device, I should get the right ad for the right product or service at the right moment and location.

Lenaghan: That's exactly right.

Gutierrez: Could you describe a particular product you helped build?

Lenaghan: The first product we launched into the ad tech space was called Audience Now. It contextualized what was happening within a single 100-meter-square tile so that our database could identify its primary demographic and psychographic characteristics with high confidence—so we could say, for example, "This tile index is very high for the 'shopper mom' type." Then, if your mobile device moved into that tile, it would be subject to "shopper mom" targeting. Audience Now was purely location-based. We didn't do any device tracking or anything like that.

Gutierrez: And how has this product evolved?

Lenaghan: Subsequently we started to ingest a large amount of ad-request data tagged with location, time, and device ID. Device IDs are obfuscated through persistent hashes—so even though we can't tie back a given device to any PII [Personally Identifiable Information], we can contextualize the movement data. We can see that a particular pseudonymous device ID has been on a golf course a couple of times in the past month and tends to dwell in tiles that have a high score for affluence. So we might tag this device ID as "golfer" type. The big push over the past year at PlaceIQ has been to target devices not only by location but also by the contextualized location histories.

Gutierrez: This seems very different from algorithmic trading. Why did it interest you?

Lenaghan: When I first came to PlaceIQ, I wasn't necessarily interested in the ad tech space per se, and my background in equity trading and physics had given me no specific expertise in mobile advertising. The mobile ad market does display certain systemic similarities to the equity market. Both involve low-latency, high-throughput systems, networks, and exchanges. I expected on entry that my work in the ad-tech and equity market ecosystems would be very similar, but the analogy broke down almost immediately. Instead of working on a lot of real-time bidding at PlaceIQ, I actually work more on geospatial location analytics that use machine learning and heuristics to build audiences to target mobile ads.

Initially I came to PlaceIQ more for the challenge of working with all of the location data, movement data, and ad requests than to specifically work in ad tech. But now that I've been exposed to more of the ad tech industry, I find it much more interesting than I thought it was going to be.

Gutierrez: In theoretical terms, how do you describe the work you do to other data scientists or to people who work in quantitative fields?

Lenaghan: In a word, I'd tell them that what we do is *ingest, transformation, and contextualization.* We ingest enormous amounts of location data—on the order of 50 billion records a month. Then we essentially do large joins against our geospatial layer. We run many different types of classifiers on the geospatial layer to determine whether or not a particular tile is residential or retail or mixed-use. Then we apply a domain-specific language we've developed to build these audiences into profiles that not only contextualize the places but also add more insight into the patterns we see in the aggregate movement data.

Gutierrez: How would you explain what you do in more qualitative terms to a five-year-old?

Lenaghan: Five-year-olds are increasingly sophisticated about mobile devices these days. I'd say to a five-year-old that we are mappers of the world. We're using the signals that come out of people's phone to better understand the types of people in any given place in the world.

Gutierrez: That's an explanation I could tell my mom: "Jonathan is using the phone signals to map the flow of different types of people in our world." "Okay, that makes sense, Sebastian."

When you did you realize that you wanted to work with data as a career?

Lenaghan: When I was an undergraduate, I majored in physics. I knew that I wanted to become a physicist—more precisely, a professor of physics. I really looked up to all of the professors I had as an undergraduate, and I was very excited about graduate school. You might suppose from my future career trajectory that I must have been immersed in experimental data when I was working in physics, but I was actually working more on the formal theoretical side of particle physics.

My study was the formal side of the strong interactions described by the theory of quantum chromodynamics: how quarks and gluons interact with each other and form neutrons and protons and things like that. Even if we can't solve the governing equation from quantum chromodynamics, we can still run simulations on it and we can solve it within certain limits.

At the time, I didn't really care what any of the experiments relating to the strong interactions were telling us. I was just interested in answering the question, "What can I learn from this defining equation using all the constraints of mathematical consistency?" My wife happens to work in the same field, but on the experimental side. She was working at accelerators while I was off in the academic clouds.

After I left academics, I worked for an enjoyable couple of years as an assistant editor for the *Physical Review* in Brookhaven. It was very interesting to see how the review and referee process works, and how the sausage is made in the academic publishing world. It set me to thinking a lot more about computation and data. And I was in the vicinity of New York and the financial industry.

Going to work in the quantitative finance industry was definitely my first jump into a very data-intensive world. Quantitative finance broadly divides into two fields. In the first field, people work with very complicated financial instruments, such as over-the-counter derivatives, which are traded very simply, usually over the phone. There are mathematical frameworks around these complicated instruments, which have theoretical, analytical, and computational aspects. The second field is high-frequency algorithmic trading, in which you have very simple financial instruments, such as equities, which are traded in very complicated ways. This field is much more computational and algorithm-driven. There's no analytical structure around this type of trading; it's essentially driven by experimental data. I originally wanted to go into quantitative finance to work in the first field dominated by analysis of complicated derivative products, but I ended up in the second field dominated by algorithmic trading of simple financial instruments.

Gutierrez: What made you change your plan from doing purely analytical work to data-driven work?

Lenaghan: I quickly learned that the empirical style of work in algorithmic trading suited me very well. I liked the experimental design—figuring out what works, what doesn't work, and how not to trick yourself with data. There are lots of challenges working with data in the algorithmic world. One of them is that thousands and thousands of other people are looking at the exact same data sets, and basically they're all just squeezing everything they possibly can out of it. Another challenge is that people under pressure to find patterns are prone to fall into the common human fallacies of overfitting models with insufficient data and overreading correlation as causation.

Gutierrez: How do the data challenges you faced in the algorithmic trading world compare to the data challenges you face at PlaceIQ?

Lenaghan: The initial data challenge when I came to PlaceIQ was that geospatial data was a data type that I had never worked with. The second challenge was that the data volume was scaled up by a couple of orders of magnitude. The volume of data in the algorithmic trading I was doing was quite large—say, a terabyte a year. But the PlaceIQ environment generates hundreds and hundreds of terabytes a year. Making these adjustments were exciting challenges for me. Another thing that excited me about PlaceIQ was being able to work in a data-intensive world where not everything is a time series. It was a little overwhelming at first, but I learned and continue to learn new techniques.

Gutierrez: Any other challenges you encountered moving from finance to ad tech?

Lenaghan: One of the differences I noticed when moving from algorithmic trading to mobile advertising was the infrastructure. The infrastructure in finance is much better. The financial industry has been around much longer, so the infrastructure is much more developed and built out and is much more intolerant to failures. In the ad tech industry the standards are a little bit looser.

Gutierrez: In finance, if you make an error, such as overfitting a model, then you lose money. Is there the same sense of urgency or pressure in your work at PlaceIQ?

Lenaghan: No, there's not the same sense of urgency or fear. If you overfit your trading model, it loses money. When making money is your job, losing somebody else's money is one of the most horrible sinking feelings in the world. So you are very incentivized to not overfit your models. If you overfit your ad tech model, the repercussions are less dire.

When an ad misserves, it's bad professionally, but the revenue hit is more incremental. I do try to instill in my ad-tech data science team a sense of—to put it not entirely facetiously—healthy paranoia about the quality of our data and the robustness of our models. But the feedback on my ad-tech team when something goes wrong is not, "Oh my God, what a disaster!"—but instead, "Let's fix it."

So in terms of the penalties for error and the allowances for correction, the two worlds are somewhat different. Internally, the urgency at PlaceIQ is not as relentlessly instantaneous as it was at the trading desk. Externally, we still have to be very careful to get whatever we publish right, because if something in a mobile campaign goes wrong it could be a very large revenue hit. At PlaceIQ we work mostly on large, direct deals rather than a lot of small channel deals, so losing one partner is always a big deal. A serious error can incur the very serious cost of losing a large client.

Gutierrez: What was the first large data set you worked with at PlaceIQ?

Lenaghan: The very first big data set that I worked with was anonymized location histories from an ambient background location app. We selected this very large data set as a test set to see how well we could actually contextualize movement. We wanted to understand at a very high level where people were living and what types of behaviors we could correlate with in-home demographics of social data.

Gutierrez: Was there an aha! moment that "This is powerful"?

Lenaghan: Our "this-is-powerful" moment came when we saw the predictability of human behavior. Location histories tend to cluster very tightly, so it was fascinating how, with a small amount of data, you could build interesting profiles of devices. Most people are at home or at work most of the time. So in that sense, it is not terribly difficult to infer high-level demographic information and associate it with a device, even when you know nothing else about that device.

Gutierrez: How did you learn and get up to speed on the geospatial analytics industry?

Lenaghan: When I joined the industry, it was still more or less a very new field. On one hand, geography is a very old field. On the other hand, the scale on which we were approaching it was fairly new, so there was not much literature to help with the learning process. A number of other companies are working in the location space, although they are not doing exactly what we do. So I started learning by looking at some of them, such as Factual—which is a data provider for locations.

Research-wise, Google was publishing a lot of white papers around latitude studies based on the experiments that they were running on people's locations. So the good and bad of it was that—aside from the Google white papers and a few blog posts—it was completely wide open, so we had to figure things out as we went along.

Gutierrez: Has the literature and industry expanded?

Lenaghan: Though the industry has expanded a great deal in terms of competitors, the literature on location histories is still fairly sparse. There has been some interesting statistical work Albert-László Barabási, who has done some things on location and, in particular, predicting people's habits from mobile data. The academic literature is focused on such questions as: What is the minimum set of location histories that you could join with another data set to infer information about that device?

Another research area is looking at the power law distributions of people's aggregate behaviors. In terms of academic research that is specific to location and applicable to what we do, there is not a great deal being produced. That being said, we have a lot of competitors now, who, in the past year, are claiming to do the things that we do. This is both good and bad.

Gutierrez: Generally, it is viewed as good to have competition.

Lenaghan: Exactly. Having no competitors is bad.

Gutierrez: What does your typical day-to-day work as an expert in the geospatial analytics industry look like?

Lenaghan: I would say my typical days in 2012 and in 2013 have been fairly different. In 2012, we were a very small company. I was the seventh employee. By the end of the year, we had doubled in size to fourteen employees. Even though we were growing quickly, we were still a very small company and we only had a few people who had experience in the ad-tech industry. In that sense, my typical day-to-day was very much like any other startup story: I was involved in doing and helping with everything.

I was building models and developing heuristics to build audiences in the cases where we did not have data or we had very sparse data. In addition to that, I was also doing a lot of the engineering and back-end work for our campaigns, checking the health of our campaigns, doing ad operations, and developing some new products. Everybody in the company was helping in all of those projects as well.

Gutierrez: Were you on a particular team?

Lenaghan: I was on a data science team, though it was a bit nebulous given all the other work I was doing. Then we had explosive growth in 2013, starting from the very beginning of the year to near the end of the year. We are now almost seventy people. In 2012 we doubled in size, and then in 2013 we quadrupled in size from where we ended in 2012.

Gutierrez: Has your role become more defined as the company continues to grow?

Lenaghan: My role has now gotten much more codified. First, we can look at it in terms of the data science team. In the first half of the year, the engineering team and the data science team worked together extremely closely. On most of the projects, you could not tell who was on the engineering team and who was on data science team. As the company has grown, more distance between the two teams has been established. That said, we are not completely split because we still work on mixed functional teams, so we still work very closely with engineering.

Second, we can look at it in terms of my personal work. In the beginning of the year, I was still writing a lot of code and developing a lot of algorithms. Now, as the team and company have been expanding, in addition to day-to-day management, I have been moving more toward a lot of more of the thinking behind the architecture and long-term planning.

Gutierrez: Do you still get your hands dirty?

Lenaghan: I like to be very hands-on, and I feel like the only time I can really be hands-on is very early in the morning. Most days I get to the office before anybody else to make sure I can get a few things done before the office fills up. Whatever project I am working on, I do my best to get that done in the early morning hours.

Gutierrez: How do you work with your team members?

Lenaghan: Once the teams arrive we have a morning standup with the engineering and the data science teams around 10…10:15. Everybody talks about what they are doing. I then sync up with a few team members throughout the day and lead or attend a variety of meetings. The meetings are usually product meetings or troubleshooting meetings. The troubleshooting meetings come up if we are having an issue with a particular campaign we are serving. These one-on-one and small group meetings go on throughout the day.

In between meetings, I am usually in discussions with account managers. These discussions center around which campaigns are feasible and which are not. What sort of targeting could be used? And is this type of campaign and targeting even possible? I will get questions whether a particular audience—even though it sounds very interesting—is going to have the necessary scale to make an impact. I find myself having to make a lot of higher-level strategic decisions. I try to keep most members of my team as separate from those as possible so that they can really focus on the clients, projects, and data. Then around 6 o'clock my day winds down and I go home.

Gutierrez: When thinking through and making these higher-level decisions, how do you think about whether you and your team are solving the right problems?

Lenaghan: I always try to look at the problem from the end. So I think about what is the final output and functionality that we want after all the days or weeks of work have been put into solving the problem. Is the final output a particular audience? Or do we want a classifier to perform much better? Or perhaps we already have a process, but the machine learning component of it is not performing as well as we would like and we want to improve it. I always start from the end.

What I have found—not only from working in industry, but academics as well—is that when you start from the beginning and everything is blue sky, there are hundreds of ideas to chase as well as thousands of ideas to try and, since everything is possible, nothing ever gets done. It can and has happened that things eventually get done, but running a company by serendipity is begging to fail. So I always focus on looking towards the end result.

Of course, many times throughout the course of solving the problem, you end up at a different place. Sometimes it is better; other times you just have to scrap the project. Keeping your eyes on the final deliverable is essential to solving the right problems.

Gutierrez: There is an idea in engineering that you build one to throw away. Do you find there is a sense of that in the data modeling work that you do?

Lenaghan: I definitely think so.

Gutierrez: What is the hit rate of things that work the first time?

Lenaghan: Around 50 percent. Our teams operate by always trying to build a prototype first. On the data science side, this initial prototype is usually a mixture of Java and/or Python and/or R. Again, we always try to keep our eye on what the final piece is going to be. If we know that performance is going to be a problem, we may start in Java from the very beginning. If we do build a prototype, we usually make it as lightweight as possible.

Gutierrez: Why as lightweight as possible?

Lenaghan: I do not like writing a lot of code or doing a lot of work for something I do not know is going to succeed. So we build the prototype and start working on it with small data sets first. One of the first tests that we do is a scaling test. Even if the prototype is not super-performant, we want to make sure that it is capable of processing all of the data. Even if our prototype code is six times slower than the production code we are eventually going to write, we do want to be sure that it is capable of processing terabytes of data.

Gutierrez: If the prototype performs well, what happens next?

Lenaghan: If the prototype performs well on the scaling test, then we move to the production phase. I would say that about 60 percent of the time we involve engineering, and about 40 percent of the time we do it ourselves. If we need something really performant and it is complicated and involves a lot of configuration, then we always involve engineering there. Eventually there is a process to migrate the prototype to production code. Engineering will push our combined work to the dev ops group, which is where it is moved into production. Then we monitor it and hopefully never touch it again.

Gutierrez: How do you do the scaling test?

Lenaghan: We slowly step up the scale of data we run through the prototype in two dimensions. We have the geospatial dimension, which is large, but not extremely large. There we are talking about hundreds of millions of entities, let's say, in the United States. We also have the second dimension, which we think of as the movement side. This is the data coming from the ad-request side. This data is on the order of tens of billions of data points per month. We want to understand how well the prototype scales up in the two dimensions—the spatial dimension and the movement dimension. Usually, we start on the geospatial side and apply our analysis to just one metro area. For various reasons, we always use San Francisco. We could use New York City, but Manhattan is too anomalous.

Gutierrez: San Francisco is the base metro area for the spatial dimension testing?

Lenaghan: Exactly. We set the initial geographic scale starting with the metro, and then on the movement side, we will start with a day's worth of data. Then we scale the data to a week's worth of data. Then we scale up the data to a month's worth. At each step we are testing to see how the prototype is performing. Depending on the project or the product, varying lengths of history are required for further testing. Interestingly, it is definitely the case that more data is not always better. It depends on the product.

Gutierrez: The testing depends on the product, data history, and what you are modeling. If it is a prototype for the December holiday season, you do not want to use data from the middle of the summer.

Lenaghan: That's exactly right.

Gutierrez: Are there any interesting aspects of the data sets outside of the most obvious information content?

Lenaghan: It turns out that a lot of the biases in the data appear from the fact that all of the movement data comes from smartphones. This means you are completely biased toward people who own smartphones. This is a large population, as there are about 110 million smartphones in the USA right now. Although this represents a large swath of the US population, it is still a biased sample. So we have to deal with that bias in the data.

Gutierrez: Are there other large biases that you need to take into account?

Lenaghan: The movement histories that we see also have a large bias, as these phones don't drop 5-minute breadcrumbs all the time. They are only engaged when someone is using an ad-supported app, for example—so you also have a bias there, which means you end up biasing toward people who use ad-supported apps. In fact, biases pop up for different ad-supported apps people use all the time, such as texting apps, Words with Friends, or other apps. So free texting apps tend to skew in one direction. Words with Friends–type apps—even my mom uses Words with Friends—tend to skew in another direction. In interpreting our data, we have to correct for these and many other sorts of biases all the time.

Gutierrez: Let's dig a little deeper into the biases. Did you and your colleagues figure them out, or are these biases industry-known demographic, sociographic, and/or psychographic heuristics?

Lenaghan: That it is something we have figured out internally. Something we're always very cognizant of is that we don't want to be an undifferentiated black-box machine learning platform. So a very large component of the bias-correcting work we do is based on social anthropology. We look at the movement data and ask people in our office with a background in anthropology or sociology questions to gain further understanding.

We want really to understand: "How do we interpret this data? It's biased in this way. Why is that?" A great deal of the time the data is not going to answer these questions. A key thing is to never underestimate the power of domain-specific knowledge.

Gutierrez: What do you consider to be the most helpful domain-specific knowledge at PlaceIQ?

Lenaghan: The domain here at PlaceIQ is people. What are people doing? Who are these people? What kinds of heuristics can we layer over our data, either intuitively or anthropologically, for it to be true? That gives us tremendous mileage.

We do have a lot of analysts who short-circuit some of the algorithmic work we do just because they know that something may or may not be true just based upon human experience.

Gutierrez: You keep in mind that it is real people behind the massive amounts of data.

Lenaghan: Correct—we want to build an analytics platform rather than give our customers black-box answers. We really want to be used and viewed as an augmented intelligence service for analysts. These are people who are running campaigns. These are people in the consumer insights business.

Gutierrez: There is an idea of the data exhaust. In the operation of ingesting and analyzing your data, you are also generating data that could be useful for other people. How do you think about this secondary data? Do you think about monetizing, giving it back to the community, or a combination?

Lenaghan: That is a very good question. Right now, we are not doing that. We are laser-focused on consumer insights and especially on mobile advertising. That said, I think the long-term vision of the company is the platform. It is the platform we can license to other people. It is a platform with potential APIs to give this contextualized information back to the community. I think that is really the direction of the company. It is not what we are going to be doing in the next bit, but over the next few years.

Gutierrez: Speaking of communities, you mentioned earlier that you are using R, Python, and Java, which are tools built by open source software communities. What tools do you use and how has that changed in your career?

Lenaghan: When I was working in trading, I worked mainly in C++ and Perl. It makes me feel very old when I say that. Now we hire young engineers, and they have never used C++ or Perl, and that sounds crazy to me.

Then moving into this world, I do most of my work in Python. The number of very useful libraries and frameworks in Python seems to be growing every day. Another benefit of Python is that you can also write prototypes very, very quickly. Since performance is not a super big issue from the perspective of building prototypes, I always go to Python. I would say for up until about the beginning of the summer of 2013, I was writing a lot in Java. That was also when I was writing a lot of the back-end code for the data science group. But now it is all Python.

Gutierrez: Does everybody use Python for prototyping?

Lenaghan: On the prototype building side, we use Python and scikit-learn, the Python machine learning library, a great deal. A lot of the other guys on the team use R, especially those that come from more of a statistics background, as they are very proficient in R. Then we also have the guys who came from more of the finance side, so they still write a lot of Java.

Gutierrez: Is data munging a big part of your work, and if so, what tools do you use?

Lenaghan: When it comes to munging, it is definitely true, even for me, that 80 percent of the work I do is munging data. When I worked in finance, I learned to do that very quickly and efficiently in Perl. Since I started at PlaceIQ I have not used Perl. Now I do all of the data munging in Python.

Gutierrez: Is data visualization a big part of your work and, if so, what tools do you use?

Lenaghan: Even though I use Python for pretty much everything, I do not use any visualization tools in Python. I know that matplotlib is great and it looks great. It is just that I haven't invested the time so that it just sort of flows out of my fingers. So to visualize data, we use a variety of other tools.

Geospatial visualization is a giant, hairy, terrible problem. We do have our own geospatial visualization program that we use internally, which works well. But for anything else that is not geospatial, I use R and ggplot2. I use R for everything else because it is what I am familiar with, everything looks beautiful, it works very well, and it is extremely functional. I can show it to people on the sales side and they like it. Amusingly, they still take the data, put it into Excel, and make their own plots with it.

Gutierrez: Tell me about a specific project that you have worked on. Take me through the thinking behind the project, how you built it, and what lessons you learned.

Lenaghan: First, let's talk about the location targeting before we cover the project, so we have a base of understanding. Before I came to the PlaceIQ, the geospatial layer had been built out fairly well. Duncan McCall and Steve Milton, as well as the early employees of the company, had very clear and very good ideas about how to tackle geospatial at scale. The big idea was that you wanted to tame the spatial dimension by keying everything in terms of the 100-meter-by-100-meter tiles. No matter what data you have, it had to be attributed to a tile.

Gutierrez: Every kind of data had to be keyed into these tiles?

Lenaghan: Every kind of spatial data. For temporal data, we divide the week up into 26 time periods that are culturally relevant, so that allows us to not have to worry about the clock time. For instance, your Tuesday A.M.

commute is contextually the same as your Thursday A.M. commute, and Sunday lunch is always Sunday lunch.

We also have a very sophisticated ontology/taxonomy that we use internally. All of our data and all of our categories of this data get mapped to this ontology. So this framework that was built out is very sophisticated. It actually makes scaling much easier to do because you are not trying to boil the whole ocean.

Gutierrez: So this is the background to the project.

Lenaghan: Correct: this was our location targeting. The big project I want to talk about, which was important to the company, was what we call our Audience product line. I briefly covered this earlier. The Audience product line is our device targeting offering, as opposed to our location targeting.

When I came here, we started to think, "So we're targeting location, which is great. Location histories are going to be even better." And so this was taking the ad-request logs and joining them with the geospatial layer that had already been built.

Gutierrez: What was the first step in this project?

Lenaghan: We started by writing a query language that allowed us to create profiles and audiences out of the ad-request logs joined with the geospatial data layer. The first Audience we wanted to build was air travelers, which meant we wanted to be able to look at all the location histories of devices that had been observed in an airport. This was actually an enormous project. It started off in fits, and there were a lot of things that did not scale so well.

We started off trying to build an Air Traveler audience by finding points in polygons across the United States. As a first step, we started off by using the polygons of airports. It is a very complicated computational geometry problem to find points in polygons mathematically [point-in-polygon problem]. There are fast ways to do it, but sort of generically. The canned ways you find to do it are extremely slow. This approach just did not scale, it was really slow, and it produced terrible results.

Gutierrez: How did you solve it?

Lenaghan: We solved it by tilizing our polygons. You still capture and map data to these tiles. It's just that—especially for larger polygons, like Walmart and airports and similar giant structure—the error that you have is small once you tilize it. Once you work at the tile level, everything becomes kind of abstract again. You have all these keys, and you are doing large key-value joints. I wrote the first framework to do that work.

Once we had the audience, the next part of the project was figuring out the demographics of that audience. You are able to make particular anonymized inferences about the demographics according to where people happen to be.

However, in-home and out-of-home locations are very different and will give different demographic results. So in order to get this right, we had to build a classifier for "what does it mean for a tile to be residential."

Gutierrez: Sounds deceptively simple. There must have been more than a few stumbling blocks. What were they?

Lenaghan: It does sound really easy. You look at the map and search for a house. Once you see a house, you know the tile is residential, so you are able to get demographic results. However, doing this across the one billion tiles in the United States means that you have to do that programmatically somehow. The power of the classifier comes from being able to designate a tile as residential or nonresidential. So this was an important step to figure out. Unfortunately, there is not a good data set that says, "This particular tile is residential."

Gutierrez: How did you develop the data set to tell you if a tile was residential?

Lenaghan: We used a lot of different data sets, including a lot of ad-request data, and tried a lot of different features to figure out where the residences were. Again, sounds straightforward, but it was not straightforward at all. As an example of why we had to use multiple data sets, the census data does not work because the census data is defined in terms of census blocks, which are enormous. So if you were to just use census data as your residential signal, you would have a residential signal essentially everywhere in the United States.

Gutierrez: Tell me about the classifier you developed.

Lenaghan: The classifier we came up with had about sixteen features that indicated whether or not the tile was residential. We then had to finish building out this very high-quality residential classifier. Once we had that, we could figure out from all these location histories what demographic attributes to give the Air Traveler audience.

Now we have these in-home and out-of-home components of the audience, which give us a base data layer for building any sort of movement profile that we would want. So we can now combine "a device that tends to be in households with this particular demographic" with "a device tends to dwell in coffee shops and has been observed on an auto lot for a particular brand."

Gutierrez: Is this where the query language comes in?

Lenaghan: Yes. Now that we have the data and the classifier, we then have to build up the query language to help us create the types of audiences we wanted. This means the query language has to be able to write these rules and has to be able to hook into the geospatial base data layer to pull out these audiences.

Overall, the project was a number of steps with problems and solutions along the way. The problems tended to get smaller and smaller and smaller as we made progress, but at the end of the day it was still a very large problem to solve. This is a project I am proud of, as I built most of it and this is something we now run on a daily basis—although at first it took a long time to run.

Gutierrez: Did it take a long time to run because it was a prototype?

Lenaghan: It was written in Java because we knew that it would be something that would have to be very performant. So we built the prototype to show that it worked and scaled. Once we showed that it worked even though it took a long time to run, we handed that over to engineering, because there is a lot of configuration that is involved in that as well. At that point, because it was not as performant as it could have been, we had one of our young rock-star engineers make it very fast and efficient.

Gutierrez: How is the data stored?

Lenaghan: This is a very important two-pronged question for PlaceIQ. The first prong and priority is to store the very sensitive location data in a way that maintains as much privacy for people as possible. The last thing we want to do is have a scandal. When we talk about this large join between location history and our geospatial layer, we never actually store the device IDs. Even though the device IDs are already obfuscated and hashed when we use them, we are super careful to never actually store them.

When ingesting data, we get the location and device ID from ad-request logs. However, once we join it against our base data layer, we drop the location. So it is stored in the format of obfuscated device ID, context, and timestamp. So it will be device123/Walmart/Wednesday, December 17, 3 P.M. Note that in this format we do not specify which Walmart it is, just that it is a Walmart. We never store any information about which Walmart it was; so we do not know if the Walmart is a San Francisco-area Walmart, a New York-area Walmart, or a Walmart somewhere else.

We are always very careful with any of our derived data that we never store any type of identifier—device ID, IP address, or similar data—and any sort of raw location data. We keep a very strict information wall between those data sets. So our data is stored as device ID and the context in which the device was, but not exactly where the device was. Our rules are built out specifically so that we only query on context and times.

Gutierrez: And the second prong?

Lenaghan: The second prong is technical in nature because of the size of data we are using. So it important to us to think about how to store, retrieve, and analyze this data. Right now, our entire infrastructure is hosted on Amazon's S3 service. Within a month, we will have moved to a colocation data center facility. The colo will help in storing location data that is very sensitive. Technically, all of the data will be stored in Apache's Hadoop Distributed File System [HDFS].

Gutierrez: As your team expands, what types of people are you looking for and how do you actually know that they are good?

Lenaghan: When we are looking for people, we are looking for very passionate people who are quantitatively minded. Even though we use Hadoop a lot here, being an expert in Hadoop is not a job requirement. We want people who can think logically, scientifically, and quantitatively about problems. We want them to be able to accurately identify what works and does not work. We also want them to know why things do not work, even though they thought they were going to work. Being self-critical is important.

Our interview process consists more of probing to understand how they think rather than, "How would you do this particular graph algorithm in a map-reduce framework?" We are interested more in raw skills than in particular skills for our data science team. Whether we are hiring a junior hire or a senior hire, we are looking for that quantitative piece. We have hired people on the junior level who have very little programming/software engineering experience. They had to learn those skills on the job and now they are writing fantastic code. So hiring based on raw ability rather than specific experience has not been a problem at all. That said, we occasionally need a very specialized person for a very specialized task, but that is the exception to our usual hiring practices.

Gutierrez: Are there any tools not currently in your workflow that you are excited about?

Lenaghan: One of the technologies we are looking at is Julia. One of the projects a particular guy on the data science team is working on is figuring out where we can use Julia in our workflow. Right now, because we are on Amazon, we pay for the compute time. So we definitely want to cut down our compute costs as much as possible. Once we move into the colo, it will be less of a concern, but we still want to cut down our compute times.

We run many processes hundreds of billions of times a month. When you are running algorithms on ad-request logs, even something as simple as converting from a latitude and longitude to a tile makes a big difference in compute times and costs. Making these types of very small changes is important in our work, so we are always looking for more performant numerical techniques. Julia looks very promising in this area, so that is why we have a person working on figuring out how to include it in our workflow.

I would also like to learn more about Clojure. I think the fewer lines of code that you have to write, the better. Just looking at some Clojure projects, it seems very promising to me. Functional programming languages lend themselves very well to things we do a great deal of—such as distributed computing, multithreaded computing, and MapReduce patterns. Being able to do that in the smallest amount of code possible is essential. Our code base is too big. It needs to get smaller. I am very excited about that.

Gutierrez: Outside of programming languages, what other tools or processes are you excited about?

Lenaghan: I am very excited about real-time processing and real-time computation systems like Storm, even though Storm is not exactly nascent. Real-time processing and computation affects us in a few places of our data/product pipeline. It affects us at the beginning of the pipeline where we are ingesting, processing, and analyzing ad-request logs, as well as writing the results to Amazon's S3 service. It also affects us at the end of the pipeline, where we do a lot of batch processing to build audiences and serve ads.

This is especially relevant in the environment in which we serve ads because it is a high-QPS [Queries Per Second], low-latency environment. I would like to move a lot of our batch processing to more real-time, on the back end. So that would mean we can find problems much earlier. We are now moving towards that.

Gutierrez: What does the future look like to you?

Lenaghan: A welcome trend we have seen more of in the last year and a half has been the consolidation of programming libraries and packages. The big push towards further consolidation of—and abstraction away from—packages and libraries is fantastic. It definitely allows more people to do interesting work without having to spend years and years in a PhD program to understand which algorithms you can apply a stochastic gradient descent to and with what convergence.

Along similar lines, people and startups are starting to try to democratize data science and analytics. I am all in favor of this move, as well. While it will make our life easier having these better tools, it will never obviate the need for somebody to make and use these tools. You will always need data scientists, even with these consolidated and democratized tools. Just because people have access to statistical tools like R, Stata, SAS, and others, it does not mean that everybody can all of a sudden run statistical analyses correctly. While you can run the statistical analyses more easily with these tools, you still have to know whether you are running the right thing or even interpreting the analyses correctly.

Real-time is also very exciting to me. As we discussed earlier, a lot of our business is built on the tons and tons of data exhaust of mobile phones and devices, so being able to make it actionable as quickly as possible really is very much the future. We are looking at a lot of interesting technologies that are making that possible. Storm, which I already mentioned, and tools that make it much easier to shard databases, so we can have horizontally scalable databases that we can actually run relational queries against. I am very excited about these types of databases because I think we wasted a lot of time doing relational queries inside Hadoop—and not only wasted time but also accumulated a lot of technical debt. I am done with that. I think the future really is an era of tools that enable more people to do interesting work faster.

Lastly, I also see a many quantitative fields forming a much more symbiotic relationship between industry and academics. Many of these quantitative fields were initially pushed by the heavy lifting of academics first and then later pushed by the work done in the industry side of things. Lately, I get the sense that the tension that there used to be between industry and academics is thawing. For example, industrial physicists and academic physicists are now working in a much more collaborative environment. You actually have conferences that everybody goes to. That is something that you did not used to see in the physics world. I think that entente will continue to spread in many other quantitative fields.

Gutierrez: What is something you see daily that you think other people do not quite get yet?

Lenaghan: When people think about the power of location, what they are really thinking about—and I get this all the time when I talk to people—is that what we do is give you an ad for Starbucks as you walk past a Starbucks. Or that what we do is give you a personalized advertisement as you walk past one of those digital out-of-home billboards at a bus station and it reads some identifier on your phone. I see that as the flying car version of the future. It is what everybody thinks it is going to be, and it is not going to be. The way the future is going to look on the outside is pretty much what the present looks like—just as the '80s looked on the outside more or less like the present. A Boeing 747 from 1968 looks just as it looks today.

But the world is very different. We have these mobile devices and access to information. I think on the location side, where you currently are is important, but where you have been is where a lot of the interesting products are going to come from in the future. So it is not a matter of instantaneously changing a billboard because you just happen to be standing there. It is that—not to sound too *Minority Report*—your life and preferences are going to be contextualized in a much richer way than they are being contextualized just from the web sites that you visit.

Web search is a great indicator. As Google has proven, it is a great indicator of intent. But even greater than that is where you are. So I may be looking for things at the Mayo Clinic because my brother is sick or something like that, but it really does not have anything to do with me. But what does have to do with me is where I have been the past six months. That is much more indicative of my tastes and interests. I think the difference is everybody knows location is going to be important. What I think people do not really see yet is that it is your location history that is important, not necessarily where you are right now.

Gutierrez: What advice do you give to junior people at your company as you and they create our future?

Lenaghan: First and foremost, it is very important to be self-critical: always question your assumptions and be paranoid about your outputs. That is the easy part. In terms of skills that people should have if they really want to succeed in the data science field, it is essential to have good software engineering skills. So even though we may hire people who come in with very little programming experience, we work very hard to instill in them very quickly the importance of engineering, engineering practices, and a lot of good agile programming practices. This is helpful to them and us, as these can all be applied almost one-to-one to data science right now.

If you look at dev ops right now, they have things such as continuous integration, continuous build, automated testing, and test harnesses—all of which map very well from the dev ops world to the data ops (a phrase I stole from Red Monk) world very easily. I think this is a very powerful notion. It is important to have testing frameworks for all of your data, so that if you make a code change, you can go back and test all of your data. Having an engineering mindset is essential to moving with high velocity in the data science world. Reading *Code Complete*[1] and *The Pragmatic Programmer*[2] is going to get you much further than reading machine learning books—although you do, of course, have to read the machine learning books, too.

Gutierrez: So knowing machine learning is the pass to get inside of the door and then, once inside the door, knowing the engineering practices is what sets you apart?

Lenaghan: Yes, in terms of the importance of everyday practice, you cannot underestimate engineering. And a lot of people do. A lot of the people we interview, even very senior people, just run some cleansed data sets that they run some R packages on. To really succeed, having an engineering mindset is important. I would say that having an analytical mindset is the most important, then having good hygienic engineering practices, and then having the tools. Where things get messed up is when you have the skillsets inverted—that is, when you just have tools that you rely on and you basically apply them blindly without good dev ops or engineering practices and without any critical thinking. The consolidation of programming libraries and practices is very good, but the tools and the packaged libraries only serve you if you first have the critical thinking skills and the engineering practices.

---

[1]Steve McConnell, *Code Complete, 2nd Ed.* (Microsoft Press, 2004).
[2]Andrew Hunt and Dave Thomas, *The Pragmatic Programmer* (Addison-Wesley, 1999).

Anna Smith

Rent the Runway

Anna Smith *is an analytics engineer at Rent the Runway, an online and offline fashion company that rents designer dresses and accessories. The company partners with famous designers to ensure every woman can have her Cinderella moment. Rent the Runway rents everything from party dresses, wedding dresses, prom dresses, and designer dresses for special occasions including Halloween, to the handbags and jewelry to accompany the outfit. Analytics is a huge part of Rent the Runway's success, which is predicated on tracking more than 50,000 unique inventory items on the website, in the mail, in cleaning, and warehouse storage, and on recommending to its customers the dress sizes and cuts that will best fit their individual profiles.*

Smith previously worked as a data scientist at Bitly, where she provided data insights to consumers and brands. Bitly lured her from the University of Oregon physics doctorate program, where her field was quantum computing. Her writing appears in publications as diverse as Forbes and Publications of the Astronomical Society of Australia. She speaks regularly at data conferences.

Smith stands out as a spirited example of how a data scientist can contribute to analytics, recommendation, experimentation, and machine learning systems while at the same time caring deeply about the people at her work and in her community. Her social commitment comes through as she talks about the lessons she's learned in being mentored and mentoring others, her advice regarding community building, and her goal of making data science less of an ego-driven field. Smith's passion for the ever-growing data science community and her thoughts about the principles that should guide its development are bright threads running through her interview.

Sebastian Gutierrez: How did you come to be an analytics engineer at Rent the Runway?

Anna Smith: The short story is that I started a PhD in physics, did an internship at Bitly, and liked it so much that I quit grad school, or, in friendlier terms, I'm in absentia indefinitely. The three things that really excited me about this transition were how I was learning so much so quickly, interacting with real data, and trying to understand how people use the Internet. After Bitly I moved to Rent the Runway, where I focus on helping people find the right dress. I think the Rent the Runway product is really great and I find the people around me really inspiring. I couldn't be happier with where I am.

Gutierrez: What type of physics were you studying?

Smith: I went into grad school wanting to do quantum computers as I really liked the theory behind it. I had prior experience in the area from a past internship, so I wanted to continue to work in it. Unfortunately, at my grad school there wasn't enough quantum computer theory, so I ended up doing lots of experiments. This really wasn't my taste, so when there were experiments I elected to do all of the computer work. After all, I didn't want to burn my eyes out with lasers or similar accidents. I did enjoy some parts of the experience, like building an interferometer.

Then an opportunity came up with some professors to do more informatics and data science–type work. They wanted to meld the disciplines of math, computer science, and physics together. Their idea was that each discipline has their own great things going on, and by combining them, they could create the future of applied math. This is more or less what has happened with computer science in the industry. So as part of their goal, the professors were developing these great algorithms and ways of dealing with data. I started by working with some companies and other professors to do a couple of projects that were separate but related to the overall big idea.

Gutierrez: So you went from theory and experiments to data. What types of work did you end up doing?

Smith: In physics, we've had large amounts of data for a long time. A portion of physics is dedicated to trying to learn how to use collected data, how to automate data-related tasks, and how to get information from the data. A great example of this data is optical images. One of the longer-term projects focused on analyzing optical images of galaxies to calculate their surface brightness. We were looking at how bright a galaxy is from its center and how you can build a profile of the galaxy based on the optical data. Different types of galaxies have different profiles, and so we were trying to learn from the data what those profiles were and use that as a classification tool. Somewhat surprisingly, this was not what most astrophysicists at the time were doing. In the past they have used and still use general curves. First, it was one-parameter curves. Then fifty years later it was "let's put two parameters in it!" and then so on and so forth. I found this to be really interesting because of the way we were using data to compliment the work being done by other parts of physics.

Sometime after the galaxy project, I went to China to look at the transaction data of a company that's pretty much a mix of eBay and Amazon. This project focused on how to group all of their products and categories. We worked on doing PCA analysis to collapse the categories to make them more manageable and to be able to get cross-classification. That was a fun project and I feel like I didn't get enough time there.

Gutierrez: How did you go from doing academic data science to data science in the industry?

Smith: Eventually, after doing these two projects and some others, the professors advised me to go out into industry to see what other people were doing in this space. We called around and found Bitly, a link-shortening company that provides data insights to consumers and brands. I had seen talks on machine learning in a grad school computer science class by Hilary Mason, formerly of Bitly and now at Accel Ventures. The professor was like, "You should look at this. It is amazing!" So I was really excited when I started talking to Bitly.

One of the things that really spoke to me was seeing what they were doing on their blog. The blog was really cool because a generous portion of the content was research-oriented, as opposed to "Here's how we support the business." From the talk, their blog, and what they were doing, Bitly seemed like a great fit. So I joined them to do an internship and enjoyed it so much that I left my PhD program to become an employee at Bitly.

Gutierrez: What was the aha! moment where you decided you wanted to pursue working with data instead of physics?

Smith: It was definitely a slow transition, because in undergrad and grad school I didn't want to deal with data. In undergrad I had opportunities to do data modeling and all things that come with data modeling. Instead, I thought that data modeling was stupid because anyone can do that and I let the opportunities pass by me.

As I got older, I realized I liked doing more computer science–related work. I enjoy working with code, which somewhat naturally lends itself to dealing with data. It really hit me upon getting to Bitly, where I finally thought "Oh, this is what I need to be doing." There was a bit of that with my other projects in grad school, but it wasn't as clear.

In grad school I was just one person in a black box basically. I either had my one professor I could go to talk about things or I could go to the Internet. At Bitly I was suddenly surrounded by people who are all very excited about data and the different things you can do with it. Now I was able to practice the machine learning algorithms that I've learned about and see them actually do things. It was really cool and I finally felt like I had found what I wanted to do.

In addition to Bitly's extraordinary data sets, I was given all the resources that I needed. These were resources that I'd only heard about but had never seen. I went from asking what a Hadoop cluster was, as my school didn't have one, to being able to work with one. Now I could play with one to figure out how to make my work better and how to change my algorithms so they were cleaner and ran faster on Hadoop. The whole transition was eye opening. Not only that, I could also look at other people's code and play with their code and data sets as well. It was fantastic and I was convinced that working with data was my thing.

Gutierrez: Was it a big transition from the PhD program to Bitly?

Smith: It was a big transition because I moved from Oregon to New York City. Now there was stuff going on all the time—all these different meetups, all these interesting people working on fun problems, and everyone was very social, outgoing, and very willing to answer my questions. Learning from them was really exciting and fun. Workwise, at Bitly, I felt like I was the dumbest one there, and so it was really great to learn a whole new set of skills, like how to give talks and how to really bring data science to everyone inside and outside of the organization. I was also surrounded by engineers, so I learned all these helpful techniques and really built up my engineering skills. We also had a fabulous data science team, where we could learn from each other and tackle these problems. It was really exciting!

Gutierrez: Is there a specific project at Bitly that you worked on that stands out?

Smith: One of the exciting problems I worked on when I was at Bitly was being developed into a product by the time I left. It was a really simple idea: instead of only collecting the data, we were going to feed it back to our users with a little more to it. Bitly operated on the premise that we tell our users what is going on for the one link or collection of links that they personally shorten with the service. However, we had all of this data about what people are doing in aggregate, so we knew what people are going to your website and what else they are doing on the internet. So this project was focused on trying to expose information about what other sites people were visiting, what topic areas those sites represented, what keywords people were looking for, what was the actual context of those websites, and trying to divide up people based on that information. Not only that, we were going to do this across geographic regions as well as browsers. The project was really exciting and I learned a great deal from all of the trial and error that went into it.

Gutierrez: What did you learn from the project?

Smith: In addition to learning how to apply a few different techniques to the data, I learned how to work with someone else who isn't familiar with data, math, or programming. At first, the project involved a business development person and myself. We had the classic technical communication mishaps where he would ask for things and I would give them to him, and he'd be like "No, no, no. This isn't right." And I'd be like, "I don't understand, that's exactly what you asked for." Or I'd bounce ideas off of him like, "I'm thinking of using this normalization technique, what do you think we should use?" And he wouldn't be able to speak my language. Eventually, we learned to work together, how to communicate with each other, and meet deadlines with non-pretty versions of the product.

Once we had the project moving along, it was easier to understand the grand vision of what we were doing. Then we got a data artist involved and all of a sudden we could really take the data I was making and make it pretty and understandable. Through this process, we were able to solidify what we were trying to accomplish, because before that I was making graphs somewhat haphazardly to show different outcomes. Once we had a more polished project and could see how it was changing and what made it special, we were really able to focus. Then I was able to work on making the process faster and we brought in some other engineers to help me make the process run in a few hours—instead of a day—for all of our customers.

Gutierrez: How did you think about what kind of data and modeling you needed for the project?

Smith: As simple as it sounds, it was a great deal of trial and error. The idea was to first try to understand at a very basic level what people were doing. So as a first step, we had to try to get all of the data that we could into one place. Once we had this data, we could then do very simple summary statistics to get a glimpse of what the data showed.

We started with questions like: How much data do we need? And how much do we have? And then we'd go back to math to understand the statistical confidence we would be able to generate. Once we understood those numbers, we could then figure out how many links we needed to see in what amount of time, so that it would be useful for our customers. For some companies, we could do it by month, and for other companies, we could do it by week. Once we were certain about the data and had confidence in it, we had to choose the models that would generate the best and most intuitive answers to a person looking at the results. There was never one right answer. It was more of a matter of figuring out what we could communicate and with what bias.

Gutierrez: What do you mean by what you can communicate?

Smith: What I mean is that in this project we are doing a lot of communication to people who might be seeing the data for the first time and/or the summary statistics of that data. So we had to focus on what type of things we are communicating that makes sense to the average person. You can explain to them all of the math, but unless they can translate that into what's going on in the graph, they're not going to get it.

Gutierrez: What tools did you use to work with this data?

Smith: I was using Hadoop to store and compute on the data. So it was stored on Amazon S3 and then we ran it through our MapReduce program. We also used Elastic search so that we could have more data processed at once. That was the core of the processing. Then once we had all of those steps done, there was a light Python script that did all of the last-minute data manipulation and pushed it out in a JSON format. The pipeline then loaded this JSON into D3.js for the charts and presentation layer. I like to think it was a clean pipeline, as it was something we worked really hard on. Though, as these things tend to go as you learn more, it could have been better.

Gutierrez: Now you're at Rent the Runway. How does it compare to Bitly?

Smith: It's been really different, as they are two very different types of companies. At Bitly, the company dealt with a more latent data source. People use it and we see things happen. It was more about trying to capture people's behaviors and understand what was going on in the Internet. At Rent the Runway, it's a lot more of trying to support the business, so a lot of pure business intelligence and business analytics. The problem here is trying to figure out how we can put data into the product to drive business goals.

Gutierrez: How do you explain what you do to someone not familiar with computer science, or physics, or data science?

Smith: Like, how do I tell my mom what I'm doing? Well, my mom's a bad choice since she loves computers. Okay, how about—how would I tell my sister? I would approach it as I'm solving problems with anything at my disposal. It's like any job—instead of having court cases to litigate, like my sister, I have problems that I need to solve. I just happen do it with data. Often times that means I need to go to the engineers and ask them for information on what people are doing on the website, and then I need to go to our databases and find the dresses that are being rented, and then I need to combine what I found out into a more refined form. This way, I can expose what's happening in such a way that we can solve the problem we are seeking to understand.

Gutierrez: How would you describe your job to a physicist?

Smith: What I do is like solving any equation. You have inputs, you have outputs, and then there's a black box. You have to figure out the black box. I guess it would be analogous to collapsing a waveform. In physics, there's a probability of where a particle's going to be, and then what happens when you observe it is that it goes to one spot. And so in data science, you have all these different possibilities or all these different arrays of data, and you just want to collapse it into one understandable piece of information that makes sense to the rest of the world.

Gutierrez: What does your day-to-day look like?

Smith: My day-to-day varies by day, as I participate in a great deal of nonre-curring meetings. It's funny—at Rent the Runway I have a lot more meetings than I ever thought I would have, but I actually enjoy them. I like the meetings because they give me time to understand where other people are coming from and what they are trying to solve. Personality-wise, I'm very much a feeler, so I enjoy trying to understand other people's point of view.

The rest of my time is focused on problem solving and thinking. Non–meeting time is great because I can sit down and do whatever I want to do. Thinking is actually something I've been working on putting more time away for on my calendar. My thinking time consists of not looking at things, not reading articles, and not even interacting with a computer. It's very much a time to just kind of sit and work through these problems more, because otherwise it's very easy to get caught up reacting. If I don't have my thinking time, a kind of ADD can kick in, and then I stop exploring things like: What does that actually mean? What happens if I look at it from this other way?

Gutierrez: How is the data team set up at Rent the Runway?

Smith: There are eight of us and we're all domain experts on different things. There's one guy in charge of our recommendations system for the web site. There's another person who deals with the product and making sure we're getting a lot of our logging completed. I'm more on their semi-team, doing the on-the-website stuff. We also have someone focused on marketing and financial reports, as well as the CAO [Chief Analytics Officer] who provides direction for the Analytics team and maintains our visibility with the rest of the company.

Then we have another semi-team who focus on the operations side. So we have one data scientist there who built the predictor for when a dress is going to be late. Then we have another guy who builds out our Tableau reports, and we're making a big effort to try to get that off of him, as he ends up getting inundated with questions all day long. And the last person is our data engi-neer, and he's really into building the framework for how all of this is being processed. I like to work with him a lot, just because I enjoy helping lessen the work of everyone around me. By helping to improve our systems and frameworks, everyone can do more of the fun things because they have less infrastructure to code and overhead to maintain.

Gutierrez: What excites you about Rent the Runway?

Smith: The data, the people in the company, my interest in fashion, and being involved in a web company that has a physical aspect to it. First, I'll talk about the non-data pieces and then we can come back to the data. In the company, in addition to all the great people, I really love all the women. It's really different than other places because most tech companies skew heavily male, especially in their engineering departments. At Rent the Runway, we're about 25 percent women in the engineering team, and we have many women in senior positions. Our CTO is a woman, and she's amazing and truly inspiring. Everyone admires her ability to get to the brass tacks of things and get everyone working. So that's really cool.

Another reason is that I've always been interested in fashion. I actually rented semi-frequently from Rent the Runway before I worked here, so it made the transition from Bitly easy. I've always been curious about the fashion industry, and though it still is somewhat of a mystery for me, at least I can understand how different parts of it work and how they try hard to part me from my money.

Lastly, I think it's very interesting to understand how to run a business that has digital and physical aspects to it. I enjoy learning how different people run their teams. It's very captivating to watch what goes on, because every boss I've had has been very different, and it's fascinating to see what sorts of things they're pushing for and how that ends up happening.

Gutierrez: How does your boss understand and champion data and your work with it?

Smith: I think my boss has done a good job of making sure everyone wants to see the data and has really been pushing that as an agenda. This is great because everyone wants to see the numbers. They all want to see what's going on with their particular department. The difficult part is making sure that we're more like data ambassadors. We know the data better than even they do, and we know how it interacts with the rest of the company. So we don't want them coming to us and saying, "We want X, Y, and Z numbers." We want to go to them and say, "What problem are you actually trying to solve? Let's come up with these metrics together and really figure it out as a team." So my boss and our team have really been trying to make it more of a collaborative effort, as opposed to just handing things off.

Gutierrez: So, much more of a consulting-type team role.

Smith: Exactly. The problem with that is to support a business, you need a lot of data going out, and a lot of that is just, "Here are the numbers. Here is how they're changing week over week." So we constantly have to figure out automated ways to provide these reports, so that we can do the more interesting and fun problems.

Gutierrez: What kinds of data do you see at Rent the Runway?

Smith: As mentioned earlier, data really excites me about Rent the Runway. We have a variety of areas where we deal with data. I separate them into a few different important silos. First, there's the warehouse operations piece. This is related to how the dress actually moves within our warehouse and how we capture that movement. We have every dress barcoded and we have stations where they scan them every time they are moved. A year ago, just from adding barcodes to the dresses, we were able to cut down the time a dress spends in the warehouse to less than a day.

Second, there's the warehouse-to-consumer-and-back piece. We have UPS data that ties a dress to a package. This allows us to know where it is on its outbound and inbound journeys. We use this data to optimize the flow of dresses. This is really different than anything I had dealt with before, because these dresses are physical things that can deactivate and go missing. Suddenly, missing data represents a much more intriguing problem. In this part of the business we've been working on predictive modeling to figure out whether or not a dress will come back in time to be sent out for the next rental.

Third, there's the customer support piece. We focus on surfacing the right data to customer support so they can be more successful and have an easier time doing their jobs. When a client calls us stressed out about an issue with a dress, we want to make sure the right data is available to customer support so that they can solve the customer's problems—whether it's related to where a dress is, why a returned dress isn't in the system, or any other type of issue. We want to minimize the stress of the customer and customer support.

Fourth, there's the web-site operations piece. This is the area I'm mainly involved in. The web-site operations group focuses on questions like: How do people find the dress that they want? What kind of features do they want? How can we make these features even better? It's been great being the first woman on the analytics team, as I have some inside knowledge of how I would go about finding a dress and what the effort entails. This helps us with the data that we have gathered and continue to gather from outside sources and our community.

For the fifth silo, there's the customary data engineering that comes with pretty much all companies: marketing, accounting, logging, website optimization, and all those kinds of things. Data in this silo is very important and fun as well. It's just that these are outside of the scope of what I think is different about the data at Rent the Runway.

Gutierrez: How do you deal with the physical aspects of body measurements?

Smith: Body measurements and fit are very hard problems. There are many ways of trying to deal with fit. It's useful to think about it in terms of the data. The easiest data to work with are the body measurements people give to us—their height, their weight, their bust size, their body shape. And then when they come on the website, they review a dress and tell us if and how the dress fits.

A more informational and harder-to-work-with body measure data set is the review a person leaves after wearing the dress. Often, the customers write a long exposé about the dress and how it fit them. In these reviews, they offer advice to other people about the dress and why it might or might not fit them based on the size of the dress they wore and their body. This type of data is harder to use because we have to parse all that information out with natural language processing to try to expose the relevant details.

Gutierrez: How can you ensure the accuracy of this very personal data?

Smith: A really simple thing we're doing is looking at the size someone says she is and the size of dress she actually wore. Additionally, we have a question that asks, "Is the dress true-to-fit?" Though the question is ambiguous, the answers give a first-order approximation of whether the wearer found the dress size large, small, or somewhere in between.

Another way we help people with body measurement–related data is an older project we have called Our Runway, where we surface people's pictures that they've worn in dresses, and then rank them by how similar their body is to your body. Right now, putting people into buckets does that sorting, and I think we can do a lot better. For instance, we could actually use some type of distance metric between what you say your body is and what it actually is.

In terms of fit, we have someone try on the dresses when we first get them, but it's someone who's a size 0, and so we can't really tell across all sizes whether it runs large or runs small. She's only one body type, so it's hard to scale out to all the other sizes and styles of dresses.

Gutierrez: I imagine that the dresses and designers themselves show some variations that are hard to work with as well?

Smith: Yes, that's a huge issue as well. We've looked into and continue to look at the fashion designer's actual size chart, and that's just really messy to deal with, as oftentimes what they say their measurements are doesn't actually align with their own dresses. So we've looked into measuring the dresses ourselves and trying to see if that sizing gives us better information on the fit.

Then there's the issue of fabrics and how stretchy they are. If fabrics are stretchy, then they are more forgiving and can fit many different people. If the fabric isn't stretchy, then it's less forgiving and fits less people. The different fabrics make sizing an even more complicated problem. So it's a huge dynamical problem that can be frustrating.

Gutierrez: What tools do you use to store data?

Smith: Here we don't have Hadoop. Here we use databases, like HP's Vertica. We store in them the data we just talked about and also the pixel logs. The pixel logs tell us what's going on on the website—like what people are clicking on, their navigation paths, and other website-related things.

Gutierrez: What tools do you use to work with the data?

Smith: When I'm left to my own devices and I don't have to conform to anyone else's stuff, I code in Python. Bitly was an all-Python shop, so that's where I developed my Python capabilities. It was really nice to go in depth with Python, and understand its specialties and how to make it clean and efficient with its special tricks.

Here nobody else really writes in Python except for a few people on the data team. Otherwise, people use whatever they want to on our team, so most of it is SQL, just to get access to the data. And then most people outside of our group crunch the data using Excel. Our recommendation engines are built on R. The whole web site is in Java, so I've been learning a bit of that again since undergrad. Then for special projects, we use more specialized tools such as D3.js. For these projects or ideas, we balance getting results and learning new tools. I've found that to really understand something, I need to physically do it or write about it. Just reading about it doesn't quite cut it. I actually need the muscle memory of working with it.

Gutierrez: What's a recent project you've worked on?

Smith: One of the projects we've been working on lately is combining all of the different types of data that we self-collect. It involves not only combining them but also figuring out an easy, fast, and robust way to replicate the process when we want to add more data by, for instance, combining our pixel logs with Google Analytics—which, somewhat unsurprisingly, is a headache. Validating it isn't always so much fun, because I'm like, "What? It's data. It's right. It's correct"— whether you parse it one way or another. That's just a different view of it. The data is correct, it's just finding the right way to look at it and combine it.

We have a procedure that takes our pixel logs and puts them into HP's Vertica, so then I created another script that pushes all our Google analytics into Vertica. Now we can look at them together through Tableau. It's not really a data science math project as opposed to a data science data cleanup project. It's what you might call a bit more of the data engineering aspect of data science. Once we have all of this data aligned, and everyone's happy, we've been going through it and looking at the numbers. We can then focus on the next steps of what else we can do now that we have these resources. We also spend time thinking about what other data sets we can combine into this big data set to make it even more valuable.

The main reason we started this project is that we just launched a mobile app, so we've been working on understanding how our users use mobile devices, who they are, and how they compare to our web audience. We are looking to better understand questions like: How is the audience being distributed across all of the devices? What does that mean as far as how they're interacting with the web site? What are they doing on the app? Why aren't they buying on the app but buying on the web site? Is it just a mental thing or is it a functionality thing?

Through looking at the data and seeking to understand these questions, we have developed a better understanding of the different groupings of people. We have women who come to the web site often and never rent, and then we have other women who rent every weekend, as well as some groups who display behavior in between. So it's very helpful in understanding each group's dynamics. By understanding their behavior, we can tailor their experience to the right platform at the right time.

Gutierrez: What have you found to be an interesting behavior that people display on different platforms?

Smith: As a data scientist, there's two responses. On one level, I really care about the data, so I just want to see what they're doing, and I'm less concerned if they buy. Obviously, on the second level, I do care if they buy and get to rent their dream dress, so I find it really interesting how the data describes the decisions and how they get made. For instance, I think it's interesting that, no matter what we show our users, they're going to find their dress. They have their own way of getting to whatever they're renting.

I still don't quite understand the thought process. We can change one page and do tests, but it's not going to really do anything in an aggregate way. So I think that's really interesting. My assumption is that they're just doing a lot of filtering or using other pages than the main ones to get to their dress. There's not even much of a conversion difference between whether they land on a particular page that we think is most important or any other page. So I think that it's really neat to be able to look at the data in an aggregate form from disparate sources and work on understanding the behavior.

Gutierrez: What specifically excites you about understanding the behavior?

Smith: One of the things I was—and am still—very excited about when I came here was the idea of how to represent someone's taste in dresses. After all, the physicality of the dress involves so many individual tastes in fabric, color, shape, fit, style, and other attributes. If we are able to represent some-one's taste, just think of how cool it would be to understand. I think we're just getting to the point where we can expose that really easily.

We're not quite at the point where we can break it down by color or any discernible physical attribute of the dress. However, we have found that there's some type of latent variable or variables that is able to represent someone's taste in a dress. We just don't know what those variables are explicitly yet. At this point, we can't just say, "Oh, it's because they're pink and flirty-looking with ruffles." That's a bit farther away from where we are because what does "flirty-looking" or "trendy" actually mean, and how do we go about categoriz-ing them in a form that is stable over time.

Gutierrez: Sounds like you have the math behind it and are figuring out how to map it to real-world physical attributes.

Smith: Correct. We have to be able to express what these variables represent in communication with others. It's all about how we would be able to communicate users' taste variables so that the rest of the company would understand that this is actually a good, accurate representation of someone's taste.

On a personal note, I'm really excited by the idea of finding more women like me—almost like my body and taste doppelgängers. My new idea is that we should be able to provide that because, once we can do that, it's nice to know that there are other women out there who face the same issues with finding the right dress.

Because the community is so nice and supportive, some of our customers are very comfortable and expose personal details in their reviews. Things like "the dress fit me oddly because I have implants" are helpful to the set of customers that face the same issues. So being able to surface the right reviews for customers to read that are based on people with similar bodies is a great thing. I dream of customers being able to say, "Okay, I trust their input. They say it's not too short, so it will be okay." It's the whole collaborative system getting to the right thing, being able to share with other people, and being really open without any judgments.

Gutierrez: Given the collaborative supportive community, is there a social network in the product that is waiting to be teased out?

Smith: That's one of the things other people have alluded to when I've given talks. We haven't set it up that way, but I think that it could be a next step if we wanted to make it even more interactive. We just launched Shortlist, which is where you can put together a collection of dresses that you like that have a theme. You can then talk about the collection with others, and this provides a little more back and forth between people.

That said, even now, with just the reviews that people leave after wearing a dress, our community of women are doing that. You can write a review, and then someone can like it. Then you can go back and comment on it and go, "Oh, that looked amazing. You look fabulous." And they can respond back, which is great since it's disconnected from any known social network so people don't actually know who these people are. All they know of them is just the picture they've seen of them. So it's nice how friendly everyone is to each other while being anonymous.

Gutierrez: As opposed to other web companies, there's a very personal feedback mechanism that a lot of other data scientists aren't going to get.

Smith: Exactly. Since I've worked here, though, I've become more of a power user, so sometimes it's hard to imagine how the product and how we've evolved is affecting people out there. We are constantly asking ourselves questions like: Do they notice when we launch anything? Do they notice when we take away some functionality? Luckily, we have friends and fans that talk to us and are very vocal about their thoughts. While sometimes painful, it's great to hear feedback like, "That was crap. You guys need to put that back up there." But, overall, it's really easy to get excited about something when you like the service or you enjoy it. And I like the idea that it's very empowering to women. It's all about democratizing fashion and making it more accessible. It's not about celebrating the material part of it, but it's about you wanting to look good and have a great night. And everyone should be able to do that.

Gutierrez: How do you pick projects to work on?

Smith: Interest and ability to persuade others that it's a good project. A great deal of my work here has been in support for other people's projects. For instance, one thing I've worked on is research into the recommendations system. They built the recommendation system and it's been running. Now I am doing the research into how it's actually working and if it's actually working.

Many of the projects end up being formulated this way. I think of an idea or a different hypothesis or assumption than what we are currently doing, and I go and test it. Then I present the data and we discuss the findings. From there we can figure out where to go next. I've always been told that it's better to ask for forgiveness than for permission. So I think that's a lot of what data science is about. We need to have the freedom to explore on our own. I don't know if it's 80 percent free time, but maybe 40 percent, definitely. We need to have monkey time to get involved in something, get really excited about it, and then still make sure you get your other work done and deliver on time.

Gutierrez: How do you measure your success?

Smith: I tend to measure success at work on project-based metrics. Even though that sounds straightforward, I think that's very hard because it's not just whether you finish something. It's also about whether all along the way you were making sure that what was coming out was representative of what was going in. It's also about whether you did the project well and whether you were able to finish something that you could communicate to other people.

The success metric here is much more about convergence. The questions we measure ourselves against are much more in line with the business. Which leads to questions like: Did we get more people by buying or renting? Did we get more people coming through and using it? Did they actually use it? Were they having a positive experience? Those are more of the metrics that we look at here, which is very different from the ways I'm used to thinking about it, where I'm like, "Look! It's awesome!"

Gutierrez: Do you get enough mathematics in your day-to-day, and if so, do you still enjoy it?

Smith: So this is where I need change how I use my thinking time because I feel like I do more latent math in my coding. I've actually been thinking about this lately and came to the conclusion that I don't actually write down equations as much as I used to. At Bitly, I had people I could work through the math with and we could work through a problem that way. They would sit down and ask, "What does this equation look like?" They could figure out if what we were trying to solve made sense from finding the math together. And so in that case, collaboratively, you need that math to talk about the same thing.

Lately, what I've been trying to do is find more people that I can do that with, because otherwise, if you're on your own, you're not going to sit down and do a whole math exercise. I mean, I know people that do, but I'm not going to sit down and do it by myself. It's nice to work with others. So I still enjoy it and think I would enjoy it even more if I had more time to think about it and work through some math with others.

Gutierrez: What are your thoughts on hiring good data scientists?

Smith: I haven't done much hiring myself. That said, I think it's the same rubric as finding good information on the Internet. It's all about common sense, like how did they approach a problem? What ways do they think about it? Do they try to approach it from many different ways? Do they get stuck on something and then give up? I think data science is a learnable skill, and it's not something you necessarily need to come with. It's just something you need to develop.

Gutierrez: What questions were you asked when you were being interviewed at Bitly and here?

Smith: For Bitly, because I was going in as an intern, it was more about what have you done and can we see a sample of your code just to make sure you actually know what you're doing? At Rent the Runway, it was much harder. I had interviews with both the analytics team and with the engineers.

The engineers had more traditional tech questions like, "Here's a sentence. Try to reverse it." Or "Here's a binary tree, how do you traverse it, and how do you make sure it's balanced?" For the data science interviews, it was centered much more around real-world case studies. They said, "Here's a problem we have. How would you approach it?" Then they asked questions like, "What kind of algorithms would you throw at it? How would you calculate the success metric? How do you know when you've won? What do you do if you don't think you've won?" After that, we talked through the problems they were facing here at Rent the Runway and how they've approached them. For these case studies, I was then asked what I think would be ways to expand on that and where their assumptions might be wrong.

Gutierrez: You mentioned that data science was a learnable skill that people could develop into. How did you do it?

Smith: Doing the work, conferences, learning from people on Twitter and other social networks, as well as reading online tutorials. I would say the best way I learn is by actually doing things. I get inspired to do something and then I do it. I see what happens and then either iterate on it or do something else. I think it all starts with the inspiration stuff, which is what I get from meetings, from sitting in, hearing people and their problems, and kind of understanding, "Oh, what could I do to fix that? What kind of sources could I combine? How can we fix this?"

In regards to conferences, I went to maybe one or two for physics in grad school and didn't really enjoy them. Everyone was doing something so different that it really couldn't be applied to any other project. In data science, the opposite is true. You can go to a conference and get excited by what everyone's doing because you can almost always apply what you learn to your own work. Whether it's a technique that you wonder if you can use with your data set or an idea of something else that allows you to collaborate with that person. Going to sessions at conferences almost always helps me improve my work or the work of someone else, which is great because it leads to an incredibly collaborative environment. It means that you're not so in depth into one physical problem that you can't really get out of that hole. Right now, the field is still at a phase where almost everyone is still interdisciplinary. We're still well-rounded and can take applications from anywhere and still have them work. So conferences are a great treat.

The only downside to a conference is that you come back with a mile-long list of things to try, and you eventually realize that you can't try every single thing. You get a high from the people doing great things that are applicable to your work, so it's tough to come back and have to prioritize what to try out. Also, when you are at a conference, you aren't working, so there's always that to think about before deciding to attend as many conferences as you would like to attend.

In regards to Twitter, I am constantly scouting for new people to follow to add to the collection of the quality people I already follow. I love to read ideas and ask myself if they have insights that I can use in my own work. After that, it's Google and Stack Overflow. I do this because it's very different from reading a book. Most of the books I'm used to reading are textbooks, which lead to the thought pattern of, "Oh, okay, yeah. I totally get that." And then you're like, "Wait. How does my data relate to these abstractions and those equations?" And so I think it's helpful when you find tutorials on the Internet that can show you, "Okay, this is my data set. This is how I relate it to these equations, and this is how it looks in code."

Gutierrez: How do you develop a taste for good work, good people, and good tutorials?

Smith: It comes from experience and having a high bullshit radar. I find that it's very easy to tell the difference between people that actually are doing work and those that aren't. Even if it is something simple, it is usually very easy to tell. I think those are the best examples of what people are doing that really inspire me. You can also almost always disambiguate between people who are just using a whole bunch of Twitter hashtags and just trying to push their own agenda. It's a lot of experience and looking at what they're doing—understanding if it's feasible or what assumptions they're making, seeing if those are correct or not.

When I first got into data science, there was the whole problem of how much domain experience you need to have. I think you need some, and the more intuition you have, the better. Just common sense and a little thought is usually a good barrier for what's good and what's bad.

Gutierrez: If you were starting out as a new data scientist today, what would be helpful for you to really understand?

Smith: If someone is just starting out in data science, the most important thing to understand is that it's okay to ask people questions. I also think humility is very important. You've got to make sure that you're not tied up in what you're doing. You can always make changes and start over. Being able to scrap code, I think, is really hard when you're starting out, but the most important thing is to just do something.

Even if you don't have a job in data science, you can still explore data sets in your downtime and can come up with questions to ask the data. In my personal time, I've played around with Reddit data. I asked myself, "What can I explore about Reddit with the tools that I have or don't have?" This is great because once you've started, you can see how other people have approached the same problem. Just use your gut and start reading other people's articles and be like, "I can use this technique in my approach." Start out very slowly and move slowly. I tried reading a lot when I started, but I think that's not as helpful until you've actually played around with code and with data to understand how it actually works, how it moves. When people present it in books, it's all nice and pretty. In real life, it's really not.

I think trying a lot of different things is also very important. I don't think I'd ever thought that I would be here. I also have no idea where I'll be in five years. But maybe that's how I learn, by doing a bit of everything across many different disciplines to try to understand what fits me best.

Gutierrez: What projects have you worked on outside of work?

Smith: I have a lot of projects I've done and even more projects I want to do. As I said before, I've played around with different parts of Reddit's data. I did a project around Reddit headlines to see if we could find any patterns for how something gets upvoted. Other times, it's small projects like one that was just exposing images from a list of images on another website. I also spend time updating my website, trying out new HTML, CSS, and other web technologies. That said, sometimes it is tough to get motivated—after all, weekends are usually time for sleep, running, or friends.

One interesting aspect of my enjoying my work as much as I do is that I end up doing work projects outside of work. Not like overtime or projects that I have to do at home, more like I am interested in the data and want to explore it further on my own. This is my time to play with my more outrageous ideas. Sometimes it's building a mini-website to better understand how dresses go on sale and how to best expose them online. Other times, it's more data and math-heavy work.

Gutierrez: What is the future of data science at Rent the Runway?

Smith: Well, I have great things I've planned for data science here. Of course, we'll build better models, have better data, and have better processes and more insight. However, I think what will really help us to be successful is to tackle getting communication better between teams. Working with engineering is sometimes an issue because we both end up defending our turf.

To make it a much easier space to be successful, we have to make the environment much more collaborative. Right now, I think they feel like all they do are services for us because we can't do them. In return, I think we can be much more open about them helping us do some of the fun projects we get to explore. This then ensures that we both get help and learn from each other. For us, learning how to really build and maintain systems that are robust and won't fail every day would be great.

Beyond that, I'd like to start working on an internal data science blog or web site so that we can publish things we're working on that we find cool and interesting. Right now, the knowledge and insights we generate appear to be thrown out here and there somewhat haphazardly. Instead, it would be great to have a collection of them so people can be like, "What are they working on? Oh, that's awesome. I would love to come help you guys and let's work on that together."

Gutierrez: Sounds like you want to make data science part of more people's lives, even if they aren't specifically data scientists?

Smith: Yes. I know a lot of the people in engineering, and they are intrigued about the whole new data science thing. They've told me that they're interested in what we're doing and that they have some ideas. So we need to get them to share their ideas and not feel like they have to just fix bugs all the time. Then if there's some project that they see that we're doing that they're interested in, they can come join us. We can delegate responsibilities and share the work and fun with them.

Gutierrez: How do you teach and communicate with people what you and the data team are doing?

Smith: My strategy is to become friends with everyone so we can talk about the problems we are tackling in our respective teams. That way I can tell them what I'm working on, and how it helps them and what my team is thinking about, and then ask them, "What are you guys doing? What do you think of what we're doing? What do you think we should be doing? Do you even know what we're doing?"

Additionally, our group has biweekly or once-monthly meetings where it's open to anyone in the company to come and see what we've done. It's a forum where we present the latest findings that we've looked at, what areas we're pursuing, and they can give their feedback and opinions, as well as ask questions. So I think it's all about being open and being very transparent about what we're working on.

Gutierrez: How do you deal with the dichotomy of data being very proprietary and controlled, while data science techniques are being, for the most part, very widely shared?

Smith: That's something I've been working on regularly, because at Bitly I was given a lot of free rein to give talks, and here, nobody's really done that, at least not about data. So there's a lot of guarding of the data. There are written and unwritten rules about what you can and can't say about any of the numbers. You can talk about how we're doing things, but you can't say what kind of success we're having, or you can't say how much of a problem this actually is for us. I think that will lighten up, and I think it's just getting trust from other people in the organization.

The data guarding is really interesting, not only internally but at other companies as well. I feel like it's hard to get people to separate from their babies. I think a lot of that is just letting them be able to express that they're upset about it, and then being like, "It's okay. We'll make sure that everything's fine." Being very patient with people goes a very long way. One of our company's core values is that happiness and positivity are a choice, and that's a very good way to think about it. You have to be open to new ideas.

I've been reading *Team Geek*, and I really resonate with that book since it's all about engineering and how to work together well.[1] Though it's software developer–focused, I think it can be applied to everything. They have this acronym, HRT—"heart," which stands for humility, respect, and trust. I think those are things you really need to cultivate, not just within your team, but also within everyone else in the rest of the greater business. Then, once you have those, you can go and stretch a little further.

---

[1] Brian W. Fitzpatrick and Ben Collins-Sussman, *Team Geek* (O'Reilly, 2013).

Gutierrez: What philosophies have you developed through your use of data?

Smith: Something I've learned through data and moving to New York is that you're always going to have problems with people, so being patient with other people and really respecting other people is very important in succeeding. You need to have a lot less ego about yourself to really succeed. You can be very pushy and aggressive about stuff, and you'll get a ways down the road. However, I think to really succeed, you need to be authentic about who you are and very true to yourself, because otherwise, it's a charade that you have to keep up. At the end of the day, you're not going to like the charade and it's just going to make you cranky.

In New York, it always seems like everyone's mean and angry, but really, deep down, everyone just wants to help you. It's exciting and surprising how helpful people are and can be if you ask them to help. So I think that's something you need to understand. They may be angry up front, but it's not because of you—it's them reacting to something else. If it is due to you, you should just try to be as nice as you can, and it's okay to say you're sorry. I know because I've had periods of stunted productivity. It's not fun for anyone, but being able to recover from that and say, "I'm sorry I went through this. Can we just move on?" is great.

Lastly, work on positivity and patience. Patience is something I'm working on right now. Things aren't going to change right away, but you can work on things, even if they move very slowly. Being able to have that patience and deal with people—and all their different emotions and reactions—leads to a very stable work and nonwork life.

Gutierrez: Going forward, what would you like to see happen with data science?

Smith: I'd like to see data science become less of an ego-driven field. I'm really excited for this to happen. I feel like we're past the rising spike of excitement around it and I'm really excited about it evolving into becoming more practical and approachable. It's not just a hype thing anymore. It's like when quants first came to Wall Street. All the physics people wanted to be quants, whereas now it's kind of de rigueur to find physicists on Wall Street. I like the idea of every company having a data team that's interested in doing research on and for their business, rather than just pushing out key business metric reports.

As this happens, I think much more cooperation and good feelings will permeate the industry. Right now, quants and statisticians are feeling somewhat sidelined right now. I've chatted with statisticians who feel like they have to rebrand themselves as data scientists. This habit should naturally die out as the industry matures and lead to a more cohesive community with less cliquey behavior. This is great because the goal should be less to have a specific title like "analyst," "data engineer," "engineer," "data miner," or "data scientist," and more to just do great things with data that helps the company or project succeed. It's helpful to specialize in different things, but it's even more helpful to make sure everyone is getting better and getting the help they need.

In New York, we have a really good data community, and I would love to see similar communities sprout all over the world. Two years ago, the New York community was really small, and now it's grown a lot. With that growth there's been all kinds of launches of different programs and more conferences, which I think is great and really builds a more supportive community with all of this sharing what we're doing. At the end of the day, we will all be more successful as long as we all work on being happy and focusing on doing great work with data.

André
Karpištšenko

Planet OS

André Karpištšenko is Co-Founder and Research Lead at Planet OS, a data discovery engine for sensor and machine data. Planet OS's mission is to index the physical world using data coming from robotic devices located in oceans, land, air, and space. Through public and private programs, more than a million satellites, radars, and mobile and fixed sensor platforms are capturing data in a variety of formats, conventions, and time scales. Planet OS finds the data sources, connects them, and cleans the data to enable access, analysis via its cloud-based solution, yielding real-time insight in a variety of settings: government agencies, energy companies, offshore and maritime industries, weather forecasting, remote sensing, and unmanned vehicle operations. Building and maintaining this platform to connect, maintain, and make the data usable to a variety of industries poses significant challenges given the variety and size of data involved, the public/private ownership behind the data generating sensors, and the required streamlining of different standards, protocols, and data models.

Karpištšenko previously founded and led the Data Analytics Research Team at Skype. While at Skype, he also led the Engineering Infrastructure group as well as the Client Quality team. Before that, he was a co-founder of ASA Quality Services, a software quality assurance company, which he bootstrapped to profitability. Karpištšenko started out his career developing software for organizations such as the European Environmental Agency. He has many publications and patents in the fields of data processing and analysis and software process improvement and collaboration. Karpištšenko holds an MSc in Management of Production from Chalmers University of Technology, Sweden, and a BSc in Informatics (cum laude) from Tallinn University of Technology, Estonia.

Karpištšenko's career as a researcher, founder, and manager of data teams exemplifies what it means to build data-driven software, services, and products. Manifest from his early days working for the public sector to his present enterprise focused on solving problems in complex systems embedded in the physical world, Karpištšenko's dedication to building solutions, nurturing teams, making the environment better for all of humanity, and keeping values as part of the decision-making process illumine his interview.

Sebastian Gutierrez: Tell me about where you work.

André Karpištšenko: Along with two co-founders, I am building a company called Planet OS. We bring together all of the available ocean and atmosphere data to help businesses in shipping, oil, and gas, as well as federal organizations, make better decisions—investment decisions, daily operation decisions, and so on. So we're a company focused on the dynamic environment and the data the environment generates. Right now the focus is very much on the ocean, but the opportunities are much wider. Weather also affects many land-based companies. Agriculture is an example of a business highly affected by the weather. Right now we're in our third year, so it's very exciting.

Gutierrez: Did you have a background in oceanography?

Karpištšenko: I did not have a background in oceanography when we started, but I did have a background in data. The story of how I ended up here was that I founded the data research team at Skype. At Skype we were looking at how the Internet worked and how we could optimize it from a Skype perspective. Our overarching goal was to help people communicate—figuring out how we could best help them find each other and talk to each other in a meaningful manner. So we looked at data to understand things like how to fit more video calls into the limited network conditions that were available at the time. I was deeply involved with data on a daily basis.

One day I met Rainer Sternfeld, founder of the company, and we shared what we had both been doing with data. He told me a story about a buoy they had built. The customer for whom they built the buoy wanted to get insight out of that data, but, unfortunately, it took the customer three months to get usable data. For me, as I worked daily with live streams of content in the Internet, it was obvious that this was a great opportunity for technology transfer that could transform how the ocean industry worked. To me, this was another stream of data to optimize, so I actually didn't even see a physical buoy until after we had built a prototype of our service. It was only when we went to present our product in London that I saw the devices involved. Until that point, I had just been working with it just as I worked with any other data set.

Gutierrez: What did you focus on in the first year, and what have you been focusing on since then?

Karpištšenko: My focus in the company is the engineering, data, and technology side of it. In the first year we focused on making sure that we were doing the right thing before we built the product. We started by aggregating public data sets and building a community around the product. Today we have about 7,400 members in our community, and to date we have aggregated data for 33 organizations, including NOAA [National Oceanic and Atmospheric Administration] and NASA [National Aeronautics and Space Administration]. In building the community and collecting data for organizations, we learned how professionals—oceanographers, ocean experts, GIS [Geographic Information System] experts, analysts, and so on—need the data to be served to them, how they want to connect their models, how they want to connect their devices, in which way it's best to integrate with them, and how best to visualize the data. By working on public data sets, we were able to quickly iterate our platform, add APIs [Application Programming Interfaces], add visualizations, and compose different data distribution functions. This work was very much the focus in the first year.

With that knowledge, we then identified different markets, like oil, gas, shipping, insurance, federal governments, defense, and security that we knew would be interested in our data. From there, we started to focus on commercializing what we had learned and built. So the second year was about building a commercially viable product, which is now available. This has very much meant that we've had to pay extra attention to the aspects of data privacy and trust. We've spent a considerable effort on metadata handling—making sure that the data streams we provide to analysts, or data scientists, and data engineers have sufficient metadata for building trust in the data.

To provide this trust and data privacy, we've looked into different sensors and standards for handling sensor data. We also hired a US cyberinfrastructure expert, who had been focused on ocean infrastructure in the US. With him and other professionals, we built a model that now supports both private and public data streams. We finalized it in a product that allows you to manage, analyze, and distribute ocean data.

Naturally we also had to do a lot of work to add the most important data types: time-series forms of data for models, satellites, high-frequency radars, as well as data from National Data Buoy Center profilers and ADCP [Acoustic Doppler Current Profiler], which is essentially sonar data measuring ocean currents. Adding those data types allowed us to create a platform that's meaningful and useful for ocean professionals. It used to be that everyone integrated this data manually on their own in isolation. Now they have a system that actually does it for them. The intent is to continue to grow the product in a way where hardware vendors and oceanographers can contribute to the platform, as well as add more data themselves.

Gutierrez: Why is the company exciting to you and your team?

Karpištšenko: Right now we are a small tightly-knit team of twelve people and are looking to grow quite intensely from here. The main interest for people who have joined the team so far is the possibility to do something meaningful instead of working on advertising or yet another web or mobile app. They can truly impact how humanity interacts with the environment and the oceans. They can start solving the problems everyone thinks about in their spare time. So that has been a great motivation.

For some, of course, there is excitement in the tremendous technical challenges necessary to make such a big data platform happen. We have to figure out how to integrate different sensors and machine-generated data streams in a trusted manner. We also have to make sure those data streams are accessible interactively. And, of course, if you work at Planet OS for a while, you start to learn about the planet more than you ever knew. It's an opportunity to broaden your horizon by not only focusing on human activities and the impact we have, but also helping to reduce the risk we create in disregarding the ocean and the environment. Essentially, we want to build a sustainable future.

Gutierrez: When did you realize you wanted to work with data?

Karpištšenko: As usual, it was a progression. I've been in the software business for 14 years professionally. In the early Internet days, I did a lot of service development work. I used to work for a European environmental agency that worked with various governmental organizations to develop systems to handle information. Another example is a public transport ticketing system. As I built these systems, I started to see some repetition in how services and products were created.

This awareness helped to shift my interests very much toward the border where the real world meets the software world—software embedded in devices. As I worked on this border, I started to understand that there's so much complexity and dynamics in the environment that it's very hard to reflect reality well in simple software models. From this, I got very excited about complex systems and sensor networks. I even did a bit of academic work in that direction. Eventually, that progressed into some interesting projects with FuturICT—a program that brings together complex-system analysts, scientists, and professionals to build economic and social models.

As you work with complex systems, be it in the embedded world or with systems like Skype, you start noticing emergent behaviors. You start to notice that the software design, architecture, and initial assumptions you put into a system can live a very long life and have unexpected consequences and behaviors years later as the system scales. To understand what was happening and why it was happening, I was motivated to start using model-driven development. Model-driven development allows you to you describe a system at a higher abstraction level and then later have the compiler generate code into byte code or assembler code. To do that, however, you need to have some type of feedback loop. This feedback loop has to rely on data.

My progression to realizing that I really wanted to work with data was that I was working on many different software products and services and, to make better decisions, I had to look into the feedback, the logs, and measurements of what the system was doing—basically all of the data. As I looked into the data, I found that I had to start to do it in a smart way, since there was so much data. Obviously, I ended up learning statistics, machine learning methods, and different data mining methods. I also started thinking about how to make services and products intelligent enough to automatically use this information and how to transform the organization so it can apply faster the knowledge generated from these systems.

At Skype, that is what I did. I improved the engineering and calibration tools so that bugs got fixed faster. I made it so that the roadmap priorities were better aligned across the organization, so that the tools used for engineering and calibration were in tune with what was required of them in the wild. As I did this, there was a natural progression from development to data, since there are so many insights you can achieve and so many decisions that you can make if you use the all of the information available. Without it, it's just your intuition guiding you, which doesn't work as well.

This is an interesting point actually. Some say that intuition is actually a better thing to use to be able to develop something revolutionary or disruptive when trying to come up with something new, rather than looking into the past, into a mirror, or into the data. Quite often, you don't come up with something new when you look to the past. You just do incremental improvements, optimizations, and make something more robust. So this is a big challenge and question for me: how much to look into the data and prior knowledge versus just creating something on my own. So now, through this evolution, I get to live in this interesting place with my feet in two different communities: software development and data science communities.

Gutierrez: Do you remember the first data set you worked with?

Karpištšenko: There have been so many that it's hard to remember. I mean, my computers are full of different data sets. The earliest ones were the easiest ones, which, of course, I analyzed in Excel, as they just had some qualitative labels for me to use. Rather than the first one, I'll talk about the one that I think is the most meaningful early work I did for myself.

In 2008 or so, when I was looking into all my communication patterns in email and instant messaging, I was annoyed by the fact that I had so many contacts, close to 700 people, trying to connect to me. At times, I couldn't even remember who they all were or what they wanted, which meant I had a long list of unread emails and unanswered instant messages. I had to decide what was relevant and what was not. In many cases, I had just been CC'ed and I wasn't supposed to take any action. In other cases, however, I was supposed to take action.

To try to get my communications under control, I started looking at those data sets by integrating them, visualizing them, and analyzing them. Some of the analysis methods included text mining and network analytics. Then, through that process, a friend and I created a productivity tool which helped me to understand the important people in my current workflow and the important contexts from past conversations I'd had with them, as opposed to conversations in which I was just CC'ed. This made it so that when I talked with someone, there would already be something that helped me to understand what we had talked about before in relation to what we were talking about now, the people the person knew, and the projects they were involved in. This tool enabled me to have more productive and efficient conversations with people.

Conversations that happen in machines are different from the ones that happen in the physical world. In the physical world, it lasts a long time and we are able to use a lot of cues other than just text or audio. In computers, interactions are usually very short and many times there are many more people involved. I don't think our brains or behaviors have been adapted to this type of interaction, so I think we need to build a bridge so that there is less context switching, less noise, and more signal.

Gutierrez: What are the main types of problems being tackled in the environment data industry?

Karpištšenko: The world's oceans directly or indirectly affect about \$10 trillion of yearly global economic activity, so oceans have a significant impact on our lives. About 40 percent of the world's population lives within 150 kilometers of a coast. Most people don't pay much direct notice to oceans, and how climate change affects us to great extent. If we look at the air we breathe, half of it comes from living organisms in the oceans. What's going on there affects you and me indirectly each day. So in a way, the problems being tackled are ones that affect all of us.

The largest businesses that look at this data are in the shipping, oil, and gas industries. Of course, nowadays, newer industries like renewables, coastal planning, insurance, and risk management are becoming more important and relevant. The types of short-term and long-term problems these businesses tackle are investment decisions, operation decisions, and decisions that affect individuals like you or me.

For shipping, the problems being tackled with this data are ship configuration, routing of ships, and location of fleets. For instance, let's say you own ships going from one place to another. You need to know which route to take so you can benefit from ocean currents. Or you need to know which areas to avoid since there is an extreme event or a storm. All of these decisions are based on ocean and atmospheric data.

For oil and gas, you want to figure out things like the best locations for your oil platforms. Or, if your platform is already operational, you want to know what the weather conditions are going to be for decisions like: Should I stop my drilling and, if so, when? When is it too dangerous for the people involved? Or when it is too dangerous for the infrastructure involved, so that no lives are lost and expensive hardware stays intact?

For insurance companies, you want to assign the right premium for policies, so you use the data you have about the environment to make smarter investment decisions and smarter evaluation of the risks involved. The relevant data can be anything from climatological studies, which are long-term analyses of what's going to happen, to data collected in the past few weeks, months, or years. Based on that data, companies then try to make the best decisions possible.

Lastly, of course, are problems that, when solved, benefit people like us. For instance, let's say that we are going diving or sailing somewhere, or choosing a vacation spot. For that we'd like to know when to go, where to go, and if we'll be close to the sea or the ocean, it's great to know what the underwater visibility will be. And if we want to sail somewhere, it's great to know the best winds to take you from one place to another. An offshoot of this area of problem-solving is that this data also helps yachting competitions.

Overall, these are largely the most important two cases: long-term investment decisions and daily operational decisions. For the most part, oceanographers usually have a background in physics, so they use their numerical models to create simulations of the dynamic ocean environment with this data. This involves a lot of statistical analysis of past events in order to predict future events. They use highly advanced technology and methods these days, so there are many, many use cases you can think about.

Gutierrez: When you started, what books, publications, blogs, or conferences did you attend to learn more about this industry?

Karpištšenko: The first thing we did was go to the top conferences and reach out to the top organizations. We visited NOAA, we interacted with NASA, and we went to different universities, such as Cornell and Rutgers, who are now partners with Planet OS. We looked into what they were doing and quite often engaged in dialog with their experts. We've done a similar thing with customers: we've interacted, over the course of two years, with close to 100 different organizations. We make sure to talk with people at executive and board levels, as well as those people who are actually on the sea. We've worked really hard to understand how their decisions depend on the data they collect or the data they have, what their decision flows are, and what their workflows are so that we can make sure our product and platform will be much more efficient than the way they do it today.

In regards to conferences, papers, and blogs, there are so many! These days, if you build a community around yourself, the news and people start to find you. In 2013 we did two challenges—one was with Kaggle and the other one was our own self-hosted challenge. With Kaggle, we created a challenge called "Create an algorithm to detect North Atlantic right whale calls from audio recordings, to prevent collisions with shipping traffic." The Kaggle data science community improved the state of the art in existing bioacoustics models in just two weeks, a significant achievement. Thanks to this, whale voices in an economically and environmentally valuable area are now detected at a much more accurate level. This has inspired oceanographers to actually change the methods they use to analyze audio. Nowadays, there are machine learning–focused tracks in bioacoustics. For our own self-hosted challenge, the Marinexplore [Planet OS's previous name] Earth Day Data Challenge, we invited people to work with data we had and to share ideas on how to analyze and how use that analysis of the ocean in different businesses. The winner of the challenge looked at growing algae near the coastline of Brazil.

We've learned the most through active interaction with professionals, both in challenges like those I just described and as we've grown our product. We've cycled through more than 30 different releases over the two-year period. From each release we've learned something new. We make blog posts and we present and interact at conferences like Strata or the Society of Exploration Geophysicists or the American Geophysical Union.

We've also learned from the people we've hired. We've hired some oceanographers from the community to work with us, and they have taught us everything known to date about the physics of the ocean and the dynamics of the ocean. Roberto De Almeida from the Brazilian Space Institute worked with us a great deal, and we as a company learned a lot about the ocean from him. He is also the author of Pydap, the Python implementation of the OPeNDAP data exchange protocol, so that helped as well. We've also learned quite a bit from John Graybeal, a US cyberinfrastructure expert. We've also worked with Chris Clark, who was leading a large research group in bioacoustics. We've also learned a lot from the friends and colleagues of these experts.

Lastly, of course, we've learned from our customers. They are in the ocean daily, so they know best what's going on. Our customers teach us about what's relevant and which ocean parameters to focus on, such as wind–wave ocean currents, bathymetric ice, and all the other crucial parameters affecting their daily activities.

Gutierrez: What does a typical day at work look like for you?

Karpištšenko: We are a company that operates 24 hours a day and 7 days a week. We have offices in Sunnyvale, California, and in Tallinn, Estonia. I mostly work in Estonia, so my day begins by opening a laptop and checking what has happened during the night and on the other side of the planet. The day ends in a similar way, by handing over what has been done during the day. These days, I've created for myself a weekly schedule that injects discipline into my work and optimizes my sleeping patterns to fit around this 24/7 life.

In the mornings, I conduct my own personal operational things, like taking care of personal matters an hour after checking that everything is all right on the US side. Once this is done, this leaves me uninterrupted work time until 8 PM in the evening. Usually, I use different task lists, roadmaps, and to-do items to plan my week. I make sure to revisit that plan during each day. Before lunch, I work on things that do not require interruptions. After lunch, my work involves interactions either with customers, with partners, or with team members. Around 6 PM in Estonia, or 8 AM in Sunnyvale, we will have a daily stand-up where everyone shares what they did and what they're planning to do next. Every Tuesday we have a more extended version where everybody shares their week's plans so that we stay in sync.

So before lunch, there is uninterrupted work, then from lunch until dinner there is work that involves communication with others. After dinner, I usually do some work on more creative items, like looking into new technologies, looking into the data sets, thinking about how we can be more productive, or what direction should we evolve toward. I use this time to look at new proto-types as well. What I look at during this time really depends on what we are currently working on. At Skype, it was very different with my data research team. But when you build a company yourself, you have to help and work on everything as it grows.

Gutierrez: How do you view success and measure success?

Karpištšenko: Success to me is when what you do is adopted by some-one else. If you create a software service, or if you create a data model or a method for analyzing data, success happens only when someone starts work-ing with it and improving it further. That's the real success. And the second part of success for me is somehow finding a flow in myself where I don't get distracted by what-ifs and instead I am able to focus on some new idea I had or something I need to finish. The ability to actually deliver something tangible—that's the main index of success for me.

Nurturing something to grow and seeing that thing grow is also a big part of long-term success. I love watching things I've built grow and continue to live. I've built two companies and five teams. These groups have delivered things like software services, software products, and professional services that are still in active use today. It's a great feeling to see that what you built contin-ues to be used 5 years after and then 10 years after. That means that you actually did something meaningful in that moment. And when you are in that moment, you never know. There are so many uncertainties. Getting through life, through those uncertainties—in a way, when you look back and see things still connect and exist, that's the biggest measure of success.

As you grow and as you start having children, or as people start to become more important to you, then seeing their successes is a great thing. I'm personally very glad that some people I managed at Skype have now grown into managers themselves and are now growing their own managers. As I was leaving Skype, an interesting situation happened where my manager, his manager, and I were all sitting around the table. At this meeting, there was also someone I had grown into manager. Being part of that full circle was what I would define as success.

Gutierrez: How do you think about whether you're solving the right problem?

Karpištšenko: This is a big question I think about frequently. I've done so many different things. As technology evolves, as the environment we work in changes so frequently and rapidly, we need to continuously adapt what we do and how we do it. If you don't, you stagnate quite often or your skills or knowledge become obsolete. So there is a big part of intuition in choosing the most important problem. I use a simple model—you could call it a decision tree or you could call it qualitative labels for the different opportunities I have—to evaluate whether I want to do it, whether that's my passion, the different rewards it will bring me either socially, financially, or culturally, and finally, if there is something new I am going to learn. I think about where this route is going to take me in the long run. That's the first set of criteria.

Once you've made the big decision, then you have to think about what problems are going to come up inside the company, organization, or team that is going to do the thing you have decided to do. Now you have to think about the situation to decide what is important and what is not. This leads to a lot of juggling between long-term goals and short-term goals. Of course, we all know about different matrices for deciding what's important. You have CRM systems for deciding on customer importance. You have different technological portfolio managing methods and tools to decide into which part of the system to invest in. What is strategic? What just needs to be sustained? What should be outsourced? So you use different tools and methods continuously, depending on the dimension you are optimizing for.

These days, a lot of what we do is positioning in the sense that we are positioning ourselves to be lucky. We follow the adage that luck is being prepared for an opportunity and seizing it when it appears. So it's thinking about which opportunities might present themselves in a week, or a month, or a year, or in 10 years and then pushing what you are and what you do in that direction so that when the opportunity presents itself, you are ready. You don't usually know when or how the opportunity comes, but you want to recognize it. Then, if you recognize it, you want to be able to seize the moment and do it. So it's saying yes to some new opportunities but then being focused enough to say no to many things that will guide you in a completely wrong direction. No magic formula there.

Gutierrez: What tools have you chosen to use at Planet OS, and how do they compare to the tools you were using at Skype?

Karpištšenko: The scale of the problems and requirements are very different because the Industrial Internet and Consumer Internet are quite distinct things. For more information about the Industrial Internet, General Electric is talking about it a lot these days. At Skype, I was focused very much on the Consumer Internet. The tools we used were things like Greenplum, R, Python, and network analysis tools such as Gephi. We chose them depending on what problem we were working on. If it was fraud detection or marketing campaign optimization, we would use different tools compared to when it was traffic-shape detection on the network, social network analysis, or social network recommendations.

As I moved into Planet OS, the key to our success, in my mind, was our productivity and our ability to iterate fast. I knew from past experience that Python had a very strong growing community and a lot of data scientists. I had worked with Java quite a bit, so I was certain that Java was not the right choice just yet. So we decided to use Python. And as we worked, we had to look for different storage solutions, data warehousing, and analysis solutions. These we picked based on what was going to scale well, as well as what was going to perform well for us in the next six months and still be a viable solution in two years. That said, some portion of my tool choices have also been exploratory to see new promising methods and to determine if there is a problem set where you can apply these methods to see whether it's applicable. I find that this applied way of working with methods and tools is the best way to learn. I've done it this way for so long and I've ended up working with so many different technologies that it would take a few pages to list all the tools I've seen and used. Of course, these come and go, so they're highly situational and context-dependent.

Gutierrez: What lessons have you learned as you've gone through this tool-and-method exploration?

Karpištšenko: The core lesson is that there is no silver bullet. Initially, when you start in the data science or software business, you think that a new language or a new framework or a method will solve everything. For example, when I was looking into how to do better feature selection, feature engineering, and better model building without my team or me being too involved in it, I looked for automated ways of doing that. Brute-force analytics looked like the best way to do it. Well, it works fine, but you need a considerable amount of computing resources for that to work, and unfortunately, you don't always have enough resources available. The resource and time constraints that you have will tell you which tool to use.

If you are using R and you actually want to deploy the code into some live web service, then quite often you are better off doing it in some other language, depending on how big the load is going to be. You have to really think about how many transactions are going to go through your model or system. In some cases, you just have to choose C++ and its libraries because you have to optimize at a very low level. This is what we've done at Planet OS on a few occasions where there are so many computations involved that it's necessary to optimize the machine learning. In other cases, you just want a quick exploratory answer. Then you can use tools like Vowpal Wabbit, Weka, or some other library.

Gutierrez: What have you learned from building successful data teams?

Karpištšenko: You need different roles and different personalities on the team. You need statisticians, software engineers, machine learning experts, system engineers, visualization experts, interaction experts, product managers, and business development experts. Oftentimes, not all of these roles are specifically within your team. Some of these roles could be your stakeholders instead. But if you look at your extended team, you have to have all of these different perspectives involved. And if you don't, then you usually end up not seeing your result go into the live production environment or finding success that fast.

The best examples I've heard about where this was true have been when operations have been disregarded. As people deployed a new service, they found out that the network configurations of the system that they used would not fit into the existing infrastructure. As they examined this issue, they found that the rabbit hole went deeper and their results getting deployed so late that the models were already outdated and they had to start over to build new ones.

Not only do you have to have all the roles, you also have to have the right personalities. You don't want to have everyone in exploratory mode. You want some people to be very focused on development. You want some people to be very focused on quality testing. You want some people to be the cheerful team members who get you through the highs and lows. Every personality in this mix of personalities is relevant because it enables frequent regular communication patterns and allows for team building and reflection.

To build successful teams and projects, I strongly believe in the Kaizen approach. Kaizen was made famous in part by Japanese car manufacturers involved in continuous improvement. I believe you should always be looking for ways to improve things, just small things. Just try it out. You'll fail sometimes, but over time things will become better. If that involves people aspects and communication aspects instead of the normal project aspects, then that's all great. Usually, things fall apart because of not improving the people aspect.

Gutierrez: How do you think about hiring?

Karpištšenko: You have to follow the trail through people. You usually have some contacts you know and so you reach out to them. You learn about their inspirations and the professionals they think are great. This process leads you quite fast to a large set of people. Of course, in addition to going out to look for people, you also have to promote your company at conferences. One of the great hires we had working for us for a while, a top data scientist who was working on top machine learning libraries, was hired through a presentation we did at PyData and then the Kaggle competition. Getting to him took a lot of interviews with people I met at the conferences, with people I met at the Kaggle challenge, and a lot of Skype calls.

As you identify the right kinds of data scientists, they should meet your team members. If the team members are all very excited, you should hire the person. You must set the goal that new hires have to inspire the existing team and have to be better in some aspect compared to everybody else. As you do this approach, it will start to pay back, as these new people will start to bring in new people as well. I find it inspiring to work with people who are passionate about what they do or who have some other reason to work other than just financial gain. Financial gain is a second-order result: if you do the right things, everything else will follow. So I look for people who push themselves. I look for some progress toward self-fulfillment or whatever it is that one is after.

Gutierrez: What's a project that you and your team have worked on in the past year?

Karpištšenko: The most exciting project was commercially finalizing the product that we had. It took an enormous amount effort by our very small team to deliver this big data ocean platform. We had to make many tradeoff decisions. We work in a field where there are petabytes of information, so we had to look at smart ways to not overwhelm ourselves with data size. We are in a field where there is a high diversity of data formats and data types as well. Again we had to navigate the landscape in order not to overwhelm ourselves with any of those. While in a sense it was a project to finalize the product, it is also an ongoing never-ending thing because we want to continue helping our customers.

Delivering a working commercial product in a highly complex, new industry with very diverse and large data sets and getting it to a state where companies trust us enough to pay for it has been greatly rewarding. It means that you've made many right decisions and what you've created is now able to grow and become mature enough for the ocean industry to step into a new era of an easily accessible data-rich environment that will allow them to make better decisions. It's great to know that I have made an impact on those businesses to help them be more responsible with the environment around us. And from now on, it's a new era with new challenges as we work with many more customers in many more different ways. So the challenges will be different.

Gutierrez: What are the challenges going forward?

Karpištšenko: Now as businesses rely on our services and software, we will have to think about all of the service level agreement aspects of our work. We will also have to work out customer onboarding, support, and maintenance. We'll have to make sure that our roadmap priorities reflect what the most important customers require while also making sure that these decisions don't paint us into a corner. This way, the business has the ability to grow and scale beyond our wildest dreams. The challenges are going to be around scaling and operational excellence for a while, rather than innovation and development as it used to be.

Gutierrez: What do you look for in other people's work?

Karpištšenko: For work that presents novelty, I want to know how it compares to something else that I know. If someone presents a new technology or a new method, I need to know some benchmarks or some baselines against which they are comparing. Without it, I just disregard it because I don't have time to do the comparison myself. With things where someone presents a success story or a use case, I look at how they approached the problem holistically and whether they are highlighting similar problems and challenges I myself am familiar with.

In cases when I look at work related to where I see the industry moving in the future—toward sub-second analytics, toward exascale data sets, and toward higher involvement of nontechnical people in analytics—in these cases, I try to get inspired and to remember the lessons learned or the models they have used to succeed, so that, when the opportunity presents itself to me, I have something in my backpack to pull out and start iterating from.

Gutierrez: What do you think the future of data science looks like?

Karpištšenko: Automated. To me, I see many similarities with what happened in the software industry in the 2000s, when the Agile Manifesto came out, as compared to what was there before, which was this rational unified process everyone talked about.

I see many similarities with what's going on in the data science community right now. In the software industry, there used to be a lot of focus on how it was difficult to deliver on time, how there are so many uncertainties, and how so many software projects failed. Then a new method, inspired, again, by the car manufacturing industry and by old production companies, brought new ways of working on software development projects. The Agile Manifesto actually made it so that these days everyone knows how to deliver software projects. If someone doesn't deliver, then it's much easier to understand what went wrong and why they were unable to ship. Of course, it's a different story if it is a highly innovative, high-risk, high-tech project.

These days in data science, I also see that best practices are being shared. Universities have set up courses. Top universities in the US are now giving you the opportunity to get a data science degree. And when you graduate as a data scientist, you will come into an environment where data integration has already been solved by companies like Planet OS, Trifacta, Tamr, and where feature engineering has been solved by companies like SparkBeyond. Furthermore, nonparametric search methods, which James Burke and others have talked about, are going to allow computers to give you ideas on which features are the best. Other things that are going to be partially solved are visualizations tools like D3.js and Tableau, which are making it easy for anyone to make great visualizations with great insight. Also helpful will be the extensibility of the language that is used for developing the models. For instance, probabilistic programming will allow you to have variables that represent probability distributions, so you can start to develop in completely new ways, as compared to today with the old languages.

In the future, the tools around you are going to be such that you can think more about the business problem and less about the technical side. You'll have to think less about how it scales and how the data comes together, because it will be more and more automated. You as a data scientist will need to learn to understand the business you are in much better, as well as learn to interact with other domains and with people from other disciplines. I think it is Chevron who has executives paired with data scientists so that when there is any big investment decision to be made, they work together to select the best course of action.

Similarly, these days, you are starting to see chief analytical officers next to chief data officers. The chief analytical officer represents the change that any major decision throughout the organization is now getting made by everyone in the organization, with data scientists and machine learning methods being the means and tools to make those better decisions. You are now seeing data scientists being planted into product teams, instead of being an isolated island or a group of experts throwing wise words at other groups. I think eventually we'll have something similar to the Agile Manifesto and agile methodology for data science—agile data science or whatever else you want to call it—that will bring some old methods and ways for collaborating and working in this field. And then suddenly, the focus will shift from methods and tools to actual end results and the delivery of those results.

Gutierrez: What advice would you give to someone starting out?

Karpištšenko: Though somewhat generic advice, I believe you should trust yourself and follow your passion. I think it's easy to get distracted by the news in the media and the expectations presented by the media and choose a direction that you didn't want to go. So when it comes to data science, you should look at it as a starting point for your career. Having this background will be beneficial in anything you do. Having an ability to create software and the ability to work with statistics will enable you to make smarter decisions in any field you choose. For example, we can read about how an athlete's performance is improved through data, like someone becoming the gold medalist in the long jump because they optimized and practiced the angle at which they should jump. This is all led by a data-driven approach to sports.

If I were to go into more specific technical advice, then it depends on the ambitions of the person who is receiving the advice. If the person wants to create new methods and tools, then that advice would be very different. You need to persist and keep going in your direction, and you will succeed. But if your intent is to be diverse and flexible in many situations, then you want to have a big toolbox of different methods. I think the best advice given to me was given by a Stanford professor whose course I attended a while ago. He recommended having a T-shaped profile of competence but with a small second competence next to the core competence, so that you have an alternative route in life if you need it or want it. In addition to the vertical stem of single-field expertise, he recommended that you have the horizontal bar of backgrounds broad enough so that you can work with many different people in many different situations. So the while you are in a university, building a T shape with another small competence in it is probably the best thing to do.

Maybe the most important thing is to surround yourself with people greater than you are and to learn from them. That's the best advice. If you're in a university, that's the best environment to see how diverse the capabilities of people are. If you manage to work with the best people, then you will succeed at anything.

Gutierrez: What is something someone starting out should try to understand very deeply sooner than later?

Karpištšenko: You should understand that overnight success takes about ten years, so you should expect to go through a lot before success finds you. You should persist and be ready to do the hard work. The idea or the initial enthusiasm is just a small part of doing something great.

Gutierrez: What data sets or problems would you love to see tackled?

Karpištšenko: That's the big challenge of the day. I think personalized medicine holds great promise. Having great open genetic libraries and having access to the gene pool of humans will be very important. Most important will be developing these data sets, tools, and methodologies in such a way that still preserves privacy and helps us to transition into this new period of much less privacy.

There is a lot of work to be done in how we collect data and how we process it so that we don't break people's privacy. A good example of the work being done in this area is a product called Sharemind, which allows us to do statistics on data sets without breaching the privacy of the individuals involved. This can be used in medicine or overall statistics of companies. This ensures that the information cannot be abused for purposes that it wasn't intended for. So ensuring the privacy of individuals and organizations will be important.

In our work at Planet OS, I see just how little is being done about the environment. So having access to environmental data through all the sensors that are out there in an integrated form will mean a great deal to mankind for understanding the impact we have on the planet. Right now, IPCC [Intergovernmental Panel on Climate Change] reports come out at such a slow pace that it's just unacceptable compared to the rate of change we see are actually seeing and the volatilities that we see in the environment. Having great means for insight into that data and what is actually happening will be very valuable.

Another interesting area is being able to look at cities as you would look at an organism. In particular, city resource mining so that there is less waste is highly relevant and important as the world's population continues to get more concentrated into cities. I think we all want to live in an environment that is worth living. As part of that, keeping track of the garbage flows and the material flows into and out of cities so that there is recycling and reuse holds great promise.

Overall, using data to look at the impact we have on our surroundings and looking very closely at this feedback loop is relevant in many fields. I hope that data scientists collectively build a better future together with engineers, businesspeople, researchers, and artists. I hope that all these different data sets get developed to enable us to have a better future.

Also interesting are things like music generation algorithms. As these become more advanced, we will be amazed by the emotions that music generated by computers will evoke in us. Also intriguing to me are ways that people will be able to navigate the knowledge graph faster. Whether it's ways to read books faster or ways to guide us through the wealth of knowledge in a way that we are not distracted and overwhelmed, but rather enjoy the rate of change and fluid flow, I think will be a great thing. Toward this end, knowledge management, text analysis, and image and video analysis management will mean a lot for education.

Gutierrez: What is something you know that few people know but that you think everyone will know about 5 to 10 years?

Karpištšenko: For that I have to reflect back on my involvement in software in the very early days. These days everyone is creating and writing software. I think in 10 years everyone will be creating models and using them. There is no single truth. Everyone is right, depending on the situation and context. So I hope that everyone keeps an open mind regardless of what the models tell us and that we don't disconnect from our values as we make decisions—that we don't just let the whole system evolve toward a state where being a human is second. I'm talking as a transhumanist: I believe some people will have the ability use different technologies to prolong their life span. As wealth and technology continue to concentrate in the hands of those with resources, I hope that the solutions trickle down fast to everyone so that the power and knowledge is used for great societal purposes instead of selfish purposes.

Overall, I'm really excited about seeing different communities come together. I've visited many different areas and communities related to software and data. I've been very happy to see people from completely different domains in both academia and industry joining in to help one another. It's no longer an academic ivory tower versus industry, and that really motivates me. Seeing computer science and data science getting applied in the real world by real people to solve real problems is very exciting.

Amy Heineike

Quid

Amy Heineike *is the Director of Mathematics at Quid, an intelligence platform that combines natural language processing, machine learning, network science, and data visualization to aid Fortune 500 companies, hedge funds, and government agencies in answering big, high-level questions about what's happening in the world. As citizen and data journalism increase in size and scope, and social, scientific, and technological information becomes available, more content than ever is available for understanding what is happening in specific sectors around the world. Quid brings together technology and patent data, and consumes more than one and a half million articles a day to enable continuous information delivery to its clients in the form they need to understand the events, trends, and patterns relevant to their concerns. Developing this product entails challenges in terms of data processing, communication of results, and delivery of a complex data science intelligence platform that is simple and clear enough for any user.*

Heineike's career before venturing into data science spanned the study of mathematics at Cambridge University and the modeling of complex human systems and cities for businesses and governments as an economic consultant at Volterra Partners. Fascinated by the graph theory behind social networks, she was particularly interested in modeling the economic and social impacts of the interaction of cities and transportation networks, which she applied to Crossrail, the 73-mile railway line under construction across Greater London. Heineike exemplifies both the drive necessary to switch tracks into a career in data science and the curiosity about data that can lead to insights, products, and even companies. Her dynamic combination of hustle and flow animates her accounts of how her desire to explore new data sets led to her collaboration with the Quid founding team, and blossomed into a core part Quid's current product offerings. Heineike's skill at communicating complex human, business, and government networks in ways understandable to nontechnical clients is on display in her interview.

Sebastian Gutierrez: Tell me about where you work.

Amy Heineike: I work for a startup called Quid that has built an intelligence platform. The idea is that we're going out and getting lots of big, complex, unstructured, and interesting data sets, and then we're building analytics and visualization on top of them so that we can use the whole platform to enable people to connect ideas and events, tell stories, and discover what's happening in the world. I am a bit of a data geek, so I find it very exciting to get to explore these big data sets that you don't normally get to look at analytically. For instance, you normally get to read one news article at a time. Now we get to actually sit and think about what would be possible if we could read a ton of news articles at the same time. It's data science heaven.

Gutierrez: What's been your career progression and how did you come to your current position at Quid?

Heineike: I started out in mathematics, where I studied a broad range of different mathematical topics. Through my studies, I became fascinated by graph theory, by nonlinear dynamics, and by the question of what these two subjects tell us about how human and social systems work. So when I graduated, my first jobs were in an economic consultancy in London. In these roles we were tasked with thinking about questions like: How do cities evolve and grow? What happens if you're building a train line across a city? Does that affect the city in a discernable way and what is that worth in money? We got to ask some really fascinating questions.

I was very fortunate in that one of the big projects we worked on was Crossrail, which is now actually getting built. I was able to swoop in after 30 years of an ongoing debate in Britain about whether it was worth building this new train line, and be there for the day when they actually decided they were going to go ahead. In those roles I was doing a great deal of mathematical modeling and a lot of data analysis, as well as also making (and then explaining) the case for what we thought was going to happen.

What struck me at the time was how constrained we were by the data that we had. I'd come in fascinated by the mathematical models we were starting to build about how cities might evolve and what kinds of nonlinear interactions might happen. But we couldn't actually really get into that depth because, at the time, we were using mainly survey data. This means that you'd literally send out a bunch of people to stand on street corners and count the cars going past on a particular day. Or we'd get survey data from the census where people would state which locations they normally commute between. So the data was very limited in the detail and resolution of what you could do. However, it was very clear at the same time that data could be generated at scale in unusual places that would make the kind of analysis we were doing far, far better and more interesting. And so when I left that, I started thinking about how I could closer to more data.

When I got married to my very dashing American naval officer husband, moved out to the States, and left behind my whole professional network in the UK, I suddenly found myself in San Diego thinking, "What on earth do I do now?" In a way, this turned out to be really helpful, as it took a while for me to get a work permit, so I had a bit of time to sit around and think about what I really wanted to do. I reached out to a friend of ours named Bob Goodson, who was living in San Francisco, where he had just founded a tech company. At that time, the company was called YouNoodle and they were building a social network for entrepreneurs. I reached out to him and said, "Hey, Bob, I think you probably have some interesting data that you're generating. I would love to analyze it, and I've got a bunch of ideas for what we could do. And have you heard of network analysis? I will look at your data for free if necessary, just let me play with it."

I started working with Bob and it began an interesting spiral of events. I collaborated closely with Sean Gourley, who is now the CTO at Quid, and it turned out that we were able to build some really interesting things with the data we had—and were able to start collecting—on entrepreneurs and startups. What we built and our results were different enough from what they started out doing and compelling enough that Quid was born around this whole new set of analytics. At Quid we refined those initial ideas, and built out a team that built this whole intelligence platform. My career evolved as the startup evolved.

Within Quid, and at YouNoodle before that, I have had a lot of different roles, which I think points partly to the ambiguity of what data science is and partly to life in early stage companies. At times, I've prototyped and tried out new approaches to analysis. At other times, I've run teams of data analysts. At other times, I've QAed the engineering team. At other times, I've done product management. I've written production code to get things to move from prototype versions into the main product. It's been an interesting and varied ride, which is why I love data science and working at Quid.

Gutierrez: How did you arrive at the title of Director of Mathematics?

Heineike: When I got to Silicon Valley six years ago, there wasn't a name for what I do. I was just curious about getting to poke around the data and figuring out what products we could build on it. At the time, it was very weird to think of having a mathematician doing that. And so I don't think anyone knew why I was there or what to call me. Pretty soon it became apparent that combining a mathematical bent with a focus on data and product building was very useful. People now typically call this 'Data Scientist', but we came up with the Director of Mathematics title—which I'm sticking with because I like it.

Gutierrez: What team do you work in and what is the makeup of it?

Heineike: I work as part of the engineering team. We're building a Software-as-a-Service platform to let our clients analyze text-based data sets at scale in order to make important decisions. That means that are complex algorithms and visualizations running, but they are wrapped inside of a software product that clients interact with directly, and therefore has to work repeatedly under lots of conditions. One of the cool things about working in an engineering team is that there are people with skills that are very different from my own. This means that it looks like they all have super powers. So between us, we are able to build some things that are really interesting. We have UI developers who build out the interactive visualization layer, we have back-end engineers who are building the data collection, search, and analysis layer, a data science team who work out how to work with the data and figure out improvements to algorithms, and then QA and DevOps who make sure it's all running well.

One fascinating thing about Quid is that most data scientists work in companies where the main business of the company is something not related to data science, whether it's finance or advertising or a social network or something else. The data science used at these companies is there more or less to grease the wheels and add interesting products and services on the side of the main business. At Quid, everything we do is with data, so most of our team can describe themselves as some form of data scientist. The people who work here are generally very excited about this whole idea and curious about what more we could be doing.

Gutierrez: Tell me about Quid's product and how it uses network visualization to help your customers understand the world.

Heineike: For our users, the product first lets them search premium, large volume data streams, and then presents the results in an interactive visualization. We use network visualization as a way to organize the data to help people explore it. We look at large, external, unstructured data sets that provide signals about markets, consumers, and innovation, including news, patents and data on startup companies. As an example, let's say you're analyzing the news, in the visualization a node in the network might be an individual article. This news article will then be linked to other articles that are very similar. When you lay this out, you end up seeing clusters of things that are related to each other, which lets you see the topics in the data. Then, between those clusters, you will see these long range links where there are connections, but they're not super close to each other, and so you can immediately get an idea of how these topics relate to one another, and what bridges them. Then you can start playing with the related metadata and laying that over the top, for example to see the change through time. The product is very flexible from a customer's point of view regarding the data that is visible and can be used. This allows them to fully explore the space in a number of ways.

We've developed the tool to cover more and more datasets over time. Initially we'd built the product to explore venture-backed start-ups. Users could search for companies with technologies relating to, say, mobile payments, and then start to analyze them to understand the technology space. They could find the clusters of companies with similar technologies and see how mature or dynamic those clusters are. They could see how they related and find the innovators bridging different technologies. They could see the profile of investments from different VC firms, or other large companies, and compare their investment strategies and how they were changing over time. They could look for acquisition targets, or find competitors or technologies that had the potential to be disruptive to them.

News was the next major dataset we integrated. We had the realization that many of the tools we'd already built could be used to help us understand an emerging conversation in the press. Now users can explore news around a particular topic, and see the stories that are being published, who are the thought leaders driving the conversation, how it's changing over time, and how sentiment maps to different stories. They can find memes and see how they are emerging. They can look at how different publications are covering a topic. It's very powerful if you're trying to figure out how to engage with the public, or want to understand how the conversation around a major topic is playing out at scale across hundreds of thousands of different sources, or if you just want to get up to speed quickly on an issue.

One thing that Quid has been very good at is that we have people on the business side of our team who are very, very good at sitting down with decision makers, with people who have a whole process for how they decide what to do, and saying: "What do you need help with? What do you need to know in order to best make a decision?" And then once they have those answers, responding with: "Here's how you could use this tool to help you do that." And so meeting them halfway and actually making it very clear how we can be helpful and useful, and feeding back to the team what we need to build better to enable them to make use of it successful. That's been really important.

It's been critical for me to get to collaborate with people who are so good at understanding how users work and what their needs actually are. We often work with them to figure out how different groups of our customers specifically use the tool, as well as understanding the evolution of how a customer goes from having their hand held while they get started to very quickly getting to the point where they start using the product in ways we hadn't even thought about. I think especially for our use case, where we've got people doing very important decision making, it is very important for us to spend time with them and talk it through with them to make sure that they're comfortable with what they're doing, and that we know what types of questions they really need to be able to answer.

Gutierrez: What have you been working on in the last year?

Heineike: This has been an exciting year, as we're getting to the stage of really scaling up what we're delivering. This means that we've been very focused on increasing the quality of the overall experience to users, as well as leveraging more datasets. I've spent a lot of time working closely with our production software engineers to help define how we interpret new data streams—understanding what's there, what is useful, where there are data issues, and how we can address them. We've also worked to improve the quality of each of our main algorithms, and figure out if and how they need to be tailored to each dataset, and responded to feedback from our users.

The data science team has grown quickly this year too, so we've brought on board people with expertise in different parts of our stack who've been able to go deeper on each piece.

Gutierrez: Why is scaling up important for Quid?

Heineike: Getting to scale the product means the business is growing, which is obviously important. It's also just really exciting to get to put the things we've been building internally in front of a lot of people though. The genesis of our product came from our own curiosity to explore what was happening in these different domains—in emerging technologies, in important global conversations—and to push the tools to see how easy we could make this for users.

On the one hand then, it's amazing to now see other people's curiosity also be satisfied and seeing them asking and answering really diverse types of questions. Our users are often very creative and smart people, and so there is a real energy when they get into the product.

On the other hand, as a data scientist I'm fascinated by how we can democratize the feeling of having the ability to really drill into and explore data. It's one thing to be able to write a script that can pull insights from a specific data set that address a specific question, and it takes a lot of technical knowledge to do that—it's another thing altogether to productize the essence of that, so that other people can drive it. Giving that control to people with the best questions is really exhilarating.

Gutierrez: What specific news sources are you looking at?

Heineike: We are using a really wide corpus of news. We gather and index about 1,500,000 news articles a day from global sources. These news articles represent a broad selection of what is being generated, because the data goes from mainstream news to unique, really personal blogs, like people's WordPress accounts, with technical sources and everything else in between.

Gutierrez: How would you describe your work to someone who is not familiar with it but is familiar with data science?

Heineike: From a data science perspective, the way I explain it is that what we're doing is Natural Language Processing [NLP] plus machine learning plus network science plus data visualization—which is a really weird combination of things to combine, actually. I don't think there are many who would include network visualization with NLP. So there's kind of a uniquely mixed stack. And for technical people, there are a lot of different bits that they might engage with from their experience.

Gutierrez: Why is putting these techniques together powerful?

Heineike: A lot of work in the information retrieval and search space focuses on the problem of highlighting the one document that you should read. Similarly, if you look within NLP, there's a lot of work around how to extract entities or topics from a block of content. But again, the output is often simply a list of things, hopefully, sorted by relevance. The focus is on finding the single best thing. We differ in that Quid is tackling the question of "can we expose all of what's in the data and make it understandable so that anyone can interact with?" So we switched to finding unique signals that exist in the data and then figuring out how those indicators all interact with each other—giving a totally different level of perspective on markets, consumers, innovation, etc. So the questions people answer using Quid are actually posed slightly differently than the normal—it's not about a single signal, it's about a holistic and nuanced understanding. This is why we end up doing the network visualizations of topics, which isn't normal.

Gutierrez: How do you describe your work to someone who has very little knowledge of mathematics?

Heineike: There are a lot of people in the research, investment, and strategy analysis space with challenging and complex questions about what's happening or likely to happen in the world. This set of questions encompasses questions like: What do people think about my business? What is the next technology bubble? What is a major market player's long term IP strategy? What do people mean when they say terms like *happiness, family, love?* The way people have historically tried to answer these questions was that they hired consultants, some interns, they sat down themselves and googled things, or they had this epically long RSS or Twitter feed that they read and read and read, so that they could eventually summarize their findings into maybe a spreadsheet or a PowerPoint slide. This is a fundamental but inefficient process done for a huge number of big decisions that are made by organizations all over the world. That's just how the system works.

So for these people, I describe my work as that we're figuring out what tools will enable them to do that better and in a more systematic and efficient way. So we build tools that empower those people, who are at the moment reading a ton of stuff inefficiently, to proficiently consume more and better information. Normally, if I explain that to anyone who's ever done any form of consultancy, they end up with sparkly, delighted eyes, and they are blown away by the fact that anyone's building tools to actually help them do this. And they want access immediately.

Gutierrez: When did you realize you wanted to work with data as a career?

Heineike: I think I realized it when I got into the economic consultancy work, though I don't think that I really thought of it as being primarily about the data. Rather, I think I thought of it as being primarily about understanding systems through mathematical modeling. But it became very clear that the data was a really fundamental component of that. The data limits how far you can go, so I think that was when it really became apparent to me that I needed to get into where the data was.

Gutierrez: How did you get interested in math and programming?

Heineike: Growing up, my family was extremely supportive and encouraging of any enjoyment and interest I showed in mathematics. My eldest brother is actually a programmer, and he was hacking around with computers ever since he was very young and I was even younger. When I was pretty young, he tried to teach me to code. I did bits here and there, and then I really got into it when I started working. I think mathematics is a good foundation because it teaches you to think very rigorously and logically. Programming became interesting as soon as I had something I wanted to build with it.

For me, it's been very natural to go from abstract mathematics into data analysis, and then further into programming. As I've learned to do better data analysis and explore different areas of analytics, I've done more and more programming. For a long time at Quid if we decided we wanted to try out some new set of algorithms or approaches, I'd be the one to learn how to do it and try it out, in collaboration with one or two others. I learned NLP when we decided we should try analyzing large volumes of text. I learned large-scale network analysis when we wanted to analyze our networks better. I spent time learning how to structure code cleanly when I needed to bring my code into production.

It's been an interesting progression of picking up skills that are really driven by the necessity of being able to do what we want to be able to do and seeing what kind stories that we want to be able to tell with the data.

Gutierrez: What was the first data set you remember working with?

Heineike: Perhaps not my very first data set, but in my economics work, I think some of the first I played with were actually government data sets. So you would log onto a portal and download economic indicators or census data for Britain, or similar kinds of data. The data was very tightly controlled—you actually had to get vetted by an agency in the UK that would decide whether you could get access to the economic statistics. If you were approved, then you could download the data in spreadsheets, depending on your query. So you would only ever see bits and pieces of it. Those were probably the first data sets I came in contact with.

These data sets were very much small data, in a way what we would maybe now think of as older-school data. So this was data that was actually collected for the purpose that people were using it for and it could fit on a spreadsheet. And not only that, it would have a bunch of statisticians arguing about exactly how to present it before you even got to see it. This is very, very different from the kinds of data that I'm using every day and the kinds of data that we think of as "big data" now.

Gutierrez: When did you realize the power of data?

Heineike: I worked for a small company called Volterra Consulting in the UK, which has done some really interesting analysis and some really interesting, different kinds of work. There were several fascinating studies that that company worked on while I was there. As one example, Paul Ormerod, who was one of the two directors, published papers on his finding that if you look at whether or not people have bank accounts in the UK, it's actually very hard to predict whether someone will have a bank account based on just their economic status, their income, or how much they earn. It's actually a very important thing for us to understand, because people who don't have bank accounts are typically excluded from financial systems and have a much harder time plugging into how our economy works. Through his research, Paul found that you could predict much more effectively whether or not someone has a bank account if you have information about whether their friends have bank accounts. As soon as you know that, then you realize this is obviously a kind of a network effect. Bank accounts are one of those things where people need to understand, "Oh, I should get one of these things. And this is how I would use it, and I shouldn't be scared of getting a bank account. I can go to a bank." And so then it makes you realize their communities really matter for transmitting this knowledge.

However, you just can't get this insight from high-level statistics. You need the data on connections between people. To understand this issue, you've got to figure out which data to look at. You might have to get it from surveys, or else be very creative about where to hunt signals down. People's interactions drive economic activity in profound ways.

Bridget Rosewell, the other director, led a lot of the city analysis work, which was really fascinating. She really pushed back against using overly complex models and focused instead on pushing to ask the right questions, and getting the data together to be able to tell the story that needed to be told. You had to be asking the right question and you had to be getting the right data to be able to tell the story.

This made me realize that rather than thinking I should figure out a much more sophisticated way of analyzing the same macroeconomic trend data to understand the economy, I should instead be thinking about how we could bring other data to bear that would give us a deeper insight on the dynamics of the economy. Could we find out about people's social connections or people's professional connections and how they impacted the society? How could you even see that and what data would you need to be able to see that? And then, if we could see that, would we have a very different understanding of how our economy works and what the flaws of it are? We still have a hard time understanding really fundamental dynamics in our economies, like business cycles and like the life cycles of companies—what data would let us understand these better?

Gutierrez: How does Quid fit into the research and strategy analysis space?

Heineike: There is a business intelligence industry with a lot of people who are doing higher-level thinking research and strategy analysis jobs, where they're trying to analyze, wrestle with, and figure out what the heck is going on in their business and marketplace. We're, however, a bit of an oddball at the moment, in that we're making data analysis products for those users. It's useful to think about where we fit into the space by looking at our industry and our technology as being two separate things.

In our industry, people are largely working to understand really big, important questions, and using the answers to inform what businesses, and governments should do. People in this industry are often working in consultancies or in teams inside their organizations—in government departments or strategy groups for example. They are normally experts in their field, or work with experts. We're, however, much more technology driven.

From a technological perspective, I think we probably share a lot more in common with people building software in other industries. So the technologies are related to organizations that ask questions like: How do you analyze text at scale? How do you extract entities or meaning from corpuses? How do you visualize large numbers of data points all at the same time and make it interactive and beautiful? For example, our search technologies overlap with consumer search products, and our visualizations have parallels with tools built for bio-informaticians.

Gutierrez: Does Quid use itself internally to understand where it should grow, what the next challenges facing it will be, and what it should be developing?

Heineike: Yes, we've definitely looked at and asked a lot of questions about ourselves, as we like to be very meta. We've used our product internally to analyze the big data space and see who all the startups in our space are, and whom we are the most similar to. We've used it to research technologies we're thinking of using, and to get up to speed on new algorithms and methodologies. So yes, we have absolutely used it and have found it useful to analyze ourselves and our industry.

I have also analyzed data science in the past to find out what the news in data science talks about. What I found, which is fascinating, is that it's a lot of big vendors—vendors of data platforms, basically—are the ones who drive a lot of the conversation. And then you find a few companies who've done a really good job of positioning themselves and their products. These companies are mainly LinkedIn, Facebook, and Twitter. So there is a conversation driven by them talking about products they've built and why they're interesting.

And finally there's this hubbub in the middle of everyone else going, "What the heck is data science? What kinds of skills are needed? What kind of teams should be built?"

Gutierrez: Are there any publications, websites, conferences, or blogs someone interested in learning more about your space should look into?

Heineike: We're bringing together information on data viz, deep learning, data science, and how it is being used, on quid.com/insights, which you can look into.

Outside of Quid, I've found Twitter very useful for keeping up with what's happening. There are a lot of interesting people from a wide range of different viewpoints who tweet links to different links and references that are good to be aware of. Some of them are journalists from places like *The Guardian* and some of them are organizations like the Stanford NLP group. For example, Simon Rogers, who was with *The Guardian* and has now moved to Twitter, posts links to different visualizations. Stanford NLP, post links to recent text mining research.

There are also creative people who use data in imaginative ways. For example, Pete Warden who just released a software development toolkit that lets leverage deep learning in smartphone apps. You can basically set it up to make it easy for you to create a mobile app that uses deep learning in the background. There's a wonderful video on his blog of him making his app detect whether his cat Dude is in the picture or not.

So for me, it's been important to piece together these people from different kinds of domains, to be able to get ideas and inspiration from them.

As for conferences, the Strata conference is, I think, one of the biggest ones. The O'Reilly team does a really good job of bringing together a mix of people from different kinds of problem domains and different kinds of views of what data science is and what the issues in big data are. So they've put together some really fascinating conferences.

Data science is very broad, and so there aren't really just one or two publications that speak to it, and that's why we end up with this kind of massively complicated Twitter feed of people from all over the place chiming in with different bits and pieces. If you could meld them together, then you can get something really interesting.

Gutierrez: You mentioned that you put together your Twitter feed to encompass different viewpoints. What are some good examples?

Heineike: I've got people from different kinds of academic disciplines, which means different kinds of methodological ideas. So the viewpoints encompass "How would I go about doing this?" There are a couple of people that I like for having different viewpoints on the data industry as a whole.

One of those is Kate Crawford, who is a researcher with Microsoft Research. She did a great keynote at Strata with a follow-up essay in the *Harvard Business Review*,[1] in which she talked about making sure that you're sufficiently skeptical of your own data. They did an analysis of Hurricane Sandy and they found that—by looking at Twitter data—pretty much everything happened in Manhattan. Then they were like, "Well, no, not everything happened in Manhattan. Actually, it's just that people who live in Manhattan tweet a lot more than people who live down the coast." So the key message was to really think about your data and how it's being generated, as evidenced by what they found out about the hurricane and what happened just based on Twitter data analysis.

Another person is Kenneth Cukier, who is *The Economist*'s big data editor. He co-wrote a book called *Big Data: A Revolution That Will Transform How We Live, Work, and Think* that's given me a lot of thoughts to mull over regarding the direction that the industry's going.[2] So it's good to have these voices that challenge you a little bit.

---

[1] Kate Crawford, "The Hidden Biases of Big Data," Harvard Business Review, April 1, 2013, http://blogs.hbr.org/cs/2013/04/the_hidden_biases_in_big_data.html.
[2] Kenneth Cukier and Viktor Mayer-Schönberger, *Big Data: A Revolution That Will Transform How We Live, Work, and Think* (Houghton Mifflin Harcourt, 2013).

Gutierrez: What in your career are you most proud of so far?

Heineike: The thing that amazes me the most is how far I, and the team at Quid, have come. There was an early day at Quid when we created a visualization poster where we used a network diagram to show a startup landscape. When I first printed this thing out and showed it to everyone at an all-hands meeting, after having collaborated with a few other members of the team, it was obviously really cool, but it wasn't necessarily clear where this type of work would lead. It's been amazing, partly because I've been fortunate and partly because I've been on an amazing team, to take that initial idea and actually ask, "What would this look like if we built a real product around it?"

Zooming forward to now, it's amazing that we actually have an entire intelligence platform built around this idea, where we have people buying access to analyze the data. And we've got this paradigm that's really helping people explore data at scale, which came out of some of those earlier ideas and early analysis. That's kind of staggering, and I feel very excited and very privileged to have been part of the work to bring our poster to life.

Gutierrez: What does a typical day at work look like?

Heineike: It really depends on what kinds of things I'm working on. I'm the kind of data scientist who's thinking about how you piece together all of the bits to make a product that works. This means that sometimes I work really closely with the business side of our organization. So I'm generating analysis for them and talking to them about how they're using it. And I'm trying to empathize with the questions people are asking and trying to determine if we are doing a good job of helping them answer their questions.

On other days, I'm working more on worrying about the nitty-gritty of what's happening on the technical side and how we're actually implementing things. Today, for example, I've been worrying about schemas and metadata. So I've been downloading tons of documents and I'm running across them and checking that all the metadata lines up in a way that I'd expect them to. I've also been looking at our engineering tickets and QA'ing some of the data processing and letting the engineers know if it works correctly. So my day-to-day really depends on the kinds of projects I'm doing.

Gutierrez: How do you view and measure success at work?

Heineike: As an organization, it comes down to having delighted users—and ever more of them. This obviously trickles back into what we're building. For some data scientists, their goal is to optimize something within their system, and so it's very clear that if they can make that number get better, they all win. In our case, the tools and products we've built are for data exploration, so it's a bit more abstract.

For us, it's about making a better product, which means that we have to have lots of little ways of assessing whether that's happening. Some questions we look at are: Are there places where there are errors? Is there feedback we're constantly getting where people really want something that they don't have? And is there feedback that we're getting where they're super, super happy and they're over the moon that they can do something? So it is that bigger picture—which I think is a really important piece of data science—that we look at.

I think there's been a lot of focus on data science as kind of the optimization piece. And I think we're definitely much more on the kind of exploration and big-picture piece. So we focus on: Are we even telling the right stories? Are we even looking at the right data? You can't really optimize for that, so how we measure and view success is a bit more product-y and product design-y, where it's hard to prove apart from seeing the output of people really wanting to use the software and actually using it.

Gutierrez: How do you view and measure your own success?

Heineike: I'm very driven by wanting to build something that I think is useful and important. When I can log into the product and use it to learn something new, that's personally very satisfying. When I see our users using the tools to explore important global issues, it is really satisfying. In my day to day, I value being continually challenged and to be learning something new. I want that to translate into real improvements in what we've made. I want to see others empowered by what I've made.

Gutierrez: Where do you get ideas for things to study and analyze?

Heineike: It comes from a few different places. Some of it comes from end-use cases, where people are saying, "Oh, I really want to answer this question." And then we might be thinking, "We can let you answer that if we built an additional feature." Sometimes it might be that they're saying, "Oh, this thing doesn't really work the way I want it to," and then we might be thinking, "We could make it better if we improve the tokenization."

And then sometimes it's we're looking at the data and we're thinking, "No one else has probably thought of this, but just by doing this and this and this, we could make something that people would want." So there's this interesting balance of piecing those things together and choosing what to do. I think some of the choosing what to do comes from us as individuals, and some of that's coming very much from talking to other people in the business and figuring out what they want, and from the product management process. We have a very long list of things we would love to do.

Gutierrez: How do you think about whether you're solving the right problem?

Heineike: Well, that's a tricky question. I think that you have to iterate to get there. With Quid, we have a lot of users who are asking questions that we think are valuable ones to have answered well, so the question is then, are we really helping them answer those questions well?

In this case, it's really good to keep the users continually in mind. Its good to understand them well enough that you can then regularly use the product the way they would, and see where it falls short, and make sure you're working towards fixing that. For us that has also meant, does the product satisfy our own curiosity? Can I quickly learn about Bitcoin or the microbiome or Apple using it?

It's definitely the case that it's really easy to get into weeds with the stuff, as there are always thousands of options of different algorithms you could try and different tweaks you could do. You have to work hard to stay focused on the big picture.

Some of it is also having hunches about what's going to have the most value. I think you have to go there and make those decisions. I think we've got a really good team here, so there's a lot of people I can bounce ideas off of when I'm working on things, or who have very strong ideas about what the right direction should be. So it's really about making the most of the people around you as well.

Gutierrez: Whose work is currently inspiring you?

Heineike: My epic Twitter feed of the people that I think are interesting is always inspiring to me. It's interesting during big sporting or political events to see all the visualizations people construct to explore them. During the World Cup, for example, Google's had their page. Twitter's definitely had a few pages on it. All of those are fun to dig around and look at, for sure. Pete Warden and his deep-learning SDK are also currently inspiring me. I think he's very good at diving into some of the really hard problems and then doing something really interesting and unexpected with it. So that's often an inspiring kind of challenge—the challenge of thinking about what more could we be doing or how we could be using what we're already doing a bit more creatively.

Gutierrez: What does it take to do great data science work?

Heineike: Data science is already kind of a broad church. There are a lot of aspects that could call themselves "data science" or could fall under that label. For me, the first thing you have to do is piece together this idea that "Here's a really interesting problem, and here's data that could talk to that, and here's the methodology that would take that data and do the right thing to it to actually reach the output." To me, the key is figuring out how you get those three things—the right problem, the right data, and the right methodology—to meld. So that's the first stage—just being able to envision how they come together.

And often actually, one of the dirty secrets is that a lot of the time it's not just that there's an algorithm that solves the problem. There's also a human workflow of some kind involved. You can actually find this out if you start picking at different big algorithms that people have and use. You find that there's often a stage where you have analysts checking things or somebody is sending off a data set to a Mechanical Turk. That means there are a lot of options for how you might go about it. Maybe you have some people who could do a bit of the work, and then you have the algorithms and the data, and then you have the problems. So piecing those together and imagining how they'll come together—that to me is where you get the magic.

But then after that, within each one of those pieces, there's lots of hard work, and often you get really interesting stuff happening. So there are some fascinating algorithms to play around with and really insightful things that people do when they have brainy insights where they say, "If I did it exactly this way, something cool would come out."

Gutierrez: What advice would give to someone starting out and what should they strive to understand deeply?

Heineike: I think perhaps they would need to start by looking at themselves and figuring out what it is they really care about. What is it they want to do? Right now, data science is a bit of a hot topic, and so I think there are a lot of people who think that if they can have the "data science" label, then magic, happiness, and money will come to them. So I really suggest figuring out what bits of data science you actually care about. That is the first question you should ask yourself. And then you want to figure out how to get good at that. You also want to start thinking about what kinds of jobs are out there that really play to what you are interested in.

One strategy is to go really deep into one part of what you need to know. We have people on our team who have done PhDs in natural language processing or who got PhDs in physics, where they've used a lot of different analytical methods. So you can go really deep into an area and then find people for whom that kind of problem is important or similar problems that you can use the same kind of thinking to solve. So that's one approach.

Another approach is to just try stuff out. There are a lot of data sets out there. If you're in one job and you're trying to change jobs, try to think whether there's data you could use in your current role that you could go and get and crunch in interesting ways. Find an excuse to get to try something out and see if that's really what you want to do. Or just from home there's open data you can pull. Just poke around and see what you can find and then start playing with that. I think that's a great way to start. There are a lot of different roles that are going under the name "data science" right now, and there are also a lot of roles that are probably what you would think of data science but don't have a label yet because people aren't necessarily using it. Think about what it is that you really want.

Gutierrez: What is something a small number of people know about that you think will be huge in the future?

Heineike: I think there's been a focus on people working in tech companies that have a big website, where there's a lot of click-traffic of people moving around and they're using data science methods to optimize user experience and sell a lot of ads or products. There's been a lot of focus on using your company's own data and then optimizing on top of it. That will continue to happen.

What I actually think is more exciting is that there's a lot of data available now which could tell us broader stories about what's really going on in everything from global economic or political systems, to epidemics and conflict, to cities and our physical environments. Maybe you need slightly different tools to analyze this data, but the analysis could really transform the way that we make decisions or interact with those systems. I'm fascinated with how that's going to change the way that will change those systems in general. And it might be dangerous, right, because you get data that's not really representative, and then you can come to the wrong conclusions or bias against some types of people. So there are definitely some parts we have to be very careful about.

For example, if you think about the broad range of information that can be gleaned from phone sensor data, that could inform us about the way that people interact with cities, and how this could change the way that our urban systems work. Or if you think about the masses of data on what our governments are legislating and how they are financed, that are really hard to understand at the moment, but could change the way we think about democracy and government accountability.

We're getting all these tools available, but a lot of the information that we could be tapping into is actually—when you dig into it—very underutilized at the moment. I think that means there are a lot of opportunities to do interesting new things with it, and it's going to be fascinating to see what those things are. And, hopefully, more of them will be, "We did a better job of understanding things. We made better decisions. We built a better way for people to interact with each other." And fewer of those will be, "Oh, that's really creepy. I'm not sure I'm happy with that."

Gutierrez: How do you think about hiring people?

Heineike: An interesting thing that's happening at the moment is that the academic space is struggling because of funding cuts. And so there are actually a large number of very, very intelligent people coming out of computational PhDs and postdocs who are thinking, "I want to get into industry," and are actually looking at others doing interesting stuff with data science, and thinking to themselves that maybe they'll have a go at that. But often they don't have any experience in working in business and they don't necessarily have any ideas about what it's like to work with an engineering team, for example, or what it would look like to have a product built on top of their work.

On the other hand, I think you have a lot of people who have been working in industry for a long time, who maybe don't have as deep a technical knowledge in a certain area but have a better idea about how to work in teams and the industry, as well as what it's like to have a product built on top of their work.

I think, in general, it's very hard to hire people who are a complete package, who know what to do and how to do it. It's very challenging, so for the hiring we do, we kind of take bets on a bit of everything, or mixing those together, or looking at the people who just have excitement and enthusiasm and who will learn what they don't know. I think probably going forward, this kind of career is going to be very much one of not being afraid to keep learning a huge amount. So that kind of aptitude and attitude is really important.

Gutierrez: What specific tools or techniques do you use?

Heineike: We use Python extensively to do computations. Python is a really nice language, which is relatively easy to learn and quite elegant to work with. Within the data science work, there's a lot of natural language processing, which there are toolkits for, and we end up writing quite a bit of our own code, too, to make sure it does exactly what we want it to do. We worry about entity extraction, tokenization, and normalization. We worry about different ways of doing dimensionality reduction. We worry about all kinds of issues that come up with text.

As for the network work we do, I think the network science space is interesting because it's a much smaller community. Probably fewer people know about that. There's been a lot of very cool work done over the last 20 years. Graph theory's been going on for ages, but it's been much more recently that people have actually had really large network data sets where they've been able to study the structure of the network and what it means. There's very active research into how to identify an interesting node in a network, how to find a community within a network, or what properties of networks are meaningful. So that's a really fun community to keep interacting with and an important source of new techniques for us.

One thing that's maybe a little surprising is that we've found some of the closest parallels to what we do are actually being done in bioinformatics. For example, Patsy Babbitt at UCSF [University of California, San Francisco] has a lab that's running analysis of proteins, where they look at large numbers of proteins, compare them all to each other, use network visualizations to examine them, and then, through analyzing those proteins at scale, find leads for what science should be done. Their results allow them to tell other scientists, "Probably one of these proteins will be doing something interesting," or "Maybe you should go and look at this," or "This protein might tell us about the evolutionary history of these proteins because it bridges them," or "This result is actually very surprising." They're able to give context to decisions about what science to do. Actually there's a lot of parallels, which is kind of surprising because they're completely different, but it's kind of fun when you find these parallels in unexpected places like that. It's also handy when it comes to recruiting people who have relevant knowledge.

Gutierrez: What's something interesting that you and your data science colleagues are currently talking about?

Heineike: One thing that's been really interesting is how much our conversations keep coming back to ethics and concerns over use of data. I think that reflects the fact that a lot of the data science that's being done is really about analyzing people, and maybe about analyzing people when they don't want to be analyzed or don't know that they're being analyzed. I'm quite a private person in general, and the thought of having my browser history tracked is slightly unnerving to me, and I think it's probably true for a lot of people as well. I think there are definitely things that are possible to do now that are getting kind of creepy. And I think that there's an onus on us and on this community to really think carefully about what we're doing and about being wise about the problems we choose to solve and the consequences of what we do.

I do think that there are a lot of problems to solve, which do not involve tracking people and getting them to buy stuff that perhaps they don't want to buy. There are a lot of problems that are very important to solve that will help people do more of what they want and need. When you think about it this way, it's getting harder and harder to understand what's happening globally—specifically, things that might affect major decisions that politicians are making or things that help us understand big social changes that we should already be on top of and care about. There are big global issues that we should worry about for which there's data that would help us unlock and understand a bit better what's going on, so we can actually tackle them more effectively.

There are some great groups out there who are encouraging people to use data science skills for solving problems for government or for nonprofits problems where, when we think about what the impact could and should be, we can feel much happier about it. And so I think that there are three parts to this. One, don't assume that everything in data science is about tracking you and being kind of nefarious. There are other uses. Two, if you're getting a job, expect that there are going to be some really cool use cases out there for using data, and so don't settle for something where you don't feel comfortable with what you're doing. Finally, keep questioning and considering the consequences of what you're doing.

Victor Hu

Next Big Sound

Victor Hu *is the Chief Data Scientist at Next Big Sound, an online music industry platform that tracks artist popularity and profitability and fan behavior across social media, radio, and traditional sales channels as reported at a granular level by record labels. Next Big Sound currently follows hundreds of thousands of artists across more than thirty Internet and social media platforms, assessing current popularity and future trajectory. Reaching that dual assessment is not an easy task. The growth and fragmentation of music audience communities, the increased speed and immediacy with which fans engage with artists via social media, and the intricacies of sales channel data all pose challenges in terms of how to model missing data, how to solve named entity recognition, and how to ensure accuracy in predicting burgeoning talent.*

Hu's career before joining Next Big Sound encompassed work for the US Department of Defense, actuarial consulting at Milliman, and statistical analysis for the New York Yankees. While still at Harvard, Hu sent his résumé and baseball analysis unsolicited to the Yankees, and they swiftly picked him up. His work for the Yankees was featured by Sports Illustrated, *which named him one of their "25 under 25," and* New York Magazine. *Hu's subsequent work leveraging social media to predict music trends and artists' sales has been featured in* Forbes, South by Southwest, *and elsewhere. Hu holds an MS in Statistics and a BS in Applied Mathematics from Harvard University.*

Hu demonstrates how a data scientist can apply math and statistics skills across fields as diverse as sports, actuarial finance, defense, and music. The freedom to pursue a broad set of interests and dreams is the backdrop to Hu's remarks about how the intelligent application of data can benefit every industry, how he chooses which projects to work on, and the importance of cooperating, collaborating, and establishing coalitions with the people who use his work, to make sure he has a lasting impact. Hu's belief in the power of data science to transform even the most traditional industries perfuses his interview.

Sebastian Gutierrez: Why does data science interest you?

Victor Hu: I think there is such an explosion of data in the world today, and our capabilities of storing and analyzing that data are rapidly expanding. The analytics that accompany data expansion are not quite developing at the same speed, so there is an explosion of the potential of what products people can work on, what questions people can answer with that data, and what industries can be affected. That is very exciting. There are so many industries where data can make a big difference. That is what inspires me more than anything else—the way that very traditional industries—finance, health, public policy, music, sports, just about anything—can benefit from the intelligent application of data.

Gutierrez: Tell me about the journey from college to major league baseball.

Hu: I studied math and statistics at the undergraduate and graduate levels at Harvard. For a long time, I wasn't sure what I wanted to do with this skillset. People always told me, "You can do anything with a degree in math"—which I think is really funny. I do not know if that is necessarily true. I think you can apply mathematics in any number of industries, but the approaches are often similar.

One of my earliest inspirations was reading *Moneyball* by Michael Lewis.[1] He brought to the forefront the concept that data was transforming the baseball industry. That was one of the earliest instances where I really saw how powerful intelligent data analysis can be. This was, I think, even before the term "data science" was really in play. Yet all these people in baseball operations were coming in and providing very valuable insights that maybe went against the norm. And, obviously, anytime you try to do that, there is some resistance. Ultimately, because it was so successful and because there was this incontrovertible truth that more data leads to better insights, it has now become a big part of how decisions are made in baseball. Hearing and reading about that, and seeing how big of an impact you can make in an industry that I never imagined you could do that in, was very inspirational.

That is how I got into sports initially. I wanted to do a lot of that work and get in on the ground floor with it. I became an intern with the Yankees, as one of the two first people they hired to do this type of analysis. It was really exciting because it felt like the Wild West. Anything was in play—anything that I wanted to do or they wanted to do was a possibility, and we tried so many different things. I think that is one of the most exciting things about data science today.

---

[1] Michael Lewis, *Moneyball* (W.W. Norton & Co., 2004).

Gutierrez: What was a key lesson you learned from your experience with the Yankees?

Hu: I think one of the most important lessons I learned was how critical it is to persuade other people. One of the big challenges of being a data scientist—that people might not usually think about—is that the results or the insights you come up with have to make sense and be convincing. The more intelligible you can make them, the more likely it is that your recommendations will be put into effect. That is very much something that you have to pay attention to. You cannot live in a cave and generate all these really cool things that nobody understands and nobody can put to use.

We had a lot of situations where I would make recommendations that would sort of go against what conventional wisdom prescribed. In these situations, I had to make compromises between the data and the real-life applications.

Gutierrez: Tell me more about one of these situations.

Hu: One example is when I was researching batting lineup decisions, trying to figure out the most efficient lineup for the Yankees. One of my suggestions was to put a more on-base heavy hitter—a big-name slugger at the time—into an earlier spot because it would yield more runs. However, this very much goes against the conventional wisdom of who should go early in the batting lineup. Conventional wisdom says that fast, base-stealing players who will not hit into double plays that should go into an earlier spot. A slow-footed slugger will hit into a lot of double plays, so conventional wisdom said to put him in a later spot. Also, there was a big concern that he would not accept the move up in the lineup because he saw himself as a power hitter, which meant hitting later in the batting lineup. So there were a lot of real-life concerns that I did not factor in when designing the model.

Eventually, we reached a compromise that involved using a different big hitter in place of the slugger. This other big hitter was on the roster at the time and he also had a good on-base percentage. We ended up moving him up into the second spot in the lineup because he was much more amenable to the move. I think this was possible because of the culture that he grew up in—where there was not as much of a focus or stigma based on where you hit in the lineup. It was a very important lesson in terms of balancing the data with the people. It is always important to make sure that you keep the people that you are making recommendations to in mind.

Gutierrez: As you have worked in different organizations such as the Yankees, the US Department of Defense, and now Next Big Sound, what have been key areas of focus for you?

Hu: Skill acquisition is always on my mind—I am always learning new things. It is interesting because I feel there are definitely different tracks of skills you need as a data scientist that all intermingle with each other, but when you are learning them, they are almost independent fields. For example, there is the communication aspect: How do you communicate your findings effectively and how do you persuade people to take them on? A great deal of that has to do with effective writing and effective data visualization. That is one very important skill that I think is just going to continue to improve with visualization software.

There is also the track of machine learning techniques, which is constantly evolving. There have been a lot of techniques that have become more widely accepted over the years. That said, every day people are coming up with new and exciting tweaks or innovations on those techniques. Especially on more complicated problems, such as text analytics, the techniques are continuously evolving—so it is important to keep up-to-date with the most advanced machine learning techniques that can be applied and what implementations are better than others.

There is also the data management track, which deals with the entire back end that goes into storing and later retrieving the data. I did not fully comprehend this track when I was dealing with smaller sets of data, such as in baseball or in the actuarial fields. This was because the amount of data that we used was relatively manageable on one machine or a couple of machines. But now as we are starting to deal with bigger and bigger sets of data, having to deal with all the back end of storing huge quantities of data, being able to access the data later, and then being able to run algorithms on it is definitely a big challenge. I think the parallelization of traditional machine learning algorithms is still something we are struggling with, especially with the most efficient ways to do it. I'm excited to see how it continues to develop.

Gutierrez: How do you learn new skills in the various tracks?

Hu: I think both speaking at and attending conferences is a fantastic way to keep up-to-date. I chaired one of the days at the Predictive Analytics Innovation Summit Chicago 2013 conference, and I got a chance to chat briefly with each of the speakers. Hearing what they have learned and what they are on the forefront of is always very exciting because you cannot keep track of everything at once. There is not enough time for that. I wrote down so many things that I learned and heard about from other speakers that I am very excited to incorporate into our workflow.

Outside of conferences, meetups, books, and discussion groups are good ways to stay up-to-date as well. There are a lot of meetups in New York City. It is hard to keep track of what is going on in all of them, but anytime there is an interesting talk or book that I hear about, I definitely try to attend or read it.

There are a lot of discussion groups that lead to a lot of free-flowing discussions. I think the New York City data science community is pretty tight. For example, I was at a data dive for the DataKind organization this past weekend and I ran into a lot of people that I have run into in the last three years who are active in this community. It is great to see that sort of overlap. Anytime you run into data scientists, you can talk shop and you can hear about what is going on in their respective fields. This is great because we all overlap in many ways, so that is also a great way to keep up with all that is going on.

Gutierrez: Tell me about the social and music industry data that Next Big Sound looks at.

Hu: Next Big Sound provides analytics and insights for the music industry by tracking different data signals to help record labels, artists, and band managers make better decisions. To that end, the data that we focus on is combining social media data with sales data, radio airplay data, events data, and any other proprietary data we can get to provide context and cross-sectional insights to the music industry. We keep track of social media on at least thirty different sources—Spotify, Facebook, YouTube, Twitter, you name it—and combine that with sales data from the record labels. The data is relatively granular. We are talking about album and track sales for each particular artist on a daily basis, broken down a lot of the time by different geographical regions and demographics.

Gutierrez: What do you seek to answer with the collected data?

Hu: The original vision of Next Big Sound was: How do we find the next big sound? Our core belief is that you can do this better with data, specifically social data. So we take the various data sets and analyze a couple of different things. Number one, what is the impact of social media on sales? Or what is the impact of radio on concerts? What is the impact of events on streaming? And then, what is the capability for forecasting album sales or future engagements in social media? So when an artist plays a concert or makes a TV appearance, what is the impact on their social media and sales? How much can we measure that? Then how can we find up-and-coming artists?

What is interesting is that you can get indications of who is becoming popular, who is going to be the next Justin Bieber, Katy Perry, Kanye West from the earliest seedings of a couple hundred people listening to one of their tracks on SoundCloud. You can see the growth of that artist. For example, Gotye had a big song, "Somebody That I Used to Know" which came out in 2011. Before the song exploded, he was relatively unknown outside of Australia. Yet, you could see right when that song came out, if you were looking at SoundCloud, that you were looking at the fastest-growing artist. Before he was signed to a major label, before anybody knew about him, he was number-one for months just based on that metric. So if you were tracking the data, you would know that he was going to become big, even though he was not signed until a few months after that. So one of the tools that we built was essentially a reverse lookup on these metrics. Who has the most SoundCloud plays in a given week, in a given month, in a given year? And that is based on all the data that we have. We can actually build a database that allows us to make that type of grid.

Gutierrez: Who finds this valuable?

Hu: The entire music industry. The major labels are our biggest clients. There is definitely a lot of value in providing these types of insights to the major labels in terms of whom they should sign. We recently signed a record label deal where we will provide specific artist recommendations and predictions of who is going to be big in the next one to two years, and if they sign them, we get percentage points on the record deal. Besides record labels, we also work with promoters, band managers, and the artists themselves. So it is really across the whole spectrum. I think it is all about being a thought leader for the music industry.

For example, last year we published a post about our research on how social media impacts sales. We ranked all of the networks based on how much of an impact it actually has on your sales, and how you can predict future album sales based on how well you are doing on any one of the different networks. We found a rather surprising insight that Wikipedia has an amazing predictive effect. We have to be careful not to say, "You need to drive people to your Wikipedia page," although that is somewhat true in the sense that Wikipedia is a proxy for a deeper interest in you. So if somebody actually cares enough about you to look up your background, what songs you have released in the past, and who you have worked with—if she wants to know more about you than just that one hit song that she's heard—then that person is much more likely to buy your album. So that insight, I think, was very unexpected in the music industry. And after we posted that, we have heard the research and results cited all over the music industry.

Recently we also expanded into new verticals within the entertainment industry. For instance, we have launched a book division and are now doing a similar type of research around how social activity correlates to sales for publishing.

Gutierrez: What was the first project you worked on after you joined Next Big Sound?

Hu: The first project I worked on was actually trying to clean up our data. We have data from all different sources, from different social media sites and different APIs. We have data that feeds in from our customers and from data providers, and these are all very different sources that are being integrated into the dashboard. The issue we run into is that we always have to deal with missing and/or incorrect data. That is, I think, a problem that every company deals with.

One of the first things I tried to do was to see if we could use machine learning to predict what values were clearly wrong. One issue we found is that there are a lot of cases where profiles are incorrectly connected in our system. So we would see that Justin Bieber, for example, is the biggest artist on Twitter as well as on a couple different networks, which is expected.

However, because he is so popular, a lot of other artists would get his Twitter profile incorrectly associated with them. Trying to catch these kinds of important issues systematically was one of the first projects I worked on.

Gutierrez: How is the data team structured and how do you work together?

Hu: Different people on our team work on different things at different times. We have data engineers. We have data scientists. We have front-end engineers. We have people up and down the stack. I think most people would say that they work across the entire stack. There is not as much segmentation of roles as there might be at a larger company. So it's great. We self-organize ourselves based on what project we are working on in that cycle, rather than sticking to specific roles. Because we are a startup—although a relatively late startup—it's a great company environment because everybody works on everything. There is very little restriction in terms of hierarchy.

Our work is based on two-week "cycles." We all get together every two weeks and we choose what projects we want to work on. And then two weeks later, we do a demo day where we present all the projects we have worked on, as well as all the progress that we made. We assess where we go next, and then we choose projects again. So we work on these very short iterations. This is based on a theory that our VP of engineering has really been a champion of—this notion of failing quickly.

Gutierrez: Should others move to two-week project cycles?

Hu: I think it makes a lot of sense if you have the capability of doing it, simply because the merits are evident—you get improved morale and you get improved dedication. People are more interested in their projects because they have selected them. Part of that might be unique to the industry that we work, in, but choosing what to work on yourself and feeling invested in your company, definitely improves motivation across the board.

Obviously, we have dealt with a lot of the struggles of maintaining such a cycle system. The main issues are around: How do you keep people accountable? And how do you keep people informed so that they can select their projects to the best ends of the company? It can be challenging, but if you have the capability of implementing this structure, it really does provide a more agile work environment, because you can incorporate things that you have recently learned at a conference. Or if you had an idea when you were in the shower, you can implement it almost immediately. You do not get caught in these long cycles where you are working on a big project without feedback–and feedback is essential. I definitely subscribe to the ideas of quick iterations and failing quickly.

Gutierrez: How do you choose which projects to propose and/or join?

Hu: This is a question that I think about a lot, and I think any data scientist wrestles with this problem. My theory on this is it is all about producing something that your audience or customer will find useful or actually needs. This means, in general, that I place much less focus on theoretical techniques. I think that is just a necessary component of working at a startup or a fast-paced environment. So in this role, I think mostly about what our customer or our theoretical future customer wants and needs most immediately.

This involves working really closely with the product team and the customer team in terms of figuring out what insights that they actually care about. We track all our products on Trello. Our team puts together an overview of all the data science questions that have come from customers over the years. We are adding new questions to the board all of the time. The key to what projects to suggest or choose is the prioritization of these questions. We are always trying to surface the projects that our customers care about at that moment in time.

Of course, something that we are always wrestling with is what is interesting theoretically versus what is useful but more mundane. There are a lot of theoretical things that I want to work on and that I find really fundamentally interesting, but when our product or customer teams hear about it, it's like, "Oh, God—why are you spending time on this? That's stupid." Once we have a conversation about it, I take that information and blend interesting theoretical things with the needs of a customer.

Gutierrez: Take me through a recent project you worked on and what insights you discovered.

Hu: One of the most interesting and impactful projects that I worked on in the last year was essentially the prediction of up-and-coming artists. This is something that we have wanted to do since I started, basically since the beginning of the company. What really made it possible was the shaping of the data and figuring out what we actually had. Of course, there had been all of the projects that came before it, where we had tried different things. What was different about this project was the aha! moment. The aha! moment came when we realized that what we had been lacking in previous projects was choosing the correct success metric.

In music, there are many ways you can define success, many different stages in an artist's career, and many different milestones that you might think of as important. But in order to really build a model and to make predictions, you have to define what your output metric is that you care about. So one day we were having discussions about how would we do this, and the success metric that we ultimately came up with was the Billboard 200. If we define what we care about in this particular question as the artists who are making hit albums as defined by reaching the Billboard 200, it gives us a very useful metric.

This metric is useful for three reasons. First, it is consistent over time. Second, it is error-independent for the rankings over time for the relevant artists we are concerned with. And three, the data is available – both historically and going forward. Though the data is available, it did take a lot of wrangling to get that data. Once we have the data and success metric, we can then make a prediction. All of the feature selection that went into modeling was also a unique challenge within itself. But the key that really set us off on that path was being able to define what we wanted to predict.

Gutierrez: What work did you have to do before you were able to create the model?

Hu: Leading up to the project, we had done a lot of work producing the backend, loading all of our time-series data into a database that can be queried for who the top artists are in terms of plays, in terms of growths of plays, in terms of totals across the networks, and other similar queries. So you can quickly pull up the top ten artists, the top million artists, ranked in order, rather than having to go artist by artist, which is how our data is stored. Once we had that in place and we had the framework for what we wanted to predict, then you could get to the juicy stuff.

I want to emphasize how much setup is involved in getting to the point where you can actually do the modeling. You have to think about what question you want to answer, as well as what question you can answer with the data. So many people, I think, neglect to think about how long that takes and what industry-specific knowledge, as well as knowledge of your own data that this takes. So that was an important lesson.

Gutierrez: Once you arrived at the modeling stage, what was the process like?

Hu: The modeling was definitely an iterative process. We started off with throwing theoretical models at it, and quickly realized that there were a lot of things we had not accounted for in the initial thinking. For example, most artists do not have all the social media networks set up and connected. So you get this unusual data artifact that, for each row of data about an artist, you only have a couple of metrics for that artist, and it varies across the whole universe of artists.

Further, it is a little bit unclear whether that is systematic or not, whether that is indicative of anything, or simply that the artist has not gone on that network yet, so that is why they do not have any data. So that was definitely an unusual aspect of the data. I realized it when I ran the model, and all of a sudden, all of these artists who did not have certain networks connected were showing up really low—like Kanye West did not have Facebook or a similar network connected, so his predictions were really low, and that obviously did not make any sense.

We had to go back and figure out how to deal with that, so it was very much an iterative process. That was where a lot of the statistical testing comes in, and you can see that the fact that someone does not have a network connected actually does provide a lot of information. Eventually, I had to code that in—the presence of a network is one of the predictor variables. So that is one interesting and kind of unusual aspect to the music data that we discovered during the modeling process.

Gutierrez: What kind of tools do you use in your data stack?

Hu: We are primarily an R shop in terms of the data analysis. Our fullstack is mostly in Java and PHP, though the modeling is done in R or Python. Our data is stored in HBase, and then we pull it out with Pig and store it usually in Mongo databases for events. A lot of times we will do SQL databases for time series, just to make the data science easier. And then the visualization is done in different things. Sometimes we will use R, and other times we will use D3.js. Actually, one of our big pushes right now is to do more D3.js visualizations.

What tools we use evolves very quickly. Just a couple of months ago, all of our data was stored in Cassandra. We made the shift to HBase literally in this last month or so. I have now been using Pig and Hive with our more Hadoop-oriented data backend. I am sure next year we will be using something different or the tools will have evolved into something different. So the speed at which new technology is coming out is really astounding.

At a conference I recently went to, PrestoDB was one of the new technologies, widely touted as an even faster version of Hive. We always struggle with connecting R or Python with the Java back end and the PHP front end. So there are different ways to do that based on different technologies that are coming out. It's all about using what works, what you need at that moment in time, and not necessarily worrying about two years down the road because everything will have shifted by then.

Gutierrez: Given the fast pace of change, how do you think about hiring or integrating someone new into the team?

Hu: Hiring data scientists is very exciting at this time because in some ways there are no established guidelines on how to do it. People have skills in so many different areas. I know when we were hiring our second data scientist I had specific things that I was looking for. My philosophy at the time was to hire someone who could do things that I could not do, or at least had a big spectrum of knowledge that I have very little in, so that I and the entire team could learn and benefit. Together we would be complementary pieces.

I think that is what we always strive for when we hire somebody, data scientists or not—we look for people who are very intelligent and can learn on the fly. I think that is a big component of data science today, because nobody knows all the answers. There is not necessarily an established playbook for things, so you really have to be able to incorporate whatever is new and interesting at the time. Those are the big things we look for when we hire: someone who is smart, who knows the basics of math, statistics, and programming, who can learn really quickly and incorporate new technologies, and who has knowledge that we do not already have.

Gutierrez: What advice would you have for people who are hiring their first data scientist?

Hu: When I was hired as the first data scientist at Next Big Sound, there were a lot of people on the team who were already very data-savvy and maybe did not have the bandwidth or the specific statistical background to do some of the work. And so they were capable of assessing the data science candidate's quality. If you do not have that background, I think it would be tough to really discern between different qualities of data scientists, because there is really a huge spectrum of people out there.

If you do not have some institutional knowledge, it would be difficult to tell who is qualified or not. This is especially true because there are not very established programs in most universities. That said, there are a lot of new programs that are just now coming into play, so this could change in the future. Currently there are a lot of people learning from online courses and from just doing it themselves, which I think is great. What I look for when hiring is people who have done projects, specifically who have delivered concrete insights using machine learning and who know how to communicate that.

Gutierrez: What does delivering concrete insights mean to you?

Hu: At the end of the day, if you are not changing the behavior of the customer or the industry that you are in, then it is hard to assess the value of your insights or your work. So that is what I think about every day in the research that I am doing. I ask myself: is the new product capability that I am thinking of actually going to change the workflow of someone in the music industry? Is the research going to change how they find artists, or how they market their artists, or how they decide to tour or release albums, or something that actually affects their decisions? This line of questions applies to any industry and to any field of data science.

Volunteering at the DataKind Data Dive recently was really illustrative of that because so much of what we are trying to do there is to affect nonprofits that really touch thousands, if not millions, of people. We were working with places like the UN, the World Bank, and Amnesty International. I was helping lead the Amnesty project, and we were able accomplish a lot in a short amount of time, because they were there to provide immediate feedback on what specific products or insights would actually affect their day-to-day. Having the NGO representatives and the people who have that industry-specific knowledge working closely with the data scientists the whole weekend was, I think, what made it a lot more impactful and successful. Having that quick feedback is key.

The longer you wait to figure out whether what you are doing is correct or not or helpful or not, the more time you could potentially be doing the wrong thing.

Gutierrez: So having data, a success metric, and a model is not enough—you also need impactful feedback.

Hu: Precisely. Again, I think that impactful feedback is a very overlooked thing and is something I have really learned to look for over time. You really need that feedback from whomever you are providing insights for or whom you are interacting with, because it is easy to get burrowed in your cave. It is easy to move towards providing this beautiful mathematical insight or applying this really sophisticated algorithm, and then later to realize you're providing something that does not have any real impact. This is especially important because a lot of the machine learning techniques do not lend themselves to interpretability. I would say a large percentage of the time, you would rather have an easily interpretable algorithm and results than a slightly more accurate one.

Gutierrez: What technologies or techniques do you see as the future of data science?

Hu: So definitely, the number one thing is natural language processing. Everyone thinks it is interesting, everyone cares about it, and everyone thinks there is a lot of potential there. Yet, no one has really done it effectively. I believe in it and its future. I work on NLP projects whenever I can, both at work and in my free time. When—not if—we solve the problem of understanding sentiment and being able to extract meaning from large bodies of text, I think that will really change the reach of data science in basically any field.

Gutierrez: What nonwork data sets have you worked with recently?

Hu: As I mentioned before, one of the big nonwork projects that I have been working on recently is DataKind. Their mission is essentially "data science for good," so they connect data technology people with NGOs that have good data or interesting data, and then we work together to provide insights for them. It is a great mission. One of the projects that I worked on recently was with a non-profit that focuses on trying to catch child predators with data from online message boards. It is a great cause and this project involved a lot of text processing. What is remarkable is that there is just so little done with this data from NGOs at this point, that DataKind can really help.

What is powerful is that we were able to provide simple insights like, "These are the topics that people are discussing in your data and this is how you can identify whom you can target from that." We can then provide this code to the organization and provide them an easy way for them to run this over time. So this idea of being able to apply a clustering algorithm fairly quickly—something that you can do in a couple of hours and that is reproducible—is very powerful. The lesson is that relatively quick simple things can really provide powerful insights both to nonprofits and to any industry where data is not being used to its full potential. What is ultimately inspiring at the end of the day is that together we really can make an impact.

Gutierrez: For people who are up-and-coming, what is something they should really understand?

Hu: A couple things. First is that you definitely have to tell a story. At the end of the day, what you are doing is really digging into the fundamentals of how a system or an organization or an industry works. But for it be useful and understandable to people, you have to tell a story.

Being able to write about what you do and being able to speak about your work is very critical. Also worth understanding is that you should maybe worry less about what algorithm you are using. More data or better data beats a better algorithm, so if you can set up a way for you to analyze and get a lot of good, clean, useful data—great!

Gutierrez: For an organization that is just starting the data collection, how would you convey to them how to capture this data correctly or how to encode it correctly?

Hu: That is a really tricky question. I think intelligent storing of data is something that maybe people do not think about, because when they're acquiring this data, they're not anticipating how the data science components will fit and what they can do with it. So as data science becomes more popular and important, having those institutions in place whereby you are getting the data you need is really critical. Unfortunately, I think it is hard to know what you really need until you dig into it. So I do not fault anybody for realizing halfway through a project that they do not have the data they need. That happens a large chunk of the time, even dealing with what we do at Next Big Sound. But it is about recognizing that as early as you can, so maybe going through and actually thinking about what you want to do and figuring out whether you have that data and acquiring it of you don't.

At Next Big Sound, we've been tracking social media since 2009, and the only reason why we can do the types of cross-sectional analyses that we do is because we have that data. Nobody else has several years of daily data recordings of each artist, or the number of Facebook fans they are getting on a particular day, or how many YouTube plays they are getting. You really just have to think carefully about what you want to do ahead of time.

Gutierrez: Just the data set alone is a very valuable asset?

Hu: Correct. A great deal of the value that we provided is based on the data we have. Further, we have a dashboard that allows you to see all of your data all in one place. That is something that I have learned as well—simply having all of your data in one place, easily accessible, easily queryable about different questions—that in and of itself is extremely valuable to a large majority of people. In a lot of cases, you do not even need advanced machine learning to do a lot of these things. A simple "group by" query can get you so far. Never forget that.

Gutierrez: Why do you personally believe in data and data science?

Hu: First and foremost, I believe in the power of data to transform industries. That, ultimately, is what drives me every day, what attracts me to Next Big Sound, and what has me very positive about the future. We can improve not just how we understand music and the ways the music industry works but also improve people's lives in all areas—such as saving children from being targeted, or protecting humanitarian efforts, or having a better baseball player. Every field can and, I think, will be revolutionized by data to differing degrees.

Working closely with people and getting to the bottom of what they need, telling stories about what you do that really excite people and convey your message well, and then thinking big—those are the important things. I am really excited to see how data is going to touch different fields and different industries, and I am excited to be a part of that future, helping to improve people's lives.

Kira Radinsky

SalesPredict

Kira Radinsky *is the CTO and Co-Founder of SalesPredict, a company using machine learning and predictive analytics to provide Customer Lifecycle Intelligence. SalesPredict increases a company's revenue by helping sales people identify their real potential buyers and by providing insights for winning their business. Through the use of sales data, customer relationship management software, and signals from the web and other data sources, SalesPredict builds a model of a company's specific high-value customers and then utilizes predictive analytics to increase leads and conversion rates, as well as accelerate the sales cycle and reduce churn. Because datasets of successful customer purchase histories are often sparse and scenarios are highly-specific, it is very challenging to find the right signals in the noise.*

Radinsky is the right person to build this technology and company. Her work sits at the intersection of constructing algorithms that leverage web-based information, developing predictive data mining tools, and utilizing data about world dynamics to predict future events. In fact, during her doctoral research at the Technio–Israel Institute of Technology and at Microsoft Research, Radinsky gained international acclaim for developing predictive algorithms that provided early warning signs of major global events, including riots and disease epidemics. In 2014, Forbes named Radinsky one of the "50 Most Influential Women in Israel." In 2013, she was named to the "35 Young Innovators Under 35" list by the MIT Technology Review. Radinsky is a regular contributor and thought leader at conferences such as O'Reilly Strata, TED talks, and WWW.

Radinsky exemplifies the data scientist who strives to forecast the future and has a track record of doing so successfully. Her sense of the vatic power of data comes through as she discusses her research on predicting cholera outbreaks, her examination of what causes riots in developing countries, and her findings on fish kills and oil spills. It underlies her description of why SalesPredict matters and her prediction that it will eventually help shape decision making at all levels in companies. Radinsky's interview brims with her passion for crunching human knowledge and experience data to suggest, predict, and tackle future events.

Sebastian Gutierrez: Tell me about where you work.

Kira Radinsky: I'm the CTO and co-founder of SalesPredict. At SalesPredict, we are working on changing the way companies do business with each other. Though almost every company today has large amounts of data about how they sell and how they've managed to sell in the past, they have little scientific capabilities to extrapolate useful information from their data. This is where we come in. Our algorithms analyze companies' historical data and tell them: "These are the people you need to sell to, this is the best way to approach them, and these are the companies you need to engage." This is why I'm so excited about our work at SalesPredict, because we have the ability to take data science and change the entire way that our economy works by adding a layer of predictability to it.

We've been around for three years. In the beginning, we started with a pilot that took less than a month to run and we helped our pilot customer achieve great results. Through the use of our product, their pipeline doubled and qualification times were reduced by 90%. Since then, we've been scaling up the technology and our customer base.

Gutierrez: What is your vision for SalesPredict?

Radinsky: The way I see it, our vision is guided by wanting to change the way companies do business. This way we can help companies connect with each other in much better ways through making better decisions. I want to see them have much better data than what they have right now.

I want to see SalesPredict evolve to make that happen. Right now we're starting with building a model of the company for each company we work with. We start with prediction in sales and marketing and guide them to understand better which people they should talk to, where to stop spending money, where to start spending money, and so on. Sales are, after all, the lifeblood of a business. After we finish that step, our next step will be to help the CEO make decisions. This is just expanding the model of the business to account for and make predictions for another part. Next, we tackle the HR department, where our goal is to have the business know which person to match to which job. This helps to actually change the entire way the business is doing what it does because, again, it supplies more data for decision makers. This is what we do and will do—slowly build a more encompassing model of a business.

Five years from now, we will have enough understanding of how companies do business to give them more insights across an entire vertical. We want to be able to tell them: "This is the golden range of where you should operate. This is what we've seen companies do in the past, and this is what you should do."

Gutierrez: What is the makeup of the SalesPredict team?

Radinsky: We're a company whose engineering is in Israel and whose sales and marketing team is in San Francisco. This means that I, as the CTO, am in Israel and our CEO is in San Francisco. The CEO and I are the co-founders and we previously worked together at Microsoft. This shared history enables us to work together closely, even though we're in different parts of the world, because we've been working together so many years that we're at the point we each know what the other one is thinking.

The reason our engineering team is here in Israel is that the engineering talent here is amazing. There is a bond between many people that help to achieve higher goals—we grow up together and we go to the army together. The army is the most amazing incubator for technological people. So we have a really fruitful environment here. However, I believe sales and marketing should always be where your target market is. Currently, our target market is the United States, so that's where the sales team for SalesPredict resides.

We're a startup, so we're only fifteen people. Of the fifteen people, we have ten people on the engineering team. Everyone is a data scientist and engineer in the engineering team. Of course, though everyone is a data scientist, some come with a more theoretical background, some come with a more data science–centric background, and some come with more hard-core engineering backgrounds, where they have scaled entire systems before.

I enjoy both aspects—the engineering and the data science. My passion, of course, definitely has to do with data. However, this work is not worth anything if I don't actually make it work, so I really like the engineering aspects of implementation and seeing that happen. This is why I think a data scientist should be, first and foremost, a great engineer. My role is about thinking how can we take this technology and move it forward more and more. So I work with the algorithms on the business side of the application, define how we're going to make it all work, and then work together with engineering. We're a small team, so I try to do less managing and more working.

Gutierrez: How does SalesPredict work?

Radinsky: We are a cloud-based solution that, after authorization, connects to our customers' CRM (e.g. SalesForce) to access their sales data. Our algorithms look at where the company managed to sell to in the past and who are the people that managed to make those sales. We collect all the data they have about those people from their CRM. When I say we collect all the data,

I mean all of it—what they told them over the phone, which sales scripts, if any, were used, what they told them, how many items they bought, and so forth. Basically, anything that has been input into the CRM is useful for us.

Once we have this data, it is passed to the web crawlers we built. These crawlers crawl the web and get as much information possible about the persons being sold to and the companies that those people work in. For the people, we want to get a better picture of who they are and how they're represented on the web. For example, it would be helpful to know if they are a technical or nontechnical people.

For the companies, we want to get a better picture of who they are—not only how they're represented on the web but also the more relevant financial and operating information. For this analysis, we crawl their web pages, we do a tax analysis on them, we categorize what the company specializes in, and we buy traffic data about a company, so we have data on how many people are going to the potential customer's website.

We also get access to our customer's website data and how their potential customers have interacted with it. For example, did a person download their white paper, what pages did they view, and so on. We have a large amount of behavioral data about a potential sales interaction before the salesperson is in a conversation with his or her prospect.

Then, after having aggregated data on the prospect, the prospect's company, and behavioral data from our customer's website, we go to the math, machine learning algorithms, and models. We want two things from this data. The first is to understand deeply how sales transactions have happened in the past. The second is to be able to calculate, based on the historical data we've seen in the past, the probability of a new person becoming a customer and the actions we should take to make them a real customer.

On our side, we build these models automatically. This includes adding in the data. The entire process from the customer's point of view is done in less than ten minutes. Everything else is done completely automatically on our servers, like the crunching of the data, gathering of the data, feature engineering, and so forth, and it poses a lot of challenges. So that is a simple description of the scheme of how SalesPredict works.

Gutierrez: What challenges have you faced as you've grown the customer base, technology, and data?

Radinsky: The challenges in these types of models emerge from moving from a simple scheme to the real world. First of all, our market is very demanding. Our customer's historical data and distributions change all the time—their perfect customers today may not be who it was, say, two months ago. As everything changes, this model has to be rebuilt automatically as the data changes—approximately every week. More than that, we have to supply the decision-making salespeople with approximately real-time data. This is because if somebody just downloaded the white paper, or somebody changed their title, or something happened to affect the sales process, or you have a new person come into the sales funnel that you want to approach, the model has to take into account this new information. Regardless of whether people came in through the website, or were on a list that was bought in or came in from some marketing material, we have to score them immediately. So it has been a learning opportunity to deal with all of these moving variables.

Gutierrez: Have you faced any data challenges?

Radinsky: A data-specific challenge we've faced had to do with the size of the data. There's big hype today around big data, but we are actually a small data company. Sure, we collect data from a vast amount of sources, but customers have a very small amount of information that is relevant for their potential customers. So they can come with, let's say, two hundred potential customers, though they have seen tens of thousands of noncustomers. These noncustomers are still relevant for the business because they didn't buy and that is a signal. Building a statistical model from the data of these two sets of groups is very hard. More than that, it's completely unbalanced.

Gutierrez: Have you faced any modeling challenges?

Radinsky: There's a prevalent paradigm that data scientists take the data and build a classifier and that it's just going to work. But it doesn't work that way, and it's important for non–data scientists to realize that. I'm going to give you the simplest example we've observed. In this example, we're trying to mimic a sales process, which can be 6 to 12 stages, depending on our customer.

Of course, each of these steps in the sales process involves data. When we receive the data, we are just getting a snapshot. For example, say you're trying to sell to somebody and this person has answered a question regarding whether he is happy or not after he became a customer. You're going to have a few repeat customers, so you'll see them. So for this question of whether the customer happy or not, it will have a value of yes or no for people who are customers, and it's going to be empty for somebody who's not a customer.

So we get this data, we build this statistical model from historical past data, and then the model focuses on this simple single rule. But then when we actually applied the algorithm in real time, it just said no about all these potential people because they still didn't have this field filled in. What's going on here? There's a process, and we're learning from a completely different stage of the process. Not only that, we have to take into account that people don't always fill out the forms correctly or even in a consistent manner. What we ended up doing is mimicking completely the process of our customers and trying to rebuild it from their data and building different types of classifiers for each step.

Gutierrez: Have you faced any challenges with giving your customers the model results?

Radinsky: After we gather the data and construct the various classifiers, what we give to our customers are scores about people to whom they are selling. I come from a search engine background, as I previously worked on building different algorithms for Bing. And there it wasn't that big of a problem. You have a query and you just give back to the user a ranked result. We do this as well—our customers also get ranked list of people to call. The challenge we faced here is that sometimes the salesperson doesn't want to call whom I'm telling him to call. They'll push back and ask, "Why is this person an A and this person a B?" You can't just create this statistical model and just give it to the salesperson and expect them to say, "Oh, yeah, that's right"—because they want to understand why certain people are ranked a certain way.

Also, the salesperson will have some kind of prior knowledge that they want to see included in the model. It's a big difference between what is statistically right and what is perceived right. I actually call it emotional artificial intelligence because we work with people. Our algorithm and the results it produces have to work with people. So we ended up doing a clustering that is not only statistically right, but also takes into account all of the emotional feelings that the person who is working with the algorithm is taking into account.

One of our algorithms to address this issue is as follows. We score of a lot of prospective people for the salesperson to call on from A to D. Then we monitor whether somebody we've scored as an A is skipped for somebody who's a B. If they were skipped, then I would say that the explanation for the A classification was probably wrong. This new insight is then going to be fed back to our classifier. The model is then going to retrain itself and build a better snippet or a better explanation.

More than that, most of our customers just want to give feedback to the system, so we've been using a lot of reinforcement learning algorithms. Salespeople sometimes tell me, "This person is not an A. I know they're not an A. This person's definitely a B." And even if we know it's wrong based on historical data, the salesperson knows something, right? So we have to work with people when we build our classifiers. Our challenge is to achieve this interaction in an automated way—and this is an interesting challenge.

Gutierrez: What do you find exciting about this?

Radinsky: The part of it that makes me excited is that we can help companies perform much better because we see their sales operation from a macro point of view. This is a completely different view than they have because we see it from the data point of view, which helps us to help them make better decisions. In the last six months, we've mostly been building and scaling because we have more and more customers coming in, and the system has to be built and working for all of them all the time.

As more and more people have started using the system and following our scores, we've had to think about how to score new customers who are people our customers have never seen in the past. We also want to think about how to have salespeople do exploration. We've actually started taking that into account as well. More than that, with more customers, we've started getting more information from several customers that work together. For example, we have data from two distributors of the same company and we can see that both distributors are trying to sell to the same person. This then lets us turn around and ask important questions such as, "Why are you doing that?" So for us, what's amazing is that because we have all this sales data, we see insights and things in the business world that wouldn't be available in a different way. We're actually combining all of them together and mining them.

Gutierrez: What are the most surprising things you've found from one of these models?

Radinsky: The most surprising element of the work we are doing deals with the issue of our customer's perception of our model results and then how we work with that perception. Everyone talks about human–computer interaction. I think we're actually implementing it here. So for me, this is the next step. We have to figure out how to produce great results based on our understanding of our customers, as well as our understanding of our customer's customers, so that we can combine them effectively. This is really exciting to me.

For example, a type of thing our system will see is that a salesperson is looking to sell to a senior engineer. That senior engineer will have the characteristics of having had their career grow very quickly and have their company growing financially as well. That is the perception that a salesperson will have of a senior engineer they want to sell to. In order to find that person, we have to figure out a few things behind the scenes, like how do we know that a company's growing financially. We want to figure that out in order to be able to add the figures as a feature of our models for the classifiers. So we have to take in a lot of data to be able to match the perceptions of how a salesperson thinks of a senior engineer they would want to sell to.

Another thing that's been pretty interesting and surprising for us is the varying levels of knowledge our different customers want about a new person who comes into the sales funnel. Because we automatically install for each type of customer and we don't know our customer's data at that point, we build something called a "buyer persona" for each of their potential customers. But different customers of ours care about different granularities of this buyer persona.

The way it works is that a potential customer goes into their system and they will want us to tell them who this person is. They want to know whether that person is a technical person, or a marketing person, or something else. One of our customers cares about the difference between this person being a direct salesperson or a channel salesperson. While a different customer would probably just care to know that this is a technical marketing salesperson, which is a completely different granularity. How we approached solving this problem is that we ended up building an ontology and finding algorithms that work with this ontology. This way we can find out the value of the network granularity layer for each one of our customers. It seemed like such a simple problem from the beginning, and the more we ran, it just became more and more riddles to make it all work together like a well-oiled machine.

The other tricky thing about this ontology and getting value from the network granularity layer is that we are rebuilding the models automatically on an almost weekly basis. There are a lot of issues with that as well. As time moves forward, the models change the distribution of how someone is scored as people act and are seen acting. Even if the information about a specific prospect didn't change, an A could turn into a B rank because we trained the model differently based on global data changes. And because we deal with small data, it can happen even on a regular basis, which means salespeople will look at the new results and ask us, "What did you change?" So we have to take into account the fact that we cannot change the distribution in a fast manner because it has to make sense to the business side, which brings us back again to the emotional AI problem.

Gutierrez: How did you get involved with computer science?

Radinsky: I came to Israel from Ukraine when I was 4. My mom bought me a computer when I was around 5 or 6. She wanted me to learn and practice the Russian language through the use of language learning programs on the computer. She also bought me a lot of computer games. Some of these computer games had math-related exercises. At one point, I had a trouble getting past a really, really difficult level. I really wanted to move on to the next level, so my aunt showed me how to write a "for loop" to solve the problem. In this way, I learned to solve some of the exercises in an exhaustive kind of way instead of doing the manual calculator way all of the time. That was the first time I thought how amazing the ability to code is, "That's awesome. I can move to the next level of the game faster by just using the computer." This was what started my fascination with computer science.

In addition to computers, as a kid I was also really excited about bioinformatics. I was really interested in the fact that you could take all genetic data, actually fit it into computers, and solve many problems that look unsolvable before that, and reach medical discoveries. You could potentially build a combination of a human and a computer together. I took part in the Technion External Studies program when I was 15 years old, which allowed me to start taking college level classes while still in high school. And once I started studying at the Technion University, this was what I wanted to do—study bioinformatics. Before my studies started, at the age of 14, I went to a research camp. At the camp, each one of us selected research he or she wanted to lead—I chose to perform a research on how natural compounds affect the proliferation of cancer cells, specifically prostate cancer cells. We studied a lot of cells and tried to identify how they affect cancer. And working with this data, we had a lot of images of how those cancer cells grow. This was the first time I did image processing, because I didn't want to count the cells by hand, as there were something like five thousand cells to count in each image. So what I did was write a system to do this image processing for me, which gave me some free time to go to the pool instead of doing tedious work.

While doing this work at camp, and then my studies in the Technion External Studies program, I started reading more and more about artificial intelligence. I thought to myself, "Wow, that's amazing! We have so much data around us that we can actually leverage." I finished high school and part of my bachelor's degree, and then I went into the army. There I was a software engineer for three years, mostly doing security-oriented work. Then I came back to the university to finish my bachelor's degree.

Gutierrez: When did you realize the power of data?

Radinsky: In 2011 there was this event in Beebe, Arkansas, where something like five thousand dead birds fell out of the sky. It was very interesting because it was close to the dates of the end of the Mayan calendar, so everybody was thinking that this was a sign of the end of the world. Not only that, just a few days before, hundreds of thousands of fish washed up dead on the shore in a nearby part of Arkansas. And all the newspapers were reporting that there is no relation between the two incidents. Nobody could understand why the birds died or the fish.

I wouldn't say that this was my first passion for data, but this was the first time I actually understood that working with data was like an adventure. For me it was, the "well, what's going on here?" moment that pushed me to do additional investigation of that. So I started using Google Trends to look for peaks of searches for bird death and fish death to see when they tend to happen. I noticed that they tend to happen at approximately the same time. The data I was using at that time was the images of Google Trends, because they didn't have an API yet. I actually had to take the image and apply image processing to them in order to scan all of the lines involved so that I could have the data behind the charts. It was a lot of work to do something pretty simple. After extracting all the data, I wrote a system that looks for correlations of events that tend to happen before people search for birds and fish death in an unusual way.

What the system automatically found out is that in the location where people search for oil spills, even up to six months before, those two queries tend to peak together. The thing with oil spills is that they cause oxygen depletion, which is the number-one cause of fish deaths. I thought to myself, wow, interesting—that might be relevant. I don't know if this was actually what was going on there, I only know that one of the main oil pipelines in the US is going through Beebe, Arkansas, but for me, this was the first data adventure I took to understand what was happening.

Gutierrez: Where did this project take you?

Radinsky: In thinking about what I had found and how I had found it, I realized that the problem with these queries is that I only see what people want to show me. That is, I only see the data for what people thought was interesting enough to search for. I wouldn't see things like "oxygen depletion is causing all the fish deaths." From this I started thinking about how I could take all the news that people ever wrote, to look for causality.

But first, I had to figure out how to define causality. I first started with things like "x causes y." I looked at newspaper articles and how somebody would write something like "oil spills are caused by accidents." So I started by looking for all of the phrases that showed causality in that particular manner. Then when I had those phrases, I started doing semantic analysis at a pragmatic level to figure out who did the action, who it was done to, and so on. I scanned all the newspapers I could find since 1851 until today. And from all of those, I built what I called the "causality graph." It's all the events and how they connected based on the phrases.

From there, I quickly realized that it wasn't enough, because I needed to add to it some kind of layer of abstraction. For instance, let's say you have an earthquake in Australia. If you look at the past, you've never had an earthquake in Australia, so what is it going to do? You cannot predict from that. However, there have been past earthquakes in Turkey, so the first thing I need to do is know that an earthquake in Turkey and an earthquake in Australia are both earthquakes and both are countries. How do I know that and have my models know that? What I did is go to Wikipedia and look for different ways of abstracting of entities. In this case, I would know that you could say both of these places are countries. With this added layer of abstraction, we could now say I was closer to having solved this problem.

Next, let's say that I do have an earthquake in Australia. It finds a similar pattern in the past, like an earthquake in Turkey. The system would look at what that particular event caused in the past, and it would find something somebody wrote in the news, like: "Red Cross help sent to Ankara after an earthquake in Turkey." So the system would look at that and would output "Red Cross help sent to Ankara after an earthquake in Australia," which is a problem, because it's not the truth and it's not something that would actually happen.

The next thing I did was to adapt the prediction to what was going on in the current event that we were looking at. This meant the system had to understand that, in the past, the cause and effect happened because Ankara is the capital of Turkey. Once it understood that, then it would apply this function to what was going on now. Now the system would say, "Earthquake in Australia.

Red Cross help sent to Canberra." I looked for these types of relations—capital and country—and others using not only Wikipedia, but hundreds of other data sets from a project called Linked Data. The biggest one of these connected data sets was, of course, Wikipedia, which is just structured information from Wikipedia.

With this causality graph I could now ask it anything I wanted. An interesting example I used to give is what it taught me when I wanted to buy an iPad. I asked the system, "How much does an iPad cost? Tell me what's going to be." The system then told me that prices were going to go up. I was curious why it thought that, so I had it backtrack how it went through this causality graph. It told me now prices were going to go up because of the tsunami in Japan. I then asked how the price of an Apple product in the United States was going to go up because of the tsunami in Japan. It then showed me the relation. The chain was basically: a tsunami occurred in Japan, there were factories on the shore, some of those factories make some of the substances needed for a chip factory in China, which makes iPads for the United States. So the system had calculated that if factories on the shore were affected, it could lead to a shortage of materials. It had seen in the past that when you have a shortage of something, prices go up. It was an interesting observation to be able to deduce.

However, the problem with looking at causation is that sometimes trivial things are generated. This was one of the things that surprised me, because it could take several hops of causality to reach something trivial. For instance, I would give it the following scenario, "Israeli professor killed after bombing," and ask it what would happen next. I would also ask people the same question to be able to compare answers. The people I asked would say that it would eventually lead to protests and other bad things happening. In comparison, the system would say, "a funeral will be held." And the funny thing is that the system would actually generate an entire text of that, which is great because it's true, but it is pretty much trivial. The reason for that is because we trained the system on what people were going to view as causes and effects. This made me realize I need to move to the next level.

The next step was then to look for important correlations. The way I approached this was to look for storylines or patterns of history. A storyline would come from something like several news articles discussing approximately the same entities for a news item. This could be something like somebody gets shot, somebody gets arrested, there is a trial, someone gets acquitted, and so on. So this would be a storyline. Once we had lots and lots of storylines, I then looked for patterns in those storylines. When a pattern emerged, I would do a semantic analysis of those news articles, extract different entities, and find out what was going on between one to another, including the abstractions I mentioned earlier—country and its capital and so forth, as well as the relationship between entities that came from the causality graph.

Gutierrez: What problems did you apply the system to?

Radinsky: At this point, I was working with Eric Horvitz who has an MD and is the head of the Redmond Lab at Microsoft Research. He is extremely interested in disease prediction and algorithms. Once the system was taking into account correlations, we decided to see if the system could predict cholera outbreaks. We wrote some code and made it so that we could tell the system "cholera outbreak" and the system would give me the probability every day of that event happening in different places in the world. When we looked at historical data, it was right 80-something percent of the time. And more than that, in 2012, we predicted the first cholera outbreak in Cuba in 130 years.

What happens is that cholera is a waterborne disease, so you wouldn't be surprised that the system found out that it tends to happen after floods. But the pattern that the system found was that if two years before those floods you have a drought, the probability of those floods leading to a cholera outbreak is much, much higher. And not only that—it usually tends to happen in countries with low GDP and low concentrations of water. Why the low concentration of water? Eric told me that cholera is treated very easily by having clean water. Clean water drops the mortality rate from 50% to less than 1%. So people who had access to clean water just don't have outbreaks, which is why being a low-GDP country matters.

Having achieved this breakthrough in cholera, we then started looking at predicting riots. For example, the system predicted the latest riots in Sudan. What we found out is that if you have a basic product in the country and the price of this product starts going up, then you're going to have student riots. Then, if in those student riots, a policeman kills one of the students, this will lead to much bigger riots that can even affect the government. This was what happened in Egypt with the bread prices. The system inferred from Egypt that the same thing was going to happen in Sudan when the gas price started to go up. In Sudan, the government had subsidized the price of gas for many years. The events in Sudan unfolded just like the system predicted—in December there were student riots, then a policeman killed one of the students, which in turn caused huge riots.

For me, it was exciting that the system predicted the cholera outbreak and the riots; it was the first time I'd seen a system do that. The system can change the way we see the world, because it can find patterns that we've never seen before. This is the dream of every scientist—a better understanding of the world we live in. This is a really great tool for decision makers, who are making decisions in the dark that affect the lives of all of us. I just want to give them the scientific tools to help them make better decisions, because we already have tons of data. We can help them so much. We've recently started working with the Bill Gates Foundation to build a much more granular cholera predictor to try to predict cholera outbreaks all over the world.

Gutierrez: How have you learned about different subject areas as you've gone from academia to research and then to sales?

Radinsky: In the past, when I was working in Microsoft Research on search engines, I would go to conferences about information retrieval. There I would learn about what people see as problems. We would also see many of those problems inside our query logs, so we could put the knowledge of what was and wasn't working together.

I also follow people on Twitter to see the development in the field. I just like thinking about different ideas. I like revisiting different workshops, even smaller ones, where people can learn about up-and-coming things. For instance, I was really excited for a long time about bio machine interfaces. My husband and I did a small project where we utilized an EEG helmet to build a music recommendation system from the feedback the helmet gave. It actually connected the electrodes to your head to identify your mood and, based on that, it would give you a better music selection. It wasn't that precise and it took a lot of time to get it on, but that was pretty exciting. So for me, each time I hear from somebody who is rooted somewhere outside of my field saying, "Oh, there's this new thing," it makes me want to play and explore the new thing.

For SalesPredict, my co-founder comes from a business and sales background, so he's been telling me about his problems. The more I work with salespeople, the more I identify the problems they have with sales systems and where the failure points are in those interactions.

For the SalesPredict technology, I also read a lot of papers on artificial intelligence. I have a lot of colleagues in the space, so they tell me about the problems they encounter. We talk frequently about whether someone read this or that about this and that. I have a few friends here in Israel with whom we've organized a reading group. So from time to time, we read things from different topics that are raised. It could be something from a conference or something that comes from somewhere else. We make sure to keep it intimate and fun, so we generally meet over dinner to discuss the latest paper. As long as it's fun, it's always easy.

Gutierrez: What does a typical workday look like for you?

Radinsky: My day is really diverse because I'm in a startup. It roughly breaks down into solving problems, coding myself, or helping our team with their tasks and learning new topics, as well as meeting with people outside of the company. These things don't happen every day, but those are the things I spend my time on. In regards to solving problems, I would say that I have a set of problems that I need to solve each day. I see a lot of problems with data that come from our customers saying things like, "I don't understand why we ranked this person as an A, because the salesperson has not been able to close them after a few months." I also look at the marketing system. Sometimes this involves building different tools to actually try to find out what's going on. Sometimes I say, "Oh, I don't have those tools," so I have to find the tools. Either I build them myself or we have to find another away. This means we can go out and find one or put it into the pool of tasks for our engineers. One of the engineers may then take it and build it, or come back and tell me that somebody must have solved it before, so then we'll dive into the literature to start looking for a paper.

In regards to helping our team learn things, we engage in a great deal of lectures. This involves me giving lectures to our group so that everybody can learn more about machine learning. Some of the days I'm giving presentations, whereas other days I'm in the audience learning from someone else. Sometimes it's theoretical topics and other times it's somebody more on the engineering side of things talking about something he did or a new technology he brought in. Here's the thing—everybody's learning something new every day. And these lectures are fun for everyone. People start thinking in the new terms they are learning when they try to solve their problems and use the different tools they care about. So that's good.

In regards to meeting people, I also meet with VCs, customers, people we are looking to hire, and people interested in data, data science, and big data, and so on. It depends on the day and it can be quite diverse. Just a few days ago I met the Israeli British ambassador, as well as the Israeli president, at a dinner I was invited to about big data. Earlier this year, I went to the conferences at MIT and Strata. So part of my job is to meet people around this topic. Funnily enough, my husband is also in data science, as he has a data science startup. In the evenings, when we relax, we talk about our life like regular couples, as well as about our data science problems. It's a pretty regular life with a lot of action and I love it.

Gutierrez: What specific tools and techniques do you use at work?

Radinsky: Sounds funny to say, but I use the tools and techniques needed to solve our problems. Our production code is in Java and Scala. Most of the stuff I do is written in Scala. When it goes into production, we usually write it in Java because it's more maintainable. The prototypes I build will be written in Scala. For our stack, we work on Amazon completely. If somebody wants to try something new, I just start a server and it finds those things and scales those things out there. For our data storage, we use MySQL because we're pretty small and we have enough relational data. That said, for some of our biggest data sources, we use NoSQL databases, such as Couchbase.

The rest of tool set is all of the different algorithms that we have developed. Many of our machine learning algorithms were written in the company as some of the known algorithms were not appropriate for the types of problems we were handling. But we still play around with it a little bit. I even have people writing in R. R is not very popular here because we write all of it in Scala, so there are a lot of Scala toolkits for the things R or Weka are capable of doing. We do a lot of NLP, so we use the Stanford Parser.

Gutierrez: How do you think about solving the right problem?

Radinsky: I do a lot of analysis before I start solving the problem because I think the most important thing is to solve the right problem. So when I approach a problem, I need to make sure that this is the problem that needs solving. The next thing I do is to make the smallest possible prototype to make sure that my solution actually addresses the problem. We learn in one environment and we test in a completely different environment because we have a temporal problem, so we keep a lot of historical data. The first thing I do is to take a lot of historical data and create a staging environment for us to do all of the relevant experiments without waiting a week. I do a lot of experimentation because this is the way to work with data.

If there's an issue from a perception kind of problem, I talk to our customers. I'll ask them, "What's going on?" We do a lot of logging in our system, so I see the usage data and I speculate on different things based on why they did or didn't click on something. Because we have so many clicks, eventually I'll build classifiers to help me understand where to focus. I want to understand what the differentiation is between places where they do click versus places they don't click. Based on that, I try to explore where to get our hypothesis. So, with most of the ideas, I just make hypotheses and check them.

Gutierrez: How do you evaluate the work you and you group doing?

Radinsky: Well, engineering has very specific guidelines, so if somebody's doing scaling, we can just look at the success criteria for that problem. We assign success criteria to every engineering task before we assign the task. This helps us understand the problem and ensure we know what it means for it to be a successful task. This is how I measure the success from an engineering point of view. In data science problems like, "We currently have a 30 percent decision rule. We need to go up." And if does, it does. Eventually, the numbers don't lie. So we look at success on the data science side through numbers as well.

Gutierrez: What do you look for when hiring people?

Radinsky: First of all, I always want good engineers. I ask questions like, "In this problem, with this architecture, what are the tools you're going to use? How are you going to solve it?" I also ask them to write code. Eventually, and more than that, it's important for me to know that they know the tools they're using in depth. So if it's someone from an engineering background, I'll ask about Java or JVM, or ask something about how does memory work—something very deep to show me that this person is willing to go deep into things. This is important for me.

If it's in data science, I'll just show them one of the problems we have and ask how they would go about solving it. For me, it's very important for them to actually ask the right questions, to see that they are not stuck on some idea, that they know how to generate the process, and then talk through the entire process of data science. I have a lot of questions around the tools and techniques they would use and why they would use them. I want to hear the depth of knowledge here, as well what works better, when it works better, and why it works better. Afterwards, I usually tend to choose an algorithm they say they know very well and ask them to dive in and explain the math.

Another thing that I am deeply interested in is people knowing data structures, like basic computer science. So I have started asking questions about very basic data structure problems and then trying to move on to questions around other data structures they don't know to see how they deal with it. The last problem I give them is usually an out-of-the-box kind of question that is impossible to solve. I want to see how creative they are in their approach. This is important because I think a data scientist has to be creative in how they solve things. I just want to hear how they solve something new and what questions they ask.

Gutierrez: What advice do you have for companies that are looking to hire data scientists that don't have a data science team yet?

Radinsky: The person you hire has to understand the business. I wouldn't hire somebody who's a junior in the beginning. It has to be somebody who not only knows the toolkits and techniques, but somebody who knows how to build a team and knows the problem domain. They have to understand the toolkits and techniques very, very well and know the math behind them. So my advice, which is applicable in many places, is to make sure the first hire is super strong and just really cling to them, because you have no alternative but to identify people like them. And it doesn't have to be somebody who's from your field, but somebody who can actually explain and show that they understand your particular business and what you think. They should not only be excited about the new technology, but they should also be equally excited about making the business thrive.

Gutierrez: Are data science skills from one industry applicable to another industry?

Radinsky: Yes. I moved from one industry to another many times, so it's doable. But, again, each time you move, you've got to work with somebody who really understands the problem. You have to understand the problem to be able to apply the tools that you've already seen, and use techniques based on the data you've investigated.

Gutierrez: What's a big thing you've changed your mind about using data?

Radinsky: I think it's that it's not enough to have a statistical background to understand data and how others see data. The problem that most people have is mostly about the perception problem. It's more about how humans perceive it to be correct. I think this was the most surprising thing for me.

In the past, it was really surprising to me how much correlation you have from different data points with events that are happening in the world and that all the data is already there.

Gutierrez: How did you develop your personal philosophy on working with data?

Radinsky: Basically, I've by working with a lot of very talented people. I think the main person who affected my views was my advisor at Technion, Professor Shaul Markovitch. He gave me most of his habits about how to approach problems. For him, most of the things he was talking about were: Is it creative enough? Is it an interesting problem nobody's solved before? The way he would approach that was, first of all, "Let's have data. Let's understand that we know how to test the novel things. And after that, let's try a different hypothesis."

The next person who really affected me is Susan Dumais, a Distinguished Scientist at Microsoft. What I really love about the way she thinks is that she isn't about "Let's do something creative." Instead, she's about "Let's make a hypothesis that is interesting" or even a psychological hypothesis that's based in data to see whether it works or not—to try to find out insights. My advisor was more of an algorithm person: "Oh, let's make a cool algorithm." And he wasn't purely problem-driven. Susan Dumais was the first person I worked with who was really problem-driven. She would say, "We develop insights from the world around us. We are scientists. Even if we're working in computer science, we're still scientists. Somebody who's working in physics has hypotheses and tests the data. This is the same thing that we do. We have hypotheses and we test the data." So she really pushed me to be problem-driven.

And at the next level would be Eric Horvitz, who is more about making something of value that's going to change big things, especially in the medical community. Use those hypotheses to make something that is actionable. I'm really inspired by Eric Horvitz. We've worked together a lot and I really love his work in the medical domain. Especially, I was really impressed with how he took all the creative thoughts that people have at Microsoft and used them to try to predict the different side effects that drugs can have. This is great research for drugs and the different side effects that they are having. For me, it was pretty amazing how you can use all these things to actually help medicine today. More than that, I'm pretty inspired by the fact he has done a lot of work about decision making, which I like, especially as I'm still interested in how decision makers make decisions.

Gutierrez: What does the future of data science look like?

Radinsky: I think that a lot of things in data science have been commoditized into very simple tools where you don't have to understand them in depth to use them. This is how I think about how our slice of the world has developed. At the beginning, we had hardware and a lot of people worked in hardware.

Eventually, fewer people worked in hardware as they started working on software on top of it. We are seeing the same transition starting to happen with data. A lot of people are working with data, handling the big data, and solving those kinds of problems. Today I see more and more people moving to working with the algorithms on top of this entire data stack that's being built.

This is where I see the future—much less in the data stack and much more in working with the algorithms on top of it. For me, it's more about how the human and computer can become one. For instance, think about all the sensors we're starting to build and starting to wear. We have to think about how we can utilize them to make our lives much better. And it's not that far-fetched. Think about someone with a smartphone that has sensors. How many times does a person actually go out without their smartphone? Almost never, right? This means you already have something on board that's very close to a medical device. Well, I wouldn't say "medical," but you already have some device going with you all the time that is capturing more and more data about you and how you live. So think about later, when we're going to have more and more devices with us. How can we use that to leverage how we have assistance from computerized systems? So I would say that there are algorithms and we are going to have more and more sensors going in, giving us more input to do more analysis on top of that.

Gutierrez: You talk about hardware to software and data to algorithms—what comes after algorithms?

Radinsky: I would say self-building algorithms. Today there is a lot of manual labor in getting the data and building a specific model. I believe that the next step will be algorithms that select the data they need based on the missing variables they have seen in the past to help in those specific kind of problems.

Gutierrez: What data sets you would love to see developed?

Radinsky: I would love to see better medical data sets and better government-backed research data sets. It's very interesting to see medical data sets be developed while taking into account the privacy and health of people. For me, it's very useful to be in a path that we have all the genetic data from all the people, so we are naturally trying to find solutions to all the diseases that we have. I think we need to cure diseases. I think it's pretty amazing that today's computer science can come close to curing diseases. To reach that goal, we need the data about how these diseases behave.

To get there, I would love to see governments release much more of the data that they are funding and collecting. This data belongs to the public. And if we can analyze that, we can make decisions in the political space in a much better way. I'm just going to quote Shimon Peres on that from a meeting I had with him. He thinks that countries are going to go away and instead we're going to have big corporates ruling the world in the next wave of globalization. He just thinks there's no place for politicians. In the next political era, we're going to have a lot of businesses that control many of the things we use, though we're still going to have a lot of governments essentially doing the real mundane things like collecting taxes. It's going to be a completely different interaction. So in order to make decisions about those interactions, we have to take an interest in how governments work together, what types of relationships are available to them, what diverse steps they're taking, and what data exists and shows these relationships. If we gather all this data, we'll be able to analyze it and make it into a much more scientific process that is a much more thoughtful way about decisions and how they affect citizens.

Gutierrez: What is something you think will be huge in the future?

Radinsky: I think that eventually we're going to have some sort of genetic hardware made from real genes grown from a lab that will allow us to combine computer and genes. There's a lot of work already in this area of people trying to do that. But I still think that trying to find a way of combining genes and computers will generate new kinds of problems, new kinds of data, and new kinds of hardware for us, which goes back to the stacks I was talking about—hardware to software and data to algorithms. It's going to be a completely new stack.

Gutierrez: What advice would you give to someone just starting out?

Radinsky: Find a problem you're excited about. For me, every time I started something new, it's really boring to just study without a having a problem I'm trying to solve. Start reading material and as soon as you can, start working with it and your problem. You'll start to see problems as you go. This will lead you to other learning resources, whether they are books, papers, or people. So spend time with the problem and people, and you'll be fine.

Gutierrez: What is something someone starting out should understand deeply sooner than later?

Radinsky: Understand the basics really deeply. Understand some basic data structures and computer science. Understand the basis of the tools you use and understand the math behind them, not just how to use them. Understand the inputs and the outputs and what is actually going on inside, because otherwise you won't know when to apply it. Also, it depends on the problem you're tackling. There are many different tools for so many different problems. You've got to know what each tool can do and you've got to know the problem that you're doing really well to know which tools and techniques to apply.

Gutierrez: What does it take to do great work?

Radinsky: You have to be passionate about what you're doing and get things done. I really believe in the Done Manifesto,[1] which says that people who do things are right. People who just talk about things take longer to finish because they either don't do it or don't feel it, so they are wrong. You need to touch the iron. So you've got to understand that and do really good work. You've got to really understand things deeply. You've got to love what you do. And eventually, you've got to make it happen and not just talk about it.

When you're getting things done, don't to be afraid to fail. Failing is just part of the process. It's fine. There's a famous Edison quote: "I have not failed. I've just found 10,000 ways that won't work." It's the same thing in data science. First of all, you have to understand the problem you tackle. Second, you have to be not only excited about the technicality of the solution, but also be really excited about the business involved, the problem the business is facing, and the tools you're going to use to solve the problem. You actually have to do the work yourself or at least mentor the people solving it. You have to be really hands-on, in other words. After all, what could be more fun? You are solving problems you are passionate about.

---

[1]www.brepettis.com/blog/2009/3/3/the-cult-of-done-manifesto.html.

Eric Jonas

Neuroscience Research

Eric Jonas *is a research scientist in computational neuroscience and signal processing, pursuing two main lines of investigation. The first focuses on how learning and memory work: how does the brain take in new information, build internal models, and do things with those models (such as save them to the neocortex). His second line of research focuses on building new types of machine learning models using signal processing technology to detect patterns in this kind of data that could not be detected natively. In both endeavors, Jonas has to grapple with two distinct sets of problems. The first set involves finding the right models and techniques to interpret and work with incredibly noisy, high-throughput neural data. The other set of problems involves bridging the gap between the scientists running the experiments and those creating the mathematical models.*

While Jonas has been continuously fascinated with the brain since he was seven years old, he has alternated brain science with entrepreneurship. He co-founded two start-ups, both of which were backed by Peter Thiel's venture capital firm, Founders Fund. The first, Navia Systems, developed a new class of probabilistic computers to make inferences under uncertainty. He was CEO of the second, Prior Knowledge, which created a predictive database that learned the deep structure of the data it contained, and then used that knowledge to generate predictions based on probabilistic inference. Just three months after being featured as a TechCrunch Disrupt Finalist, Prior Knowledge was acquired by Salesforce.com. Jonas transitioned with the company and went on to become Salesforce.com's Chief Predictive Scientist. In early 2014, Jonas returned to pure research, an environment he knows well from his career at MIT, where he took four degrees: a BS in electrical engineering and computer science, a BS in brain and cognitive sciences, a MEng in electrical engineering and computer science, and a PhD in brain and cognitive sciences. He is currently a postdoctoral fellow in EECS at UC Berkeley.

Jonas is an exemplar of the data scientist who is building the tools of tomorrow. This comes through as he talks about creating instrumentation to record signals from brain cells, his frustration with the slow pace of tool development in the past decade, and his vision of having the right tools to analyze the deluge of brain data that will be available in the future. His desire to build tools that help scientists and non-scientists make sense of their world through data—together with his views on how quantitative and computational the life sciences will become going forward— energize his interview.

Sebastian Gutierrez: You recently co-founded a startup and then sold it to Salesforce. Tell me about this journey.

Eric Jonas: Some friends from graduate school and I started Prior Knowledge [P(K)] in August 2011 with the goal of building developer-accessible machine learning technology. Our vision was building ubiquitous machine learning. We were really inspired by companies like Heroku and Twilio and the way they had democratized access to a lot of what at the time was fairly cutting-edge technology. We felt it was crucial to preserve uncertainty—the ability to say "I don't know the answer"—when putting this technology in the hands of normal people. At the time, that was a really radical thought.

Certainly, the Bayesian statistical community had been doing this for a long time, but a lot of machine learning methods that were out there were just all about giving people an answer. It is important to ask whether you can trust this answer or not, especially since often whether or not you can trust an answer varies greatly on the question you happen to be asking. Machine learning systems will quite reliably tell you that there are two genders, male and female, but they won't always be accurate in predicting which one you're talking to. So that was really the goal.

We did the normal startup thing—we built the product, we demoed and made it to the finals at TechCrunch Disrupt, and we raised funding from venture capital firms. Peter Thiel's group, Founders Fund, backed us. We then were acquired in December of 2012 by Salesforce, where we found a great team of people who very closely agreed with the vision of a ubiquitous predictive platform, where at the end of the day saying "predict" should be as easy as saying "select." I was at Salesforce for over a year, where my team and I continued to work on machine learning technology. I was promoted to Chief Predictive Scientist of Salesforce and oversaw this area.

Gutierrez: You have now left Salesforce to do computational neuroscience research. Why neuroscience research?

Jonas: The brain is fascinating because it computes. The liver is interesting too—it's this complex metabolic soup, and when it breaks you die, so there are people studying it. However, as a CS person, it's amazing that there's this blob of goo in my head that somehow is doing all this intractable computing.

So these days I spend my time doing two things. First is working with neuroscientists and answering neuroscience questions about how learning and memory work. How does your brain take in new information, build models internally, and then do things such as saving that to the neocortex? Then second thing I spend my time doing is building new types of machine learning models to find patterns in exactly this kind of data. I think that's really where the field is headed, especially as we see a tremendous set of interests in neuroscience data now.

I have been interested in the brain for most of my life. Even before college, I was already fascinated with it. Then when I arrived at MIT, I double-majored in electrical engineering and computer science [EECS] and brain and cognitive sciences [BCS] as an undergraduate. I then did a master's in EECS, and then finished it off with a PhD in BCS. So my background is computer science and brain and cognitive sciences, and I have been studying it for many, many years. This area is something I've been thinking about even when I was working in the industry.

Going from being a startup CEO to working in a large company is a jarring transition, because suddenly you have extra time after work. In this extra time, I finished my PhD, and I continued doing some of the neuroscience research I did in my PhD. I was always very interested with the idea of building machine-learning signal processing technology to find patterns in neuroscience data that we just couldn't find natively. As undergraduates we're taught things like the Fourier transform or principal component analysis, but those are all like fifty years old. Clearly the machine learning technology that's out there has become a lot better at finding those sorts of patterns in data. So lately I've been working with researchers at UCSF, Northwestern, MIT, Harvard, and other academic institutions.

The Obama administration has spearheaded the new BRAIN initiative and there's going to be all this high-throughput neural data around soon, but frankly, not a lot of technology to even look at it—especially technology that's accessible to regular scientists. And I say "regular scientists" partly because the people running the experiments and formulating the questions are often actually quite different from the people with the ten years of computational or mathematical training necessary to build and understand the models. I am trying to carve out a space in between them to be at the impedance matching layer, serving as that buffer because I believe that you can be really productive there.

Gutierrez: Why is this important work now and ten years from now?

Jonas: One of the reasons that I left Salesforce was in some sense that my team was there, everything was in good hands, and I was looking for the next big challenge. Aaron Levie, the Box CEO, at one point made the comment, "One of the questions you have to ask yourself when doing a startup is why is now the right time to do the startup?" I think Peter Thiel told him that.

So for me, the question is: Why is right now the right time to be trying to build tools to solve these neuroscience problems? One reason is that the data are going to be available very shortly. About five years ago there was a real question of tracing out all the neurons in neural systems and how they can connect, because no one had the schematic of the brain up until then. Connectome projects are projects that are tackling this goal of creating a comprehensive map of the neural connections of the brain. So people have been building these connectomes of really dense schematics of the systems, led by groups such as Sebastian Seung's group at MIT.

However, the problem with biological circuits is that the schematic that you get out doesn't have a nice little box drawn around parts saying "This is an adder" or "This is a register." No, it's just this dense graph of crap. You can imagine what it's like by trying to figure out how a processor works by just looking at how all the transistors are connected. Obviously, that really limits your understanding. So Konrad Kording[1], a scientist at Northwestern, and I started trying to build models to discover the structure and patterns in this connectomic state. We have a paper that was just sent out for review on exactly this idea of how—given this high-throughput, ambiguous, noisy, sometimes error-filled data—you actually extract out scientific meaning.

The analogy here to bioinformatics is really strong. It used to be that a biologist was a biologist. And then we had the rise of genomics as a field, and now you have computational genomics as a field. The entire field of bioinformatics is actually a field where people who are biologists just sit at a computer. They don't actually touch a wet lab. It became a real independent field partially because of this transition toward the availability of high-quality, high-throughput data. I think neuroscience is going to see a similar transition. I like to say that neuroscience is generally ten to fifteen years behind the rest of biology because, in many ways, it's a harder problem: there's more ambiguity, and getting the data is much, much harder. So the hope is that right now is the right time to strike.

Scott Linderman[2], at Harvard, and I are organizing a workshop at the 2014 Computational and Systems Neuroscience [COSYNE] conference on discovering structure in neural data, organized around questions like: How do we find these items? And how do we build algorithms to find patterns in this data? In ten years, the data's going to be there, and if people just keep taking the Fourier transform or keep doing PCA on this data, then we're really going to be screwed. There's just no way you're going to understand these systems.

---

[1]http://www.koerding.com/.
[2]http://people.seas.harvard.edu/~slinderman/.

It takes sometimes a decade to get these sorts of ideas out into the scientific community, because in sciences there is never a kind of transparent ROI, so it's harder to expect people to be really eager to get onboard. But the hope is that if we can start building the right models to find the right patterns using the right data, then maybe we can start making progress on some of these complicated systems.

Gutierrez: When did you start wanting to study the brain? And what motivates your current work?

Jonas: When I was a kid, I used to build circuits and taught myself how to program a computer. I remember sitting in a 7th-grade class and wondering what the assembly language was for my brain, because clearly there's this computing thing going on there. When I arrived at MIT, I decided to double-major in EECS and BCS sciences partly because I just couldn't give up the engineering side and partly because I had some smart faculty members tell me that if you actually want to do this sort of work, you really need to have the hardcore quantitative background.

Then when I began graduate work at MIT building instrumentation to actually record from the cells, I realized partway through that what we really needed were the tools to understand the data that was being generated. I've come to the slow realization over the past several years that I'm not really the kind of scientist who has the patience to sit there and methodically explore a given system with existing tools. Usually, after a few weeks of working, I get frustrated with current tools, and I'm like—no, let's just build a better tool. And it's this frustration that's part of what is really motivating me to help build better tools.

The other part that drives me is being able to understand the brain faster. The area of the brain that we study is called the hippocampus: it's kind of like the RAM for your brain. There was a postdoc in a lab I worked in who discovered a phenomenon called "reverse replay"[3] basically by staring at the data for five years using traditional methods. I remember thinking I could have written an algorithm five years ago that would have just found that. It had me thinking of how much understanding is sitting on the hard drives of people in various labs just needing these sorts of analytic techniques. So that really continues to motivate me when I look at these data sets. I'm like, "Look, the answer is in here somewhere!" The challenge for us is to actually build the tools to find it. It's no different than trying to build a telescope to find something far away or build a microscope to see something really small.

---

[3]David J. Foster and Matthew A. Wilson, "Reverse replay of behavioural sequences in hippocampal place cells during the awake state," www.nature.com/nature/journal/v440/n7084/abs/nature04587.html.

Gutierrez: When did you realize you were on the right track?

Jonas: The real aha! moment for me was realizing that a lot of other people are doing computational neuroscience. Their work was just being completely ignored by the actual people who were getting papers on the cover of Nature. It's an interesting question of why we have this entire community of very smart people putting out all these papers in these low-tier journals, and over here we have the actual experimentalists, and they never seem to communicate. Is it because everyone is insular and territorial? Or is it because what's being produced by the computational neuroscientists isn't of use to the experimentalists? Why, if I'm an experimentalist, would I go and learn all these additional techniques if it's really not going to give me any sort of new capabilities? I saw that happening in 2004, and then I came back to it ten years later and still seeing the same thing is very frustrating. This has to change. Someone just has to do this.

There are more and more people waking up to this realization, and that's part of what motivated us to get a workshop going, because I'm going to be wicked bummed if these data sets come out and no one knows what to do with them. I think everyone—including the funding agencies—is going to be very frustrated. We don't want to find ourselves again in kind of the regime that genomics found itself at the end of the '90s where it's like, "Well, okay, we have this data. What do we do with it? Where's this clinical miracle that everyone was promising?" Part of that, of course, is because biology ended up being fantastically harder than we ever anticipated. But part of it was also that we just weren't really sure of the questions that we were going to ask of these large data sets.

Gutierrez: What are the main types of problems that people are working on in computational neuroscience?

Jonas: There are two main problems people are working on—one part is the wiring of the brain, and one part is the activity of the brain. The set of wiring questions people have—especially with this connectomics data—are: What are the circuits? How are they organized? And what are the kinds of modules that are connected? The set of activity questions are: What are the repeat patterns of activity? And what do they really mean? The problem with figuring out the answers to these sets of questions is that all of the data is incredibly noisy.

One way of understanding the data issue we face is by imagining that you are in a stadium and you can listen to 500 people at once. Your goal is to figure out what's going on in the game just from listening to those people. There are certainly some things you can tell, like who's winning, who's losing, and that sort of thing. But if you wanted to actually understand things play by play, it's actually much, much more difficult. Translating this example back to the brain, we can think of cells in the brain as being individual people. So there are lots of people trying to build algorithms to tease apart those signals in the brain to figure out what the different conversations are that are going on in a given time and understanding how to relate these conversations back to the larger state of the system.

Gutierrez: Are other people tackling other biological systems of the body with similar data and algorithm goals?

Jonas: Certainly. A lot of my friends do computational biology work where they're trying to understand how proteins interact and give rise to signaling networks. We used to think that each gene was turned into a single protein—that was the end of the story. Now we know that a gene gets turned into mRNA that then gets more or less sliced and diced and then turned into proteins. This kind of slicing process—called alternative splicing—is the reason why it looks like we only have like 20,000 genes in the human genome, but we have vast amounts more of all these different proteins.

There's incredible diversity in proteins. So lots of people—such as Yarden Katz[4] at MIT—are developing algorithms to take this high-throughput data and understand what's actually happening and what the generative stochastic processes are. If you take the naïve computer science view, every cell is basically a little computer, right? It has this chunk of memory, and DNA is the compressed obfuscated buggy binary that runs inside this complicated set of stochastic differential equations. If we're going to figure out how this all works, then we have to start applying better computational techniques. So, yes, it's very much the case that there are people tackling different biological systems with similar data and algorithm goals.

I make the perhaps slightly controversial statement that I don't think humans are going to be able to understand biology. I think our notion of what it means to understand something is going to have to change. We're going to have to be much more comfortable having a complicated model inside a computer, where we only understand parts of it. In some sense, we were incredibly lucky with physics. The fact that Maxwell's equations are four linear partial differential equations that explain all this behavior is amazing and magical. There's no reason to expect that these gross bags of fluids that we call our bodies, which have evolved over 4 billion years, are going to exhibit this kind of aggressive reductionism.

When you watch Steven Boyd[5] lecture, he keeps referring to the 19th-century mathematics that we all know and how this 19th-century approach to science just doesn't work. So we have to start developing algorithms and we have to be using computational tools to redefine what understanding is. In fact, I think

---

[4]http://www.mit.edu/~yarden/.
[5]http://stanford.edu/~boyd/.

industry is actually a bit ahead in this regard. An executive from Target— the proverbial or perhaps apocryphal Target predictive application—doesn't necessarily care about the underlying causal process giving rise to someone buying diapers versus beer. What the executive cares about is: "Does this model have predictive power and does it let me then go do something else?" I think you're going to see much more of that trend in the life sciences.

Gutierrez: For someone who wants to start working in this area, what material should they be consuming?

Jonas: On the computational bio side, there are lots of blogs out there actually. For instance, *Nature* and *Science* both run blogs. However, one of the most useful resources, which I didn't appreciate as an undergraduate, are review articles. A review article is an up-to-date survey of some particular subfield. This is something no one tells you about when you're 21 and struggling through some material. It would be so much easier if someone said, "Guess what? Some poor graduate student out there has written a 15-page article on the state of the art in extreme but accessible detail because it's designed for the wider scientifically literate audience. You should go and read it to understand the material." Both *Nature Reviews Neuroscience* and *Nature Reviews Genetics* are both great sources for having these sort reviews. Those are my two go-to resources.

The other thing that people really don't appreciate is that graduate students, perhaps because of an adherence to sunk cost fallacy, often write really great surveys of the field at the beginning of their PhD thesis. Often, when I want to get into a new field or an adjacent field, I go find a recent grad student's thesis and then read the first chapter. This is because they're going to talk about the review of their field in a way that keeps in mind a general audience for that section. So they're often great pedagogical tools. I have lots of printed-out PhD theses around my living room from random graduate students.

Gutierrez: Let's switch back to your startup experience with Prior Knowledge. What was it like to be a guest lecturer at Peter Thiel's startup class at Stanford?

Jonas: The whole experience was great and the Stanford students were very enthusiastic. To the class I guest lectured specifically, the experience was very fun, as the class was very much focused on AI and not data analytics. What made it a very interesting experience was that I was a guest lecturer with Scott Brown, the CEO of Vicarious, and Bob McGrew, the director of engineering of Palantir. These two companies are pretty much on opposite sides of the space of possible data/AI companies.

Palantir is a very successful company that does nothing AI-related in the slightest. In fact, they tend to be very AI-agnostic. They build tools to let human analysts do better data analysis. On the other hand, you have Vicarious, who says that in ten years we're going to solve general artificial intelligence. I don't think a lot of people in that class recognized the degree to which the things that Vicarious was saying are the things people have been saying for the last twenty years, as it's very easy to overhype and overpromise on these sorts of technologies. Then you had us, Prior Knowledge, who sat right in the middle of the spectrum.

Personally, I think that the best thing that happened in the class, which I'm not sure it ever made it into the notes, was that Peter asked, "Do you ever worry about your technology being used for evil?" The people at Palantir said, "Well, yes, in fact, we have an entire legal staff dedicated to making sure that that doesn't happen." I think that's the best spin I've ever heard on a compliance department. Having a bunch of lawyers because you like government contracts is very different from saying no, we employ John Connor.

Having Peter teach the startup class was great because Peter's extremely smart and very willing to make long-term technology bets. He's excelled in investing in science and state-of-the-art engineering as well as pure-on commercial consumer investments, like Facebook. There aren't very many VCs who make those kinds of bets, and Founders Fund has always been very much willing to take the risk here for deeper science plays. Of course, I'm obviously very grateful for that. Finally, it was fascinating to watch the class reach a certain degree of infamy due to Blake Masters taking extensive notes and posting them online.

Gutierrez: Could you turn your neuroscience research into a startup?

Jonas: I wish it were possible to make a billion dollars developing technology that would make neuroscience better. I wish it were possible to make more money building tools for science because you could hire an engineering staff and a support staff to help speed up progress. However, it's just not sustainable. The reality of the current situation is that science tool companies are a shitty business. If there's a very clear and useful clinical application for a product, then often it's more viable. Illumina, which makes systems that analyze genetic variations and biological functions, is an example of this type of company.

Unfortunately, at this time, I don't think there's a way to build a startup around neuroscience tools. Even if I could accomplish something as dramatic as building a tool to record from a billion neurons, which I think everyone recognizes would be one of the biggest breakthroughs in neuroscience in the past 50 years, the total addressable market for this tool would be around 20 labs, each of which has a couple of million dollars in funding a year from the NIH if they're lucky. It's not a market that can sustain large companies. I wish it was—but no, alas.

Gutierrez: What did your typical day look like at Prior Knowledge?

Jonas: There was no such thing as a typical day at P(K). As CEO, it meant that I worked with customers, managed the team, owned the product vision, and dealt with investors. The responsibilities and day-to-day work in each of these areas changed very rapidly as we went from idea to execution to being bought all in the span of eighteen months. Each day I would work on all these areas and, before I knew it, my entire day would be gone. Sometimes I would feel like I had done nothing. After all, as an MIT person, I was used to thinking about my day in terms of how many equations I had written and how many lines of code I had committed, so it was very tough. Some weekends I'd go to the office to code because I missed doing the actual technical work so much.

One of my extremely smart friends, Alex Jacobson, who was the entrepreneur in residence at Founders Fund, who made the original introduction for us, at one point told me, "You're never going to get to do new technical work again because now everyone knows that you can manage technical people, and in many ways that's a far more valuable skill. Most people don't know how to evaluate technical people or how to convince them to do anything. So the fact that you can manage a group of 24-year-olds, and get them to do something real is all that anyone's going to want you to do". This was somewhat disheartening to me because I don't want to be pigeonholed away from doing technical work.

The only way that I survived the CEO experience was that my team was amazing. My cofounder Beau Cronin handled the product side. My cofounder Vimal Bhalodia handled all the COO-type work, and planning, and execution. My cofounder Max Gasner handled all of the transactional work and was the person out on the road fundraising with me. Cofounder Cap Petschulat, my best friend from high school in Idaho, was our lead architect. So it was just a really great group of people that kind of helped me through that. The first two hires we had—Jonathan Glidden, a Berkeley undergrad, and Fritz Obermeyer, a PhD from CMU—were also fantastic. It was my first experience actually managing people who were obviously much smarter than me. It was always great to come back at the end of the day to find out how much technical progress they had made.

Gutierrez: What did your typical day look like at Salesforce?

Jonas: At Salesforce, a lot of our challenges revolved around overall integration, integration with the existing systems, and talking to customers. This made life much less hectic. The challenges inside of any large company are very different from startups, as the incentive structure is so different. In a startup you can move very quickly. The Facebook mantra of "move fast and break things" works at startups because when you break things, no one cares—because generally you have four customers, whom you likely met through a friend of a friend, or a friend of a VC. So if you break something, you can call up the CEO and say sorry. In a big company, you can't do that, so it becomes more of navigating those waters and understanding how to play that game. Though a different structure, it was still an important challenge for the team and me. I think we pulled it off successfully, as evidenced by the fact that most of the team is still around even after the one-year mark.

Gutierrez: What does your typical day look like now that you do research?

Jonas: I wake up most mornings and go to Philz Coffee, where I sit and code or read papers for a couple of hours. These days I'm trying even harder than ever to aggressively defend large blocks of time. I've discovered that if I want to get anything technical done, I need a four-hour contiguous block. Otherwise, work just doesn't get done. Even just one 15-minute phone call can totally mess that up.

It's been a bit of a struggle being an independent researcher without an admin or a team to work with. I come home and—like, wow—the accounting paperwork isn't all magically done for me. So the big challenge for me right now is trying to figure out how to be technically productive while still getting admin tasks done. Similarly, travel also gets in the way of being technically productive. For instance, I'm going back to MIT next week. The week after that there's the workshop that we're running at the COSYNE conference. Then two weeks after that there's a machine learning conference in Iceland and then one in Copenhagen. It's just so easy to have all these little things kind of eat away at your time as you're trying to be technically productive. I think that's often why in academia, graduate students and postdocs do all the real work. There's no way that a principal investigator, given all of their responsibilities, could possibly ever sit down and do real technical work.

Gutierrez: How did you view and measure success as a PhD student?

Jonas: "Poorly" is the most honest answer I can give. I've yet to meet a graduate student who doesn't make the mistake of confusing inputs with outputs, especially in the experimental sciences. We think that because we're there at nine in the morning and stay until midnight, where we sit in front of a computer all day, and maybe even actually coding instead of hitting reload on Reddit or Hacker News, that we're being productive. We work very hard and think that's successful. We say things like, "I haven't slept in 48 hours!" And everyone's like, "Ooooh." There's this tendency to think that's the metric that matters as opposed to the number of papers you write or how close are you to actually graduating. So in grad school, I didn't have the best view or measure of success.

Gutierrez: How did you view and measure success as a startup CEO?

Jonas: Fortunately, in industry, especially in startup world, people won't let you get away with that view or measure of success. If you have a smart group of people that you're working with, like I did, all the traditional startup vanity metrics are ignored in favor of the question of "Are dollars coming in?" That measure of success is all that matters, especially when you're VC-backed, because the dollars-going-out number is typically very large. So for us at P(K), the primary focus was: How many customers, doing real things with the system, are paying us money? Of course, when we were very early on in the process, we took the wrong view of measuring success by focusing on much more technically questions, like: What's our uptime look like? How long does it take to process a job? What's our predictive accuracy? and similar questions. However, we very quickly realized that no one gives a damn. What really matters is who's actually using and paying for it.

For example, there's all this talk about predictive analytics. Kaggle became this big thing because everyone seems to think that predictive accuracy matters. In reality, almost no one actually cares about predictive accuracy because in almost all the cases, their starting point is nothing. If you have something that gets them 80 percent of their way there, it's an infinite improvement and they will be so happy. The number of industries where the difference between 85 versus 90 percent accuracy is the rate-limiting factor is very small.

Sometime in the future, after everyone has adopted these sorts of technologies, the predictive accuracy will start to matter, but at this point it doesn't matter as much as people think it does. Sure there are some areas like quantitative hedge funds that are fighting tooth and nail over that last epsilon, but most people are not in that position. So it really comes back to the question of "What value are we providing?"

Gutierrez: How do you view and measure success now that you've transitioned back to research?

Jonas: As I've transition back to research, it's been very important for me to keep the startup experience view and measurement of success at the top of my mind. Our first employee who we hired out of Berkeley, Jonathan Glidden, wants to go back to graduate school. I'm really excited for him because I don't think he's going to make any of these cognitive errors in graduate school, having now gone through this process. He really understands, in some sense, how to ship. But it's hard because a lot of academia tends to value novelty in a way that I think is actually very counterproductive.

One of the things I struggle with is that comparative advantages are actually complicated, because they intersect with your utility curves. There are problem domains where I know that the models that I have would actually be transformative. But I feel that that's not the most important domain for me to be working on right now, or someone else will do that, or that I can come back later. And so it's hard when you've been so trained by graduate school to think that what really matters are papers. So it's novel to recognize that, yes, some papers aren't actually worth your time to write. So a lot of it right now is how can we uniquely have an impact? Where can we have these force-multiplying effects? Even for some of the things I care about, are papers the right metric?

Roughly nine months ago, I decided that what I really needed was a board for myself. So I have this group of four friends that I send weekly status reports to who monitor my progress and help me set goals. Sure it's a little Type A, but it's tremendously helpful. One of the things that we've talked about is that there are a lot of technologies that technically exist. You can read *Scientific American*, or *Newsweek*, or other publications, and you think these technologies are out there. Sadly, you can never really buy one off of the shelf. This problem comes from the fact that so much of what academia is oriented toward doing is getting a prototype that basically only works once and then getting that result out there by writing a paper and then moving on. There's a real gulf between that and actually having impact in people's lives.

I think the rise of university press offices has actually been a double-edged sword, because my mom reads an article in *Tech Review* or from the MIT news office about research and it always says, "And this may lead to something for cancer" or some other impactful result. I then have to be like, "Mom, when they say that, what they really mean is that they had to put that in there for the grants. In reality, this protein may lead to curing cancer in the same way that you living in this house may lead to you being very, very rich, through homeownership, but it's probably not going to happen."

Since I returned to science, I've really started trying to track the kind of weasel words that scientists use. Of course, no one's trying to be disingenuous; it's just that part of our jargon includes phrases like "may show a relationship to," or "may share a common cause with," or "strongly suggests that," and none of these are definitive. The general public interprets these statements as being far more certain than we, the scientists, intend. Are we actually learning anything? No. We're waving our hands around a lot. The real metric should be: "Do I know something now that I didn't know yesterday?" And a lot of times for a lot of results, the answer is, "Slightly." So for me personally, the question is still out as to whether or not that's rewarding enough to keep waking up in the morning. We'll see.

Gutierrez: How do you choose what to study and analyze?

Jonas: My list of goals is to learn everything, be able to build anything, save everyone, and have fun doing it. That's a nice simple list. It's nice to have application domains that I'm actually passionate about or questions that I'm really curious about. On the neuroscience side, I actually do care a great deal about how the system works. And so—while there are other application domains in epidemiology and genomics and other domains that are also very interesting—when times get tough, I'm not going to drag myself out of bed for those problems. So a lot of it is kind of intrinsic interest, and then part of it is also clinical impact.

I've started working with clinical schizophrenia data partly because there are people out there suffering from this disease and, in some sense, that's absurd. The fact that disease is a "thing" is absurd. We're mechanistic, we should be able to fix ourselves, and the world I want to exist in 20 years will solve that. So going back to the goals, it's important to ask if we are closer to this world or not?

Startup culture teaches you to be like Steve Jobs, in that you're right, everyone else is wrong, and your vision will power through. Academic culture teaches you that you're dumb and that you're probably wrong because most things never work, nature is very hard, and the best you can hope for is working on interesting problems and making a tiny bit of progress. Just doing that is seen as an amazing career. So the question is: How do you reconcile these kinds of things? I don't know, I struggle a lot with reconciling these two cultures in myself.

Some of the best scientists out there are the ones who are extremely opportunistic—when they see novel ideas and how things suddenly fit together, they drop everything else and work on that for a while. Others are consumed by a single, all-encompassing vision and aggressively pursue that forever. The downside, in many ways, is that the academic funding system really rewards the former, in that if you have three *Nature* papers with no clear coherent tie to them, it doesn't matter. "You have three *Nature* papers—congratulations, Professor!" Whereas if you've been working on the same problem for 10 years, but only making incremental progress—"Well, sorry, you're not getting tenure at a place like MIT. I really hope you enjoyed working on the problem."

That's one of the reasons why I'm not necessarily excited to go back into academia, because the incentive structures are so confused around this issue.

Gutierrez: What kind of tools have you used and do you use now?

Jonas: From a technical point of view, I'm almost entirely a Python and C++ person. I do C++ for the heavy numerics and Python for basically everything else. It's an extremely productive environment. It's nice because, as someone with computer science training, I can do complicated things in Python. It's not like MATLAB, where you have to jump through a million different hoops. And I can drop down in C++ when I need it for speed.

Mathematically, a lot of what I work on is Bayesian models using Markov chain Monte Carlo to try and do inference. I really like that universe because the world is so simple when you think about it probabilistically. You can think of a stochastic process that you can condition on your data and do the inference. That's great! It means the set of math I have to know is actually shockingly small, especially because often the problems that I'm working on don't have data from a billion neurons yet—we have data from 100. And so I'd much rather spend my time building complicated correct models and then, when the data gets larger, figure out how to simplify those systems; rather than start out with something simple and later rework the models when the data set size grows. That's really been my tack thus far academically and in the startup world. I think everyone at P(K) shared that bias as well.

Gutierrez: As you're building these tools for yourself, is there any chance you'll go and build another tool company?

Jonas: I've thought of it. I actually just received funding from DARPA to fund some of the construction of these tools and to hire some engineers to specifically build them, which is kind of exciting. But I don't know. The problem is that it's really hard to do a tools company, especially an open-source tools company. Tools companies don't fly anymore, especially for these sorts of things. Of course, the entire big data ecosystem is all built on top of these tools.

But how large is the space for companies like Cloudera? I'm not really sure. I think most people who buy things from Cloudera are buying them because Mike Olson did a good job selling them something and there's a CIO someplace whose CMO has turned to him and said, "I want to do what Target does. We need a Hadoop." So they call up Cloudera and they're like, "I'm buying two Hadoops." And then they're like, "Great. We just bought two Hadoops. Now we have some Hadoops." And you're just like, "There's no value being created here." I think that's how a lot of the most successful tool companies have managed to get off the ground.

The other thing I really learned at P(K) is that the right thing to do is to not build a tool company but to build a consultancy based on the tools. Identify the company, identify the market, and build a consultancy. Later, if that works, you can then pivot to being a tool company. If you're selling to the enterprise, which you should as they're the only people with money, you're never going to make headway without a substantial services arm. So start with the services arm first because it's quick revenue, it's nondilutive, and it's great. If that works and gets traction, then you can go down the standard Silicon Valley VC trajectory.

I think there are a lot of data-related startups right now that, had they started by doing that, would be in much better shape. What you don't want to be doing is burning through VC money just to figure out who your customer is going to be. It's a painful truth and it's hard work, but it's much better to approach building a company that way. We had meetings at P(K) around this question of turning to consulting or continuing to build the platform even after we had taken VC money. Ultimately we decided to shoot for the moon and it worked out very well for us. However, having gone through that, I now think the right thing to do is to start out with a consultancy.

Gutierrez: How do you know you're solving the right problem?

Jonas: It really depends on what you're going to call the "right problem." For me there are two parts to that question: "Is the problem right in some sort of global sense?" And "Is the problem going to be one that I'm going to be willing to see through?" That latter question is just as challenging as the former. So a big part of it is: "Am I intrinsically interested in the answer here, and do I think other people are going to be interested in it as well?" Then I can ask, "Am I having fun, and is this the best use of my talents?" I have a clock that shows the estimated number of days until I'm 80, which is a reasonable life expectancy. It helps to remind me that each day actually really matters.

So on the neuro side, I talk to my neuroscience friends to try really hard to make sure I'm not just doing math-wanking. Sometimes I still veer in that direction because math is fun and solving technical problems is fun. Sometimes you veer in that direction and end up finding another path back. However, science is hard, most stuff doesn't work, and you have to be willing to stare at shitty, ambiguous results and be like, "I'm going to keep doing this for another 8 hours today, anyway."

Some of the research projects I've started lately, I've worked halfway through them and realized that I'm not that into them, so I stop. People are very understanding about that sort of thing. Everyone likes to talk about how if you're not failing some fraction of the time, you're not trying hard enough, so at the very least, I console myself by thinking, "Well, I tried this thing and it didn't work, and that's okay." Megan McArdle has a new book out about the role of failure and success, and she makes the same argument that, in some sense, it's figuring out how to both fail and recover is really crucial to all of this innovation stuff that we do.[6]

Specifically to the question of "How do I know I'm currently solving the right kinds of problem right now?" is the good response for our workshop, which is a good indicator that people give a damn. Private and public funding agencies are getting really excited about funding this type of work, which also suggests I'm on the right track. On the other hand, the universe is very fad-driven. In 2008, I might have thought hacking on Hadoop was going to be this really big thing, and now I'm like, "Well, honestly, what's the real value there when most people probably could have done their data analytics on top of PostgreSQL?" You don't really know if what you're working on is the right problem to be solving sometimes until years out, but you hope you're on the right track.

Gutierrez: When you're thinking and solving problems, do you approach it modeling first or do you approach it data first?

---

[6]Megan McArdle, The Up Side of Down: Why Failing Well Is the Key to Success (Viking, 2014).

Jonas: If I'm not familiar with data, then I generally don't even start. I recently met Winfried Denk, who invented two-photon microscopy and is a very smart applied-physicist guy who's received many, many, many awards. His comment to me in this area was that the number-one thing you have to be able to do is actually know what questions to ask. And so I try not to get involved in projects where I don't know what the right questions are. And then generally, if I know the questions, I understand the data well enough to then start thinking about the modeling. The nice thing about modeling is that you can fairly rapidly turn around and try a bunch of different things. But if you haven't even looked at the data and done the most basic things, then it's very easy to be led astray.

Gutierrez: How do you look at the data?

Jonas: Matplotlib in Python. I make a bunch of initial plots and then play around with the data. A lot of the data I work with looks very different from the kinds of data that show up more on the industry side of things. No one in science really uses a relational database, because we either have time series, or graphs, or images, or all these weird things. Rarely do we get relational facts. So I don't end up using SQL that much. It's much more about writing a bunch of custom scripts to parse through 100 gigabytes of time-series data and look at different spectral bands or something similar.

Gutierrez: What do you look for in other people's work?

Jonas: On the research side, my answer is different from many of the people I work with and other people in the field. One of my colleagues told me, that I read more papers than anyone they know. I don't actually really read most of the papers. I read the title and the abstract, look at the figures, and then move on. For example, when I evaluate machine learning papers, what I am looking to find out is whether the technique worked or not. This is something that the world needs to know—most papers don't actually tell you whether the thing worked. It's really infuriating because most papers will show five dataset examples and then show that they're slightly better on two different metrics when comparing against something from 20 years ago. In academia, it's fine. In industry, it's infuriating, because you need to know what actually works and what doesn't.

So a lot of what I look for are: "Do I think that their approach was valid? Do I know them?" The degree to which I will read papers from people I know and trust far is far higher than those whom I don't know. People complain that it's hard for new people to break into fields. Well, that's partly because at any given time, 99 percent of the time people are all new and they're cranks. So a lot of it is: "Do I find the structure of this model to be interesting? Do I think they did inference properly? Did they ask the basic questions? Do I believe those results? Is the answer something that I would have believed before

I read the paper?" If the answer is yes, then I'm more likely to have believed it. "Am I surprised? What is the entropy on the result?" The more surprising results, the more thorough I want to be when evaluating the work.

On the industry side, I think that the ability to do software engineering is something that is very important, but isn't really taught. You don't actually learn it as a computer science undergraduate, and you certainly don't learn it as a graduate student. So for me it's very important that someone has learned it somehow—either by themselves or from someone else. I basically can't hire people who don't know Git.

There's a universe where companies at a certain scale can afford to hire people who have tool deficits. For almost anything I've been associated with, that's not really the case, which is unfortunate because it means that you leave out some very smart people. You want people to be productive on day 10, not on day 100. I also really think that the Bayesian approach to machine learning has incredible legs. I think that it encourages a certain kind of precise thinking. If you're not doing it, it's very easy to confuse yourself and lead yourself astray. So I generally look for people who are at least familiar with that part of the universe.

Gutierrez: What have you've changed your mind about with regard to using data?

Jonas: I started dating my best friend in undergrad around 2001. We were together for nine years and then broke up. So I found myself single at 29, and realizing that I was going to have to learn how to date. I decided that if I'm going to start dating, I should keep data on it. So I create a dating spreadsheet, and then went on something like 100 first dates and ended up with this massive chunk of data. Coming out of MIT, you think you're stupid because everyone else was smarter than you and you think all these sorts of negative things about yourself. However, after looking at the data, it turned out that I'm actually kind of a catch—which was a great thing to change my mind about.

I found that if I could get to the second date, then generally I could get to the fifth date with someone. There's this initial evaluation process, but I was generally pretty good at that. So I was actually in a much better position dating-wise than I had ever thought. You wake up and 30s are around the corner, and your friends are marrying and you're single, you're like, "Oh, my God—I missed the boat!" With the dating spreadsheet it was easy to see that no, actually the data says the opposite. This was this phenomenal. Psychologically it's so easy to fall into these kinds of anchoring effects where you always remember the last date that went poorly. You don't think of the previous 10 out of 15 that went well. If I hadn't been keeping track of that data, I think it my confidence would have taken a big hit, which interplays with all these other things in my life. So I'm dating a great person now and I don't think that would have been possible had I not had this epiphany that the data shows that some people do in fact like me.

The dating spreadsheet has become somewhat of a joke now, but it actually really helped. Everyone talks about "quantified self" and everyone wants to track themselves. But no one's writing down a lot of the interpersonal inter-actions that actually matter—who cares about how many steps you took last week, who did you kiss? So I think the dating spreadsheet is a good argument for the quantified-self approach in this kind of data.

Gutierrez: What does the future of data science or computational neuro-biology look like?

Jonas: I know that everyone wants to talk about big data. It's now this phrase that has somehow entered the lexicon in a horrible sort of way. And also that being a data scientist is "the sexiest job of the 21st century"—admittedly said by a data scientist in an article he wrote, so not really objective. Sure it was in *Harvard Business Review*…but come on! I actually think a lot of the future is in small data. Or what my friends at Bitsight call "Grande Data", as in the Starbucks cup sizes—it's neither Tall (short) nor Venti (large); it's Grande (medium). The amount of things you can discover out of a gig of data are often far more interesting than the things you can discover out of a terabyte of data, because with a gig of data, you can ask more interesting questions. You can build more interesting models. You can understand more about what's going on.

On one hand, there's the Peter Norvig philosophy that with enough data you can use simple models, which is true if you are Facebook, Google, Walmart, or companies of that size. Otherwise, most companies have a thousand, or ten thousand, or even a million customers, which is nowhere near what you actually need for Norvig's philosophy. Most people who are buying and using technologies like Hadoop are using it as a recording engine, where they comb through all this data, then stick it in an RDBMS and actually do their data analysis in R and SQL. I think that as the big data hype cycle crests, we're going to see more and more people recognizing that what they really want to be doing is asking interesting questions of smaller data sets.

On the computational neuroscience side, the data are coming and the data sets will be getting bigger over the next ten years. Right now, if we're not building the right models, I think we're going to be a little bit screwed in ten years. What are we going to do—linear regressions? I've talked to very smart, famous machine learning people at Google, and I asked them, "What do you do all day?" And they replied, "Well, you know, we do feature engineering and then run linear regressions on our largest data." "But you wrote a book!" I thought. "What's going on?"

I hate the phrase "predictive analytics." If you think that the world is all about predictive analytics, then the entire universe is in some sense solved and uninteresting. If you care about what's going on inside the box and if you want technology to let you see new things, then that's kind of a green field right now.

The data microscopes project that DARPA is funding is all about building exactly these sorts of tools to let you see things in data that you couldn't see before. A great deal of data analytics startups these days are like data visualization technologies in that it's great when you can think of the questions to ask, and then ask and visualize and plot the conditionals and these sorts of things. However, for a great deal of data regimes, we're far beyond that.

When you start looking at these kinds of Grande Data problems, and it's too much to visualize, and it's not enough data to do something Google-scale, and you don't even have the resources to do Google-scale—well, what do you do now? More linear regression? I really think that's going to end up being the future, especially for the sciences. It's inevitable. And I think you're going to eventually see these sorts of data microscope techniques being taught to undergraduates. We're going to see this real transition.

Gutierrez: What data sets would you love to see developed?

Jonas: There are two data sets. One is that I would like to have all of the connections between every cell in the brain. This would be the spatial locations and connections of every cell in the brain. Two is that I want the time series from every neuron in the hippocampus. We have to start building these sorts of high-throughput neural data sets. I'm not necessarily content with these being in animal models, as they're going to be for a while, but we'll get there eventually because the systems are there. The data is there; it's just currently inaccessible. We have to change that. Fortunately, there's more and more interest. Somewhere in my brain, there's some glob of goo that knows my phone number. We have no idea what that is, what that looks like, how it even works, and that's ridiculous because it's in there. There's some circuit in there. I want to understand that, and I want the data to exist to help me understand that.

Gutierrez: Early in your PhD you built a device to measure this data. Have you or others thought of pursuing this?

Jonas: It is something I've thought of pursuing. The question is partly one of comparative advantage. I think that this space is large enough. What I really want to get to is—if you can record all those neurons, how do you then go use that technology to make $10 billion? Because then we can let the capitalist innovation machine do its work. However, what we're currently on right now are rats. No one really wants to read the mind of a rat. Even pharma companies have no interest in reading the mind of a rat because a rat is basically too big of an animal to do large-scale experiments with. They like mice because they're small, and even then, mice studies are horribly expensive from pharma's perspective. We'll get there eventually. Hopefully, the hype doesn't kill it first. But, I can imagine going back on the instrumentation side. I've spent a little bit amount of time over to Berkeley's AMPLab, working with some people there that do compressed sensing, trying to feel out these areas. We'll see. Hopefully there are lots of opportunities there in the future.

Gutierrez: If you could give advice to someone starting out, what would you say?

Jonas: I think that if you're an undergraduate today and you don't know how to program, you're basically screwed. If you're doing anything remotely technical, especially on the biology side, you have to learn how to program. That's inevitable. For my liberal arts major friends, maybe it's less mission-critical for their career trajectory. The ability to work with data really ends up being sort of crucial, and to that end you have to know how to talk the language of the computer.

To people starting graduate school in sciences I would suggest reading Derek Lowe's blog, "In the Pipeline."[7] Derek's a medicinal chemist who's been blogging for years, and he's an amazing person of insight. He talks about working medicine, pharma, and life sciences for thirty years and how he's never had a drug he's worked on actually makes it into a patient. He talks about how that's common because the median success is zero. He also talks about how the purpose of graduate school is to get out of graduate school, as there's nothing else that matters. I think that's really true. I would also encourage people to go into the more quantitative programs because it's so much easier to later become less quantitative.

To academics, I would give the advice that startups are not a source of funding. It's surprising the large number of graduate students who approach me with something that basically has no market and say, "Well, I think I could probably get VCs to give me funding for this." My response is always, "You don't understand the game being playing here. VCs are going to want you to focus on things that you don't want to focus on, and it's not going to work." And even when you have VCs who are extremely supportive, like we did, you will eventually realize that these aren't grants they're giving you. They want you to turn around and give them a billion dollars back. If that's not your intent when you take the money, then that's fine, but you need to tell them that upfront when you start.

This whole using-VC-to-fund-science is a difficult and a duplicitous thing to do, and it's very easy for graduate students to convince themselves otherwise. It's easy to say, "I'm going to build this tools company." And you're like, "Well, no. Let's apply the same rigor to this process that I apply to my other science. How many companies are there like this? How have they been successful? What have their real trajectories been?"

It's one of the reasons I think things like pursuing nondilutive capital like DARPA or early consulting gigs for any of these hard tech problems is actually the right way to go. If you look at a lot of the really successful companies

---

[7]http://www.pipeline.corante.com/.

that have been built this way—even, to an extent, companies like Cloudera, where they had already built a lot of the hard technology when Yahoo! spun out Hadoop. Someone has to be footing that bill, and VCs do not have the risk appetite or patience to let you try something for three years. They'll let you try it for a year, and they'll probably still keep funding you over the next two years, but what's going to happen is you'll bleed through your cap table, and you're going to wake up and maybe finally have a success and realize you've sold 90 percent of the company. Oops.

Gutierrez: What do you look for when you're hiring people?

Jonas: It depends a lot on the role. The first thing to find out is if they are an asshole. Life's too short to work with assholes. At MIT we used to talk about how it would take freshman a while to de-frosh. They would come in thinking they were the smartest person ever because they grew up being the smartest person they knew. Then they get out into the real world and realize that no, they're not. They have to have that arrogance beaten out of them. There are some people who never lose that. There are some people who very much think that being smart is an excuse to not have interpersonal skills. And the world is just too collaborative for that to work anymore.

My cofounder Beau Cronin made the comment the other day when I was talking about an academic who I was working with who was a little bit difficult, and Beau said, "The nice thing about doing a startup is you get to say, 'Nope! No! No! Do not talk!'" In academia, because of the way the incentive structures are often set up, that's not as much the case, so you might end up working with difficult people.

At P(K), we evaluated a lot of really smart people that just weren't a good fit. Startups spend too much time talking about culture these days, and often culture is a euphemism for "not exactly like me." Which is a terrible way to look at culture. What really does matter, and how we looked for fit was by asking ourselves the following questions about them: Are they excited about the same technical problems as we are? Are they excited about being collaborative? Do they like sharing their successes and failures? Do they possess some degree of appropriate humility and understand why it's important? Finding the right person with the right fit is hard, especially in the machine learning and data space. But that's the most important thing—making sure they are a good fit.

Obviously, a strong math background is necessary. If I have to explain probability to someone, it's going to be a really hard slog for everyone involved. I would rather take someone in the top 20 percent of quantitative skills who also is a great software engineer over someone in the top 5 percent who doesn't know how to code. The quantitative finance model really popularized the notion that raw cognitive talent is all that matters. This is the D. E. Shaw and Renaissance Technologies model of "We're going to take people who have been doing algebraic topology for a long time, and we're going to then teach them quantitative finance, and this is going to be a good scheme." In some sense, it obviously worked out very well for them, but especially on the data side, data analysis is so much messier than actual math. I have friends who work on these topology-based approaches, and I'm like, "You realize these manifolds totally evaporate when you actually throw noise into the system. How do you think this is really going to play out here?" So I would much rather someone be computationally skilled. I'm willing to trade off what their Putnam score was for how many open source GitHub projects they've committed to in the past.

I'm also very skeptical of this notion where a data scientist comes in without the domain knowledge and starts producing work. I think you actually need to care about the domain. I do think that a lot of the interesting problems, especially those I'm interested in, necessitate that you have already been doing work in the area for a while. So rarely do I find myself hiring someone who just has data science experience.

One of the things I've seen a lot in the neuroscience community—or in industry even—is that you get people who really like math showing up and being like, "How can we apply this thing I have to your problem?" They just want to do the math and they don't really care about the application. But if you don't actually care about the underlying problem, then you're not going to be willing to make the compromises necessary to understand how to guide your own work. In academics or industry, if you're not actually speaking in a language that your customers understand, then you will have a nice time talking, but no one will really listen to you.

Gutierrez: What is something you know that you think people will be wowed by five years from now?

Jonas: Either that Bayesian nonparametric models let you see things in data that you didn't know were there or that Markov chain Monte Carlo actually scales to data at a size you care about. Being properly probabilistic solves so many of the problems we face in machine learning, like overfitting and complicated transform issues that I still don't fully understand. There's an entire set of machine learning work that starts with the predicate that your data are a fully observed, real-valued matrix where the matrix is $R^{n \times m}$. From my point of view, problems almost never look like that. This predicate forces you to do all this stuff with your data to try and force it to look like that. And then, once you have it in that form, you do a bunch of linear regressions. I'm of the opinion that it's better to do slightly more sophisticated modeling here by modeling the likelihood function and taking a generative approach. I think that in five years, that's going to be the way most people do things. I think it's inevitable. However, I think it's going to be a lot of work to get there.

The other thing I think people are going to be really surprised by is how much of a quantitative and computational science the life sciences will become. In some sense, everyone's always saying this—it's kind of a trope at this point, but it's only going to become increasingly true. Every time we look back, we're much better than we were five years ago. We always still hate ourselves though, because we're never where we want to be—but I think we'll get there.

Gutierrez: What is something someone starting out should try to understand deeply?

Jonas: They should understand probability theory forwards and backwards. I'm at the point now where everything else I learn, I then map back into probability theory. It's great because it provides this amazing, deep, rich basis set along which I can project everything else out there. There's a book by E. T. Jaynes called *Probability Theory: The Logic of Science*, and it's our bible.[8] We really buy it in some sense. The reason I like the probabilistic generative approach is you have these two orthogonal axes—the modeling axis and the inference axis. Which basically translates into how do I express my problem and how do I compute the probability of my hypothesis given the data? The nice thing I like from this Bayesian perspective is that you can engineer along each of these axes independently. Of course, they're not perfectly independent, but they can be close enough to independent that you can treat them that way.

When I look at things like deep learning or any kind of LASSO-based linear regression systems, which is so much of what counts as machine learning these days, they're engineering along either one axis or the other. They've kind of collapsed that down. Using these LASSO-based techniques as an engineer, it becomes very hard for me to think about: "If I change this parameter slightly, what does that really mean?" Linear regression as a model has a very clear linear additive Gaussian model baked into it. Well, what if I want things to look different? Suddenly all of these regularized least squares things fall apart. The inference technology just doesn't even accept that as a thing you'd want to do.

The reason my entire team and I fell in love with the probabilistic generative approach was that we could rationally engineer in an intelligent way with it. We could independently think about how make the model better or how to solve the inference problem. A lot of times you'll find that by making the model better—that is by moving along the modeling axis—that inference actually becomes easier, because you're more able to capture interesting structure in your data.

---

[8]E. T. Jaynes, *Probability Theory: The Logic of Science* (Cambridge University Press, 2003).

Another great thing about this approach is that you can have two different sets of people working on the same problem space. With the current techniques, if you want to move jointly along this space then you need people who know everything. And that's really hard when hiring. It's hard to find the person who knows optimized C++ numerical methods and really understands all these kernel tricks or similar techniques; whereas with the generative approach I can find people who are really good at modeling but only work with small data, and I can find people who are as not as up on the modeling but know how to do really efficient inference. Then I can put them together and get a lot out.

Gutierrez: What is nontechnical advice you give your friends?

Jonas: The biggest thing I think people should be working on is problems they find interesting, exciting, and meaningful. Today I saw a quote on Facebook that said that a data scientist is a scientist who wants to feed his family. This is not entirely incorrect. There are a lot of interesting problems out there that I think a lot of people can get excited about—and life is too short to not be having fun. So I hope that most people are operating in that space. For my friends who are just graduating college, I tell them, "No, don't go do finance if you're not really excited about it. There are so many other interesting things." In thirty years, you're not really going to care about that extra money. It won't be a thing if you work on problems you find interesting and meaningful.

Jake Porway

DataKind

Jake Porway is the Founder and Executive Director of DataKind, a nonprofit dedicated to using data science to tackle the world's biggest problems. DataKind convenes data scientists to help mission-driven organizations such as the World Bank, Sunlight Foundation, Grameen Foundation, Amnesty International USA, and the United Nations Global Pulse address their biggest data challenges. With an unparalleled team and advisory board, DataKind is taking the nonprofit world by storm. Supporters of DataKind and its mission include The Knight Foundation, Alfred P. Sloan Foundation, Blue Ridge Foundation, Teradata, Cloudera, Informatica, and thousands of data scientists who donate their time and skills to DataKind projects to help make the world a better place. With the launch of its latest wave of local chapters in mid-2014, DataKind is poised to help even more NGOs and other mission-driven organizations improve their data literacy and data capacity, as well as assessing how data science can help them better serve humanity. Match-making data scientists with cause-driven work and applying data science to unconventional datasets generate interesting and diverse challenges.

Porway's career before DataKind included internships at Bell Laboratories and Google, where he designed and implemented statistical models and algorithms for event detection and the identification of anomalous behavior. He went on to do his PhD at the UCLA Center for Image and Vision Sciences, where he not only conducted his own research, but also coordinated a team spread across the US and China to integrate his code and research into a unified software system. Porway then did a stint at Utopia Compression Corporation—helping the company improve machine autonomy and convert large amounts of image data into usable knowledge— before joining The New York Times, where he was a data scientist in the research and development lab, building prototypes for the future of media.

In 2011, Porway founded DataKind to connect data-rich mission-driven organizations with teams of data scientists willing to donate their time and knowledge to solve social, environmental, and community problems. DataKind has been widely featured in such publications as WIRED, The Economist, Forbes, The Washington Post, and Harvard Business Review. Porway holds a BS in Computer Science from Columbia University and a MS and PhD in Statistics from UCLA. He is the host of The Numbers Game, a National Geographic Channel television show focused on raising data literacy levels.

Porway personifies the data scientist who believes in the power of data to change the world and wants to ensure that this change benefits humanity. This ideal shines through as he talks about his motivation for founding DataKind, his reasons for making the organization a nonprofit, and his vision for creating a movement that helps the world become a place where data is both better understood and better deployed for social gains as small as improving the lot of a single family and as large as helping millions of subsistence farmers in Africa. Porway's passion for using data science to help others is consuming and contagious, compelling anyone who is data-driven to reassess where to apply their energy and skills.

Sebastian Gutierrez: Tell me about where you work.

Jake Porway: I am the founder and executive director of DataKind, a nonprofit dedicated to tackling the world's biggest problems through data science. We bring together high-impact organizations dedicated to solving the world's toughest challenges, with leading data scientists to improve the quality of, access to, and understanding of data in the social sector.

We live in exciting times. A friend of mine put it to me like this: just like in the 1990s, when every field had its computing moment, today everyone's having their data moment. It's not just data and tech companies that can benefit from data now. Everyone is benefiting in some way, because of the ubiquity of cellphones, laptops, and other digital interfaces that collect and transmit data. Not to mention emerging device sensors and many other things. Even organizations that are not traditionally data companies are suddenly inundated with data that they can use to make better decisions.

I'm really excited because we at DataKind focus on applications of data science to make the world a better place. So using the same technologies that help Netflix recommend movies you want to watch, we apply similar techniques to problems, like sourcing clean water, combating human rights violations, or addressing other pressing social issues. It really feels like a brave new world, and the chance to use data science skills to do something good at this time is just incredibly rewarding.

Gutierrez: Tell me about your team and the organization.

Porway: We're eight people now but on track to double our staff. We also have volunteer-led chapters in six cities around the world—Bangalore, Dublin, San Francisco, Singapore, the UK, and Washington DC. One of the ways that

DataKind works is by connecting volunteer data scientists with social organizations and the issues they're working on. We curate these collaborations to make a big impact—bringing together teams of expert data scientists who volunteer their time with social issue experts from mission-driven organizations—all focused on a data challenge. We think of our team as our staff and volunteer leadership, but also a deep bench of expert data scientists and a whole host of organizations, like Amnesty International, the World Bank, and the United Nations.

I love my staff immensely—not only because they're incredibly brilliant people, but also because I look up to them every day. They have such a flair for social justice, for ethics, and for thinking about the real ramifications of what we do. They're that perfect combination of technically-gifted people who are also incredibly thoughtful, well spoken, and really want to make the world a better place. We've been blessed that these great traits have also extended to the volunteers, as well as the organizations who want to do good.

We think a lot about how to use data science to make a difference while still remaining ethical. There's often a misperception that data science can be scary. That it should be feared because people who use data science are just using it blindly, using algorithms and things to manipulate our emotions on Facebook or predict whether your daughter's pregnant based on what she buys at Target. We are not like that and we all—the staff, the volunteers, and the organizations we work with—strive very hard to make sure that we are sensible, ethical, and really think about the real ramifications of what we do.

Gutierrez: How does DataKind work?

Porway: One of the ways that DataKind works is by connecting volunteer data scientists with social organizations and issues, and then designing these collaborations to make a big impact. One of the challenges in getting data scientists to work on social problems is that data scientists themselves don't always know the problem spaces. So if you're going to use data science to alleviate hunger, for example, a data scientist might not necessarily know what all the problems in hunger are. So we set up projects that are specifically designed around collaboration with, say, a group of hunger alleviation experts who've been doing this with the World Bank for fifty years, but don't have the technical data science expertise. We unite those issue experts with expert data scientists around a specific set of data challenges designed to make a big impact for the beneficiary organization.

It would be great if we could do the matchmaking and just say, "You both now know each other. Have fun." The challenge on the other side is that the people in the social sector side, who are new to data science, don't always know all the ways it can be used. They may not know the right questions to ask. On the flip side, the beneficiary organizations have the deep expertise around their issues and in the regions where their organization works. And finally, we have found that there can be communication or "translation" challenges when the two sets of experts get together.

So a lot of the work that DataKind does is not only to recruit—bringing the community together—but also to help make it as easy as possible for people to work together and scope a project that will make a big difference. We focus on ensuring that volunteer data scientists have the background and framework they need to take an ethical approach to their work. We also work closely with the beneficiary organizations, digging into the data to scope out a project that will be useful and in support of their mission. Then all along the way, DataKind stays close to the teams, as supporters, helping out when there are roadblocks, providing camaraderie, and ultimately making sure the project is a success for the volunteers and the beneficiary organization

We also see a large part of our role as the storytellers. We want to herald the results and spread the word about the power of data science for good. We hope that projects result in new tools or approaches for the beneficiary organizations, but also that the work done during the project can scale to benefit other people and organizations.

For example, we recently worked with a nonprofit organization to build a tool that looks at Google Maps to identify poverty levels in a Kenyan village based on the buildings' roof types. In this case, we knew from the organization's regional expertise that iron roofs connote more wealth than roofs that are thatched. It's exciting to think that by using a computer-aided process, coupled with knowledge about a regional culture, a nonprofit could save hundreds of hours of investigation by foot, traveling village to village, and instead spend limited resources on programs to address pockets of extreme poverty. The work was done for one organization, but the tool and approach has the potential to benefit any number of organizations.

DataKind has been working to codify our process in a "playbook" so that, ideally, anyone can replicate this type of collaborative work to harness the power of data science.

Gutierrez: Why develop a playbook for how DataKind does its work?

Porway: From the very beginning of DataKind's project work, we've been fielding interest from around the world. We have been fielding questions like: How do I do DataKind in my city? How can I be involved in the Data for Good efforts? I want to do it in my particular town—can you help me? We realized from those requests that DataKind could achieve its mission of tackling the world's toughest problems through data science by enabling others to do data science volunteer work. They just needed the tools to ease their journey. Now, three years into our operations as a stand-alone organization, we feel like we have enough experience under our belt to share our framework and process to support successful volunteer data science project work.

We're sharing that framework through a deliberately placed global chapter network. We launched our first chapter in the UK in 2012. And this summer we welcomed chapters in five additional cities—Bangalore, Dublin, San Francisco, Singapore, and Washington DC. Volunteers lead each chapter and will be serving as our ambassadors in their communities. The chapter locations were each chosen for their unique combination of local technical data science expertise and the range of opportunities to work with mission-driven organizations tackling the world's biggest problems. This means that we're building on our track record of project work, where we've been functioning like a Data for Good consultancy, and taking the first steps to build a global Data for Good movement, where people are doing the same thing around the world in their own communities, on their own, with our help and our playbook. And of course it's an iterative process. They'll learn from our experience and framework, but they'll also find ways that work better and help us improve our process. This year is going to be a really big year for us in terms of understanding this process, its impact, and helping scale it out to others.

Gutierrez: Why is it important to scale up DataKind?

Porway: We really feel that if we're going to tackle the world's biggest problems with data science, then we want as much of a movement and community as possible. To step back a second, I really feel like the world will be more effective if everyone can at least converse about data science. The more we demystify what this new "big data" resource is and take it down from being this black-box, jargon-y field, the better. I think that maybe it should be more like law. I don't have to be a lawyer and I don't even need to get a law degree to do anything with law, but I know when I need a lawyer and I know what I'm going to ask them.

So at the very least, we want to scale DataKind so we can create many similar environments where the power of data science for good can flourish and be usable by the general population. That way we can get a nonprofit doing important work to say, "Hey, we'd better get a data scientist in here." That's what's going to help us elevate the power of data science in the whole social sector and the whole world.

This goes back to some of the principles of the way we run projects. Some people come to DataKind and say, "Oh, I get it. DataKind does data science projects." Yes, we do. We do around twenty projects a year, and in the end, some social group gets a little better. For example, Amnesty International digitized thirty years' worth of their human rights urgent action alert data. This means that they can better predict the urgency and prioritize incoming action alerts based on data collected over three decades. A DataKind volunteer, Victor Hu [Chapter 13], worked on that project and helped Amnesty International build a model to enable them to understand, based on the information they had digitized, how to rank these urgent alerts as they come in, and say, "Hey, really act on this one now or something bad is going to happen." Which is great.

But the projects—they're actually Trojan horses to this greater movement. We see the mix of project work and our global chapter network as a way to help mission-driven organizations realize the power of data science in their work—and ultimately enhance data literacy in the social sector. So we've had tons of groups coming in saying, "I don't know what this is," and walking out saying, "Oh, I see how data science could help me and my organization." Once they see the power that data science can lend to their efforts to achieving their mission, advocating for a cause, or working with communities, they have the ammunition they need to make the business case for their organization to invest in data science professionals.

There's a flip side to this too. I also think that pro bono service should be built into the data science profession as it is being established. People sometimes ask us, "Why would a data scientist who makes six figures volunteer to do good?" However, no one bats an eye at pro bono legal work. It's just a part of the profession. So I think that by creating an opportunity for professional data scientists to give back and see the impact that their skillset can have on some of the most pressing issues we face allows us to set the stage for companies and others to make this a part of how we all do business. Data scientists in the business world are all generally well-compensated. Let's give back.

Gutierrez: You were at *The New York Times* and then you decided to start a nonprofit. Why the change in focus and why was a nonprofit the right vehicle for your goals?

Porway: For me, what it really came down to was—and this is going to sound corny—this has always been a dream job for me. I always wanted to find ways to use my skills for good. I had this feeling that I needed to give back. However, every time I got out of school with a degree—I started with computer science and later with statistics—most of the jobs available were not particularly socially fulfilling. I had bills to pay and it was very hard to find those fulfilling gigs that could pay those bills. This work did both. I just didn't actually expect it to come along this way. I could have gone and tried to work for Amnesty International, but it just turned out that there were other data scientists out there who wanted to do this too and so we got together and said let's build this right now. Now is the time, and everyone else is interested. This is more than me. This is a movement.

As to your other question about establishing DataKind as a nonprofit, we did so because we didn't want people questioning our motivations for solving problems and our goals. The work we do is so sensitive, because we're handling people's data and because we're getting out there to do good with that data. If we were a for-profit, I'd be very concerned that people would always wonder what our motivations were for doing things. Would we end up taking the higher-paying projects over the projects that would make a bigger difference? And so establishing DataKind as a nonprofit was a very loud and clear statement that we exist as a public good and that our data science solutions should be aligned with what will make the world a better place and what will have the most impact on helping the most people.

Gutierrez: Why should data scientists join forces with DataKind?

Porway: We want data scientists to volunteer and work with us to have a tangible impact on the world. There's a great opportunity here to directly help mission-driven organizations leverage new technologies and the power of data science to support their work to make the world a better place. In short, data scientists can apply their skills for good!

This is also an exciting time to be involved in building a Data for Good movement. I really do think people are looking for answers. I hear it all the time from the social sector: What is data science? How can we use it? Who can show us the way? DataKind and our volunteers are part of the community that can start providing answers.

Our projects offer an opportunity to work with real-world data that is horribly messy—a challenge. This work allows you to face different challenges than you would face in working with data at a company like Netflix. At Netflix, they've got a great data architecture in place. They have control of all the data collected. The difference between the data and how good it is at Netflix and an organization working to provide clean water to communities in rural Africa is night and day. The NGO in Africa has probably been recording things in Excel at best, though more likely on paper, and most likely all of the data is rolled up across various people's computers and has been input differently. It's a great challenge to come work on a project where you get to see what people are really facing in the trenches when it comes to data.

Lastly, another great reason is that you get to make a real impact while doing something really fun and enlightening. For example, nothing's cooler than walking away feeling like you found a new use for an old drug through data mining old medicine databases. Who doesn't want to feel happy about doing good for the world at the end of the day?

Gutierrez: When did you realize you wanted to work with data?

Porway: It was actually a total accident to be honest. I've always been really interested in artificial intelligence. From a young age, the idea of building thinking machines was just fascinating to me. I also really liked problem solving, as well as being creative. I always thought I was one of those left-brain and right-brain people. So through a computer, you could combine both halves—you could create whatever you could dream up.

I took a computer vision course in my last year of my undergraduate degree. My professor, Shree Nayar from Columbia, one of the best teachers I've ever had, stood in front of the room and said, "What does it mean to see?" It became a much more philosophical, even religious, conversation, and I was just drawn in. What a crazy problem. How do you replicate something that we can't understand ourselves? Leaving the philosophical answer aside, the current thinking at the time was that machine learning—writing algorithms to help computers learn models of the world from data—was the cutting edge of approaching this problem.

I was accepted into the PhD program at UCLA in Song-Chun Zhu's lab to continue working on the problem of modeling human vision. As a result, he advised me to switch my major from computer science to statistics, since the latter was going to be so much more instrumental to our work. That's when things really changed for me.

During my time in the statistics department at UCLA, bolstered by internships at Bell Labs and Google, I saw how my machine learning work could be applied to different interesting problems. I saw that the world is changing under our feet and that statistics will be needed everywhere. As one of my advisors, Mark Hansen, pointed out, every field is going to need statistics and computing. I still believe that a data scientist is just a statistician who can program well. Being able to collect data, model the data, visualize data, and draw meaning from the data allows you to see in ways that you've never seen before. And now because of the way data is coming off of everything, there are no limits to where this thinking can be applied.

Gutierrez: What publications, web sites, blogs, conferences, or books should people read to learn more about the types of problems being tackled in non-profit and social organizations?

Porway: First, I would say check out DataKind, of course. We publish all of our case studies. We're also going to come out with our first failure report on a project that didn't work, because we want to share learnings about everything—not just the good, but also the bad. We're not shy about that. So definitely come and look at our case studies.

If you want to get your imagination going, the book *The Human Face of Big Data* has really good for examples of how the latest techniques are being applied to interesting human and social problems.[1] There are a couple of other groups that are definitely worth looking into. The group Markets for Good is looking at how to design data markets to tackle social issues. Data and Society, Danah Boyd's new group, is worth looking into because they are tackling questions like the anthropological implications of big data. Also worth looking at is the United Nations Global Pulse. They are dedicated to using data science to tackle initiatives the UN identifies, such as coastal preservation informed by satellite imagery.

---

[1]Rick Smolan and Jennifer Erwitt, *The Human Face of Big Data* (Against All Odds Productions, 2012).

I'd say if you want to get a sense of what people in the nonprofit sector are talking about, then NTEN, the Nonprofit Technology Network, is a good place to learn how nonprofits can use technology. In general, the *Stanford Social Innovation Review* is fantastic because they're looking at the ways and problems social innovation is wrestling with.

Of course, I'd also recommend that people read up on other groups who have similar missions to DataKind—Data Science for Social Good out of the University of Chicago, the Data Guild, and Bayes Impact. They're all trying to make the world better through data science and have unique perspectives. I love the work they're doing.

Gutierrez: What does a typical workday for you look like?

Porway: Most people think that I do data science all day, but the first thing I learned about becoming a founder/CEO is that I hardly touch data at all anymore! My job is primarily governing the company, making sure we're funded, building strategic partnerships, and plotting out DataKind's course over the next one, five, ten years and beyond.

Of course, a big aspect of that is surveying the landscape and making sure our work aligns. I spend a lot of time meeting with nonprofits, foundations, and governments to understand their data problems so we can deploy programs to address their issues. I also spend a fair amount of time at data science conferences understanding what techniques could best be applied to social sector problems. And, of course, I'm also scanning for other data science do-gooders to join the movement.

I often check in with our chapters and project teams to understand how our processes are working and what could be improved. DataKind serves as the connector and translator for other data scientists and social organizations, so I see my job as making sure we're delivering what everyone wants to get out of this—that they're having their promise fulfilled.

Gutierrez: What project have you worked on that you think showcases the great work that DataKind does?

Porway: We did an early, high-impact project with the Grameen Foundation's Community Knowledge Program. The basic idea is that subsistence farmers in Africa have a tough job. You're a farmer in Africa. You are poor and therefore you do not have a lot of access to information. You don't know what the weather's going to be like the next day or if a storm is coming. You don't know the crop prices in the town next door. So the Grameen Foundation's premise was that if they got volunteers with cellphones in the community to go to farmers and ask them what information they wanted, look it up, and give it to them, the farmers could make better decisions. DataKind put together a team of volunteer data scientists to work with the Grameen Foundation to find out if the program was effective by using their data.

What was interesting—and the goal of our work together—was that we were able to tell if certain programs that they had instituted were actually working. For example, Grameen gave their teams bicycles to travel further, which was an expensive program to put in place. They wanted to know if and how well that investment was working. Our volunteer team discovered by looking at the GPS data from their services, that there wasn't a statistically significant increase in the range for the groups that had the bikes. Based on that insight, Grameen cut the program and chalked the learning up to an experiment that didn't work. This allowed them to put the funding to other great uses.

Gutierrez: What were some insights from this project for the data scientists?

Porway: I think an interesting thing that people may not think about when just getting into data science is that you always need to question your assumptions. I know that sounds trite, but specifically here, Grameen and the data scientists started with the premise that more searches means better performance. They built a histogram of search volume over all volunteers and found "For this month, this person did 200 searches. Everyone else did 10." And at that moment, as a data scientist or statistician, you might say, great, they're my good performers. I'm going to go back to Grameen and say, "Here are your good performers."

But a good and even better data scientist is one who doesn't just compute but thinks hard about all of the potential interactions of people and data. A good statistician really thinks about all the confounding variables, all of the things that are being said and not being said. So they didn't say, "These are the best people." They said, "Grameen, this is what we got back. Some people did 200 searches. Does this sound right? This is what we've observed." And the people at Grameen sort of scratched their heads. "Those look high." And the data scientist quickly went back—this was all happening at the same time because they were in the same room—and looked at it simply hour by hour, and it turned out that some of these people were doing 200 searches an hour. And we all said, "Wait, that's impossible." Grameen left committed to finding the issue with their data, which could have been that people were gaming the system or that there was an issue in their cellphone systems.

Another lesson from this project was that we shouldn't underestimate how much little things like that can transform an organization. And so finding out about the data was just a simple analysis that found a problem in data quality. Grameen could help so many more people by fixing their data collection, but without having a data scientist available, it just wouldn't have been possible.

Gutierrez: In addition to the great lessons from that project, what were other lessons you've been able to extend to other projects?

Porway: That data science is a lens through which we can view, understand, and change our world. I'm actually borrowing this from Moritz Stefaner, who is a data visualization guy with the best title I've ever heard of—truth and beauty operator. At the very end of the Visualized conference last year, he talked about this idea. He said, "I read a book in the 1980s by Joël de Rosnay called *The Macroscope*.[2] Joël was saying, 'Back in the old days, we invented the telescope to look at the infinitely large, infinitely far away. Fantastic, great tool. We can now see the cosmos. And then we invented the microscope. We can see infinitesimally small, right down there, little bugs and stuff. But what we need now is something that lets us look at the infinitely complex'—what he called a *macroscope*—'that lets us look at the complicated patterns of society and nature and the ways that they interact.' And at the time, there was not a really good way to do this."

To me, data science and the vast quantities of data in our world are that lens, that macroscope. Data science offers the ability to interpret this data. It's not just cool algorithms or data in the sense that you would normally think of as service, and spreadsheets, and sales pictures of binary number tunnels. Data is new eyes. And data science is a way to see the world through the lens of this new macroscope to learn the patterns of society and nature so we can all live better lives.

Gutierrez: Whose work is currently inspiring you?

Porway: The great Jer Thorp and Kim Rees are doing absolutely fantastic work regarding thoughtful, socially conscious data visualization and interactive design. Pete Warden's work never ceases to amaze me in its technical prowess yet its accessibility. That guy has an unmatched curiosity. Of course, I'm also inspired by everyone on our advisory council—Hilary Mason, Mike Olson, Drew Conway, Cheryl Heller, Mark Hansen—to name just a few.

Gutierrez: What do you look for in someone's work?

Porway: To me, someone's work has to be thoughtful and sensitive. It's easy to throw exciting tools at anything, but it's a very, very different challenge to do something thoughtfully and with sensitivity. For example, Pew Charitable Trusts just teamed up with SkyTruth to use satellite imagery to counter illegal fishing. I love these clever applications of new technology in ways that are thoughtful and helpful.

Gutierrez: What's the biggest thing you've changed your mind about?

Porway: This is very wonky, but coming from AI, I always felt like the more sophisticated the model the better. I had this idea that using massive amounts of data was somehow a crude and crass way of solving problems, like it was

---

[2]Joël de Rosnay, *The Macroscope: A New World Scientific System* (Harper& Row, 1979).

cheating somehow because there often wasn't a beautiful model of "reality" behind the calculations. But now I've come around to think that simple models are so much more useful, given all the volume data now available. I used to say, "If the model doesn't explain *how* the process works, you didn't work hard enough." Now I'm like, "Eh, it's pretty functional. Just build a discriminant function over some data. That's good enough."

I want to highlight another thing that I've changed my mind about. As much as I wanted to use my skills for good, I actually did think that data and data science were mostly going to be applicable to tech problems and tech solutions. What has really shocked and surprised me in a good way is that there's almost no limit to where data and data science can be applied.

Gutierrez: How do you measure success for yourself and for your organization?

Porway: I have tried to align DataKind with my own values. I want to create something that makes a significant improvement on the world. That's the ultimate success for me.

At DataKind, we're challenged in a very practical way in figuring out how we should measure the success of our projects. We measure everything from how much improvement we've seen in organizational efficiency because of analytics and data science, to how much more data literate an organization is. Ultimately, we better at least leave this place a little better than we found it, and that's only going to happen if we help each individual organization we work with get better at what they do.

Gutierrez: What does the future of data science look like?

Porway: The data moment is now—and data literacy in the public is going to be a requirement going forward. Right now, data literacy is like literacy in the old days. Only the monks used to be able to read and write. That's dangerous because it means the monks, even with their best intentions, are acting from just their own worldview in terms of what gets written and what gets read. It was only once everyone was taught to read that we had the beautiful range of communications where people could share ideas.

I think right now, similarly, data science is in the belfries of academia, big companies, and Wall Street. We need to smash those silos to the point that everyone else can at least converse about data science. If we don't, we as a citizenry risk losing the ability to critically assess data-driven results, which means we can be manipulated, taken advantage of, and cut out of the conversation. As governments and companies move to use data science more heavily, data literacy almost starts to feel like a civil right.

Gutierrez: How would you compare where we are now to four years ago? And where do you think we'll be four years from now?

Porway: I think science is going to be democratized. Here we've been talking about data and data science, but I think what we're really talking about is a new age of science. It's no longer just the experts in academia or government labs who have the technology or the capacity to draw in information and create conclusions that they release in papers, findings, and actions. We're all going to become scientists.

Our kids are going to laugh at us about the old days. "How did you pick your plumber? You just read about him in the yellow pages? You just went down the list calling people? What? Why didn't you use any data or evidence to make your decision?" Or, "How did you guys eat? Oh, you just followed some diet? Didn't you know your endocrine levels? Were you guys doing something like an experiment on yourselves where you just followed your gut?" I actually think that's absolutely coming and I think that's really exciting. That's definitely something that I don't know if people really grasp yet. The new age of science—of citizen science—in which we're all empowered with the data collection and analysis tools to study our world and learn about ourselves.

Gutierrez: What does it take to do great data science work?

Porway: I think a strong statistical background is a prerequisite, because you need to know what you're doing, and understand the guts of the model you build. Additionally, my statistics program also taught a lot about ethics, which is something that we think a lot about at DataKind. You always want to think about how your work is going to be applied. You can give anybody an algorithm. You can give someone a model for using stop-and-frisk data, where the police are going to make arrests, but why and to what end? It's really like building any new technology. You've got to think about the risks as well as the benefits and really weigh that because you are responsible for what you create.

No matter where you come from, as long as you understand the tools that you're using to draw conclusions, that is the best thing you can do. We are all scientists now, and I'm not just talking about designing products. We are all drawing conclusions about the world we live in. That's what statistics is— collecting data to prove a hypothesis or to create a model of the way the world works. If you just trust the results of that model blindly, that's dangerous because that's your interpretation of the world, and as flawed as it is, your understanding is how flawed the result is going to be.

In short, learn statistics and be thoughtful.

Gutierrez: Should data scientists have an ethical responsibility for their work?

Porway: Data scientists should and do have a very strong ethical responsibility in what they do. As a data scientist, you may find yourself running a version of what are essentially psychological experiments on users. That's something that people really need to think deeply about. And that decision of what to do is yours to make alone, but we can't hide from the fact that we as data scientists have a lot of influence.

People always treat data as "fact" when it is anything but, and that is a truth that we need to continually convey. It is our ethical responsibility as data scientists and data-literate people, to make data and data science understandable, and accessible, and less scary to everyone around us. If we're going to create a world where everyone can use data and data science for the better, we've got to dispel jargon and make our field more understandable. The easier it is for the general populace to understand what we do, the better conversations we'll be able to have about the ethics of what we do.

Gutierrez: How do companies fit into this picture?

Porway: I think corporations have this really great opportunity to become partners in this ethical data science movement. If you think of all the stuff that could be done with corporate data, it boggles the mind. Just through apps, all of these companies have become giant warehouses of human and natural data that we could learn so much from and that could make the world better. Robert Kirkpatrick from the UN Global Pulse talks a great deal about this, so a lot of credit to him.

For example, look at OkCupid, a seemingly noninvasive dating app. Yet they have data about what interests people in each other. They probably have one of the biggest databases of modern human relationships, taking into account the obvious caveat that it's a younger, tech-savvy crowd. Someone recently brought up the fact that they were trying to get an understanding of how American youth looked at religion in relationships. They wanted some insight into questions like: Do religious youth primarily gravitate toward other religious youth or nonreligious youth? Most researchers would be struggling with the challenge of collecting this data through complex surveys. However, OkCupid could actually easily answer that question with their data.

Every company has data that can help make the world a better place. LinkedIn could probably predict a recession before it happens. Google released flu trends, but did they have an ethical responsibility to release it? If they can predict the flu and this could save lives, should they have to release it and tell the world? I think that's actually a really interesting discussion now and in the future as we go forward as a profession and as companies continue to hone how they use data science. It's very important to think about the ethical ways to use data to make the world a better place.

Gutierrez: What personal philosophies have you developed from working with data?

Porway: Question everything, but be an optimist. I am always deeply critical of technology because of the potential for misuse. George Box has been saying for a century that all models are wrong, but some are useful. And to me, that is the optimistic part. Sure, some people might say because of data privacy issues we just shouldn't do anything with data science. We should lock data down and save ourselves from ourselves. I think that's too extreme. I think it's right to question the issues around data, data science, data privacy, data ethics, data misuse, and so forth. But it shouldn't dissuade us from finding solutions to those problems, because the benefits far outweigh the risks. And I'm optimistic. Yes, be skeptical, but also make sure to be optimistic.

I

Index

Get the eBook for only $10!

Now you can take the weightless companion with you anywhere, anytime. Your purchase of this book entitles you to 3 electronic versions for only $10.

This Apress title will prove so indispensible that you'll want to carry it with you everywhere, which is why we are offering the eBook in 3 formats for only $10 if you have already purchased the print book.

Convenient and fully searchable, the PDF version enables you to easily find and copy code—or perform examples by quickly toggling between instructions and applications. The MOBI format is ideal for your Kindle, while the ePUB can be utilized on a variety of mobile devices.

Go to www.apress.com/promo/tendollars to purchase your companion eBook.

Other Apress Titles You Will Find Useful

High Impact Data Visualization with Power View, Power Map, and Power BI
Aspin
978-1-4302-6616-7

Big Data Analytics Using Splunk
Zadrozny / Kodali
978-1-4302-5761-5

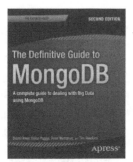

The Definitive Guide to MongoDB
Hows / Plugge / Membrey / Hawkins
978-1-4302-5821-6

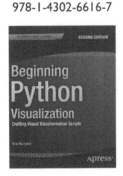

Beginning Python Visualization
Vaingast
978-1-4842-0053-7

Big Data Application Architecture Q&A
Sawant / Shah
978-1-4302-6292-3

Pro Apache Hadoop
Wadkar / Siddalingaiah / Venner
978-1-4302-4863-7

Big Data Imperatives
Mohanty / Jagadeesh / Srivatsa
978-1-4302-4872-9

Pro Data Visualization with Microsoft Business Intelligence
Stirrup
978-1-4302-3647-4

Pro Python
Browning / Alchin
978-1-4842-0335-4

Available at www.apress.com

Made in the USA
Middletown, DE
03 September 2015